BRITAIN'S CITIES

Since the pioneering work of the nineteenth century by social commentators such as Charles Dickens, reformers such as Charles Booth and Ebeneezer Howard, and political analysts such as Benjamin Disraeli and Frederick Engels, the uneven distribution of life chances has always been topical. Socio-spatial variations in human well-being remain a dominant feature of capitalist societies, with geographic clustering of disadvantage characterizing most cities of the Western world. Major planes of division are apparent in terms of income and wealth, health status, crime rates, in the quality and quantity of housing, and in the nature and availability of employment opportunities, as well as being linked to factors related to gender, age and ethnicity.

Britain's Cities offers a lucid and informative introduction to current processes of urban restructuring and geographies of division within the contemporary British city. A key principle that guides the organization of the book is that the geography of Britain's cities is the outcome of interaction of a host of public and private economic, social and political forces operating at a variety of spatial scales from the global to the local. Consequently, a proper understanding of the nature of urban division and of the problems of and prospects for local people and places in urban Britain must be grounded in an appreciation of the structural forces and processes that operate on and in combination with contextual factors to condition local urban geographies.

The book is structured in two main parts. The first examines the key structural level processes and agencies operating to influence the geography of the contemporary city, and the second part provides a set of empirically informed issue-specific studies of the major planes of division within Britain's cities. A major feature of the book is the manner in which it combines structural and local level perspectives to illuminate the complex geography of socio-spatial division within urban Britain. The book provides a comprehensive and authoritative account of key concepts underlying urban restructuring as well as up-to-date analyses of contemporary conditions within Britain's cities. Combining conceptual and empirical analyses from leading researchers in the field, *Britain's Cities* is essential reading for all interested in the urban geography of modern Britain.

Michael Pacione is Professor of Geography, University of Strathclyde.

BRITAIN'S CITIES

Geographies of division in urban Britain

Edited by Michael Pacione

London and New York

First published 1997
by Routledge
11 New Fetter Lane, London EC4P 4EE

Simultaneously published in the USA and Canada
by Routledge
29 West 35th Street, New York, NY 10001

Typeset in Garamond by Florencetype Ltd, Stoodleigh, Devon
Printed and bound in Great Britain by Biddles Ltd, Guildford and King's Lynn

British Library Cataloguing in Publication Data
A catalogue record for this book is available from the British Library

Library of Congress Cataloging in Publication Data
Britain's cities: geographies of division in urban Britain / edited by Michael Pacione.
p. cm.
Includes bibliographies and index.
1. Cities and towns–Great Britain. 2. Urban geography–Great Britain.
I. Pacione, Michael.
HT133.B717 1997
307.76'0941–dc20 96-9722

ISBN 0-415-13774-8 (hbk)
0-415-13775-6 (pbk)

Christine, Michael John and Emma Victoria

CONTENTS

Part III Prospective

PLATES

FIGURES

TABLES

CONTRIBUTORS

Dr Rob Atkinson School of Social and Historical Studies, University of Portsmouth

Professor Helen Bartlett School of Health Care Studies, Oxford Brookes University

Dr Tim Blackman School of Social Sciences, Oxford Brookes University

Dr Liz Bondi Department of Geography, University of Edinburgh

Dr David Byrne Department of Sociology, University of Durham

Ms Hazel Christie Department of Geography, University of Strathclyde, Glasgow

Dr Sarah Curtis Department of Geography, Queen Mary and Westfield College, University of London

Mr Aram Eisenschitz School of Geography and Environmental Management, Middlesex University

Dr Nicholas Fyfe Department of Geography, University of Strathclyde, Glasgow

Professor Mark Goodwin Department of Geography, St David's College, University of Wales, Aberystwyth

Dr Anne Green Institute for Employment Research, University of Warwick

Dr Rob Imrie Department of Geography, Royal Holloway College, University of London

Dr Ian Jones Department of Geography, Queen Mary and Westfield College, University of London

Professor John Lovering Department of City and Regional Planning, University of Wales, Cardiff

Professor Graham Moon School of Social and Historical Studies, University of Portsmouth

Professor Michael Pacione Department of Geography, University of Strathclyde, Glasgow

Dr Ronan Paddison Department of Geography, University of Glasgow

Professor David Phillips Department of Geography, University of Nottingham

ACKNOWLEDGEMENTS

Few can be unaware that since the mid-1970s the incidence and intensity of socio-economic disadvantage and socio-spatial division within Britain's cities have exhibited a marked increase. This book employs a combination of structural and local level perspectives to explore the geographies of division within the contemporary British city. Each chapter has been written by an acknowledged authority in the field and I should like to express my appreciation of their work and of the timely and efficient manner in which they each responded to my varied editorial requests.

All photographs were provided by individual authors with the exception of Plates 2.1a, 2.1b and 7.1b which were supplied by Michael Pacione; Plate 7.1a by M. Gould of the University of Warwick Photographic Service; Plate 8.1 by David Hoffman; Plate 8.2 by Philip Wolmuth; Plates 9.1 and 9.2 by the East London and City Health Authority Health Promotion Department, and Plates 10.1 and 10.2 by Robert Perry.

As always, I should like to express my particular gratitude to my wife Christine for her support and encouragement throughout the project. I should also like to thank my son Michael for his computing expertise, my daughter Emma for helping with the index, and Alfie the cat for his judiciously timed interruptions.

INTRODUCTION

> Blessed are the meek for they shall inherit the earth – provided its all right with everyone else
>
> (anonymous graffito)

The onset of industrial urbanism in Britain brought existing social differentials into sharp focus and created new class divisions between rich and poor. For most of the ensuing two centuries, the conditions experienced by people living at the disadvantaged end of the quality of life spectrum have attracted the attention of social commentators such as Charles Dickens, reformers such as Charles Booth (1893) and Ebeneezer Howard (1899), political analysts (Engels 1844, Disraeli 1845) and academics.

The earliest investigations were largely descriptive and set out to uncover the nature and extent of urban poverty. The work of Engels (1844) in Manchester graphically illustrated the plight of the poor working class and anticipated the research of the Chicago ecologists almost one hundred years later. In sharp contrast to the ecologists, however, Engels identified structural factors, and in particular the unequal distribution of economic power, as the principal cause of poverty and deprivation. Towards the end of the nineteenth century, Booth's magisterial study of poverty in London further illuminated the disadvantaged position of the poor. Significantly Booth, who was opposed to all forms of socialism, began his survey in the belief that personal factors lay behind poverty but he later modified his view to acknowledge the role of structural forces such as the distribution of good housing and job opportunities. Booth also recognized the analytical value of mapping the incidence of concentrations of poverty.

Since the pioneering work of the nineteenth century, the uneven distribution of life chances has remained a dominant feature of capitalist societies, with geographic clustering of disadvantage characterizing most cities of the Western world. Explicit government acknowledgement of the geography of deprivation in British cities emerged in the early 1970s following the 'rediscovery of poverty' in the 1960s and analysis of the 1971 census (Holterman

1975). The developing interest in disadvantage was supported by a Social Science Research Council's (1968) recommendation for increased academic attention to poverty research. This also lay emphasis on the importance of geography, on the grounds that certain groups within the social structure were vulnerable to the threat of poverty at certain stages of their life cycle and that these risks could be accentuated for residents in *areas* where resources and opportunities for betterment were restricted. Supported by calls from academia for a 'welfare approach' to human geography (Smith 1974), these trends stimulated the formulation of territorial social indicators (Bebbington and Davies 1980, Carley 1981), and the emergence of a range of studies designed to explore the nature, intensity and distribution of deprivation in contemporary society. This was further encouraged by the emerging policy emphasis of the Department of the Environment (1983) on the problems of the inner city, and the decision to adopt an area-based approach to the analysis of urban deprivation and to identify 'areas of need' for remedial action. Over the course of the past two decades subsequent research (Edwards 1975, Knox 1975, Millar 1980, Begg and Eversley 1986, Pacione 1986, 1987, 1989, 1993) has built on these foundations to advance our knowledge of the geography of poverty, deprivation and the uneven distribution of life chances in British society. Within this tradition of applied social research, this book provides an introduction to the major dimensions of socio-spatial division that characterize the modern British city.

Few would deny that Britain's cities are marked by deep social, economic and spatial divisions. Although this is not a new phenomenon – the urban industrial agglomerations of nineteenth-century Britain exhibited sharp variations in poverty and affluence – what commands attention is the fact that in Britain today, despite half a century of the welfare state, the gap between rich and poor is *increasing*. Futhermore, a significant minority of the population, unable to compete successfully in the arena of market capitalism, are effectively excluded from participation in mainstream society. As McCloughry (1990) observed, one of the most debilitating aspects of poverty in the UK at present is that the poor person lives in a rich society where people are valued according to what they own.

Major planes of socio-spatial division in Britain are apparent in terms of levels of income, health status, crime rates, in the quality and quantity of housing, and in the nature and availability of employment opportunities, as well as being linked to factors related to age, gender and ethnicity. Such divisions are revealed most starkly in Britain's cities by the concentration of disadvantage, both spatially in deprived areas and socially with reference to particular population groups. The vision of a divided city is encapsulated in the view from Parliament Hill:

> the six immense tower blocks of Nightingale estate (Hackney) stand . . . less than four miles from Hampstead and Highgate in space, but

> if social distances were measured in miles it would be half way round the globe.
>
> (Harrison 1985: 21)

The global metaphor employed underlines the need to look beyond the horizon in order to understand and explain the kind of socio-spatial divisions viewed from Parliament Hill. It highlights the fundamental importance of the fact that the geography of Britain's cities is the outcome of interaction between a host of public and private social, economic and political forces operating at a variety of spatial scales from global to local. Consequently, an understanding of urban division and of the problems of and prospects for local people and places in urban Britain must be grounded in an appreciation of the structural forces and processes that operate on and in combination with local contextual factors to condition local urban geographies. This key principle informs the structure of this book which comprises two complementary sections, the first outlining the main structural level processes and agencies operating to influence the geography of the contemporary city, and the second part providing a set of issue-specific studies of the major planes of division within Britain's cities. A concluding prospective chapter examines approaches to bridging the revealed urban divide.

The focus of attention on the city may be justified simply by the concentration of the British population in urban areas. The urban focus of the book, however, also reflects the restructuring of the global economy, the increasing power of multinational corporations and of internationally mobile finance capital, and the process of de-industrialization in traditional manufacturing locations, all of which have contributed to major socio-economic change between and within states. In the United Kingdom the impact of these trends has been manifested most starkly in the cities.

A dominant force behind the changes that have occurred in the nature of capitalist development during the twentieth century has been the transition, in the mid-1970s, from the post-war era of 'Fordist' production supported by a Keynesian mode of social regulation (encapsulated in the vision of a welfare state) to a 'post-Fordist' era marked by flexible production and a neo-conservative mode of social regulation (applied most vigorously by successive Conservative governments since 1979). Significantly, each of these political-economic episodes has been accompanied by a distinctive geography. The economic geography of 'post-Fordism' is generally seen to be based on new industrial spaces which, significantly for the distribution of opportunity, will almost always be located at some distance (both socially and spatially) from the previous loci of 'Fordist' production (primarily the older urban industrial centres). Equally, in social welfare terms, a strong characteristic of the mode of social regulation under flexible accumulation is a division between those who are dependent upon public provision of welfare services and those able to take advantage of new forms of private provision (such as private health care or education). These social and economic differentials reflect the growing polarization in incomes, lifestyles and access to resources between different people and places in contemporary society. Most significantly, a recurrent feature of the emerging 'post-Fordist' society is a tendency for socio-spatial differentials to increase. While such dualism is evident at all geographic scales, it is particularly evident within urban environments.

While the nature and consequences of global restructuring are well documented at the *structural* level (Burrows and Loader 1994, Amin 1995), much less attention has been focused on *local* manifestations of and responses to 'post-Fordism'. This represents a significant omission since, as we shall see, a substantial proportion of those disadvantaged by the current process of restructuring live in towns and cities, large areas of which have been devastated by the effects of global economic restructuring, the de-industrialization of the UK economy and ineffective urban policies. In view of this, and the dominant position of the city in most societies, investigation at the intra-urban scale of the problems and prospects experienced by those people and places marginal to the capitalist accumulation process is of particular value in understanding the local impacts of global restructuring. This book addresses this question directly by focusing on contemporary urban conditions and responses to the restructuring process within the British city. Within this context particular attention is directed to the position of those people and places disadvantaged by the capitalist accumulation process.

The book begins with an introductory chapter by Michael Pacione on the processes of urban restructuring and the reproduction of inequality in Britain's cities. This provides a foundation and framework for the remainder of the text which is divided into two main sections. In Part I, five key chapters provide a conceptual and policy grounding for the subsequent studies of specific aspects of urban division and exclusion. Chapter 2, by John Lovering, addresses the relationship between global restructuring and local impact. Chapter 3, by Rob Imrie, provides an overview of national economic policy and its implications for urban Britain. Chapter 4, written by David Byrne, considers the urban impact of social policy in the UK, while the following chapter by Tim Blackman examines the planning context of urban restructuring. These 'top-down' perspectives on British economic and social policy are complemented by Chapter 6, prepared by Aram Eisenschitz, which examines local level responses to urban economic and social division. By providing a balanced overview of the major debates and discussion of recent policy and practice within each subject area the first six chapters together provide an informed basis for the subsequent discussion of exclusion and disadvantage along a number of key planes of division within the British city.

While acknowledging the interrelationships between global, national and local forces in the production of urban environments, the critical scale of reference for Part II of the book is the intra-urban. In this the principal aim is to provide insight into local geographies of urban division, and the problems of and prospects for those people and places marginal to the mainstream capitalist development process. A number of key planes of division are subjected to in-depth investigation. For reasons of organization and analytical clarity each of the dimensions is discussed in a separate chapter. Clearly, however, individual planes of division, such as age and ethnicity, crime and gender, or income, housing and health may overlap to compound a situation of disadvantage for particular people and places. Bearing this in mind, eight individual chapters examine the nature of and responses to disadvantage experienced by different population groups and places within the British city, focusing on issues relating to income and wealth (Anne Green), housing (Mark Goodwin), health (Ian Jones and Sarah Curtis), crime (Nick Fyfe), ethnicity (Graham Moon and Rob Atkinson), age (David Phillips and Helen Bartlett), gender (Liz Bondi and Hazel Christie) and politics and governance (Ronan Paddison). These detailed, empirically informed investigations provide a valuable exegesis of the major planes of division in Britain's cities. In the concluding chapter, Michael Pacione adopts a prospective viewpoint to examine possible ways forward for those people and places disadvantaged by contemporary socio-spatial divisions within urban Britain.

REFERENCES

Amin, A. (1995) *Post-Fordism: A Reader*, Oxford: Basil Blackwell.
Bebbington, C. and Davies, B. (1980) 'Territorial need indicators: a new approach', *Journal of Social Policy* 9: 145–68.
Begg, I. and Eversley, D. (1986) 'Deprivation in the inner city' in V. Hausner *Critical Issues in Urban Economic Development*, Oxford: Clarendon Press: 11–49.
Booth, C. (1893) 'Life and labour of the people of London', *Journal of the Royal Statistical Society* 55: 557–91.
Burrows, R. and Loader, B. (1994) *Towards a Post-Fordist Welfare State?* London: Routledge.
Carley, M. (1981) *Social Movements and Social Indicators*, London: Allen & Unwin.
Department of the Environment (1983) *Urban Deprivation* Information Note No. 2, London: HMSO.
Disraeli, B. (1845) *Sybil*, London: Henry Colburn.
Edwards, J. (1975) 'Social indicators, urban deprivation and positive discrimination', *Journal of Social Policy* 4: 275–87.
Engels, F. (1844) *The Condition of the Working Class in England*, London: Panther Books.
Harrison, P. (1985) *Inside the Inner City*, Harmondsworth: Pelican.
Holterman, S. (1975) 'Areas of urban deprivation in Great Britain', *Social Trends* 6: 33–47.
Howard, E. (1899) *Tomorrow: A Peaceful Path to Real Reform*, London: Swann Sonnenschein.
Knox, P. (1975) *Social Well-Being: A Spatial Perspective*, Oxford: Oxford University Press.
McCloughry, R. (1990) *The Eye of the Needle*, Leicester: Inter-Varsity Press.
Millar, A. (1980) *A Study of Multiply Deprived Households in Scotland*, Edinburgh: Scottish Office.

Pacione, M. (1986) 'Quality of life in Glasgow – an applied geographical analysis', *Environment and Planning A* 18: 1499–520.
——(1987) 'Multiple deprivation and public policy in Scottish cities: an overview', *Urban Geography* 8: 550–76.
——(1989) 'The urban crisis: poverty and deprivation in the Scottish city', *Scottish Geographical Magazine* 105: 101–15.
——(1993) 'The geography of the urban crisis: some evidence from Glasgow', *Scottish Geographical Magazine* 109: 87–95.
Smith, D. (1974) 'Who gets what where and how', *Geography* 59: 289–97.
Social Science Research Council (1968) *Research on Poverty*, London: Heinemann.

1

URBAN RESTRUCTURING AND THE REPRODUCTION OF INEQUALITY IN BRITAIN'S CITIES

An overview

Michael Pacione

INTRODUCTION

Many of the world's cities are currently experiencing social, economic, political and environmental changes of unprecedented magnitude. These changes are the outcome of the interplay of a host of private and public interests operating at a variety of geographic scales. In seeking to understand the process of urban restructuring, theorists from both Left and Right have attached particular importance to the workings of the capitalist mode of production and the rise of a global economic system. Within this context, geographers have focused particular attention on the socio-spatial variations in levels in living or human well-being that arise as a result of the operation of the capitalist global economic system.

Central to this analysis is recognition of the fact that uneven development is an inherent characteristic of capitalism which stems from the propensity of capital to flow to locations that offer the greatest potential return. The differential use of space by capital in pursuit of profit creates a mosaic of inequality at all geographic levels from global to local. Consequently, at any one time certain countries, regions, cities and localities will be in the throes of decline, as a result of the retreat of capital investment, while others will be experiencing the impact of capital inflows. At the metropolitan scale, the outcome of the uneven development process is manifested in the poverty, powerlessness and polarization of disadvantaged residents. Britain's older industrial cities have suffered most from the restructuring process.

The existence of socio-spatial divisions within British cities is not a new phenomenon. Cities have always represented a mixed blessing. For some the city is the acme of civilization; for others urban life reflects the pit of human despair. Nowhere was this contrast in life quality more starkly illustrated than in the burgeoning urban-industrial agglomerations of nineteenth-century Britain. The gulf between rich and poor in cities such as Manchester and Glasgow provided ample testimony to the fact that the process of urban development in capitalist societies is inherently problematic, being accompanied by socio-spatial division and conflict. More than one hundred years later, despite the social legislation of the twentieth century, Britain's cities are still characterized by extreme variations in levels of living between different population groups and areas. Marked divisions exist between rich and poor, privileged and deprived, skilled and unskilled, employed and unemployed, healthy and ill, old and young, male and female, resident and immigrant, included and excluded, and inner city and outer city. The nature and extent of these contemporary planes of division in the British city have lent support to the concept of the 'dual city', 'polarized city' or 'two-speed city'. The existence and overlapping nature of these diverse planes of division creates complex socio-spatial patterns of advantage and disadvantage which reflect the position occupied by people and places in the hierarchy of power in which 'some decide and others are decided for'. For many observers, the scale of social division and the extent and intensity of the problems experienced by the disadvantaged residents of Britain's cities represent a contemporary urban crisis.

UNDERSTANDING URBAN RESTRUCTURING

Over the past half century, four main types of theory have been advanced to explain the structure and internal dynamics of urban areas. According to the classical ecological perspective, under free market conditions natural areas distinguished by homogeneous social characteristics arise as zones of the city are occupied by land uses that maximize the use of a particular site. Through a process of invasion and succession of land uses distinctive sub-areas, such as a central business district or ghetto, emerge as the city develops. A second main body of theory proposed to explain urban structure and change is based on neo-classical economics. In this a key role is assigned to relative location and the differing needs of activities for accessibility in determining land use patterns. Despite some success in predicting general patterns of urban land use within particular socio-historical contexts (most strikingly in the Chicago of the 1930s), the limited ability of the ecological and neo-classical economic theories to explain observed socio-spatial patterns led researchers to examine the place of the city in relation to the process of capital accumulation.

Within this political economy framework the work of Harvey (1985), Smith (1984) and Dear and Scott (1981) afforded valuable insight into the key processes and agents responsible for the production of the built environment of the capitalist city. This approach has exposed the roles of and relationships among various fractions of capital (such as property speculators, estate agents, financial institutions) as well as those between capital and the state (in the shape of the ideology, policies and planning practices of central and local government) in influencing urban change. A further advantage of the political economy approach is that it highlights the impact that economic, social and political processes located *outside* the territory of any particular city have on its internal structure and development.

The resurgence of humanism and the advent of post-modernist/post-structuralist interpretations in human geography represents the fourth main body of theory that has contributed to our understanding of urban change. This approach aids interpretation of the city by revoking Marxian reification of the market and demystifying the role and actions of the agents involved in the production and reproduction of urban environments. Post-modern critique of meta-theory and emphasis on human difference provides a useful corrective to macro-level structural analyses. By adding another level of explanation, a post-structuralist perspective augments the exigetical value of a political economy view of urban change.

Each of the four theoretical perspectives can claim to illuminate some part of the complex dynamics and structure of the city. But no single approach provides a comprehensive explanation of urban restructuring. The question of whether an accommodation is possible among the different approaches has tended to be polarized between those who accept a pluralist stance – 'agreeing to differ', on the grounds that there is no single way to gain knowledge (Couclelis 1982) – and those who insist on the need to make a unitary choice of theoretical framework due to the perceived superiority of a particular epistemology (Hudson 1983). Others have sought to combine approaches in different ways (Johnston 1980). The latter route, which incorporates a search for a middle ground between the *generalization* of positivism and the *exceptionalism* of post-modern theory, is the approach favoured in this book. The restructuring of Britain's cities is best seen as a manifestation of structural forces within a particular context. A full understanding of urban restructuring in Britain requires examination of both the *general* processes of the capitalist mode of production (by means of a political economy perspective) and an (empirically informed) appreciation of the *particular* social formations that emerge from the interaction of structural forces and local context. In this book the study of the particular – geographies of social division in the British city – is set within a structural framework comprising national and international economic and political parameters.

The importance of employing a combined perspective that encompasses global and local scales, structure and agency, and theory and empiricism in seeking to understand the process of urban restructuring in Britain is illustrated most graphically by the fact that economic and political forces operating at the global level can reach down to influence the quality of life of individuals and local communities. For cities, the impact of the emergence of a world economic system is encapsulated in the distinction between the city as an autonomous self-governing polity (which existed in medieval

Europe prior to the development of an economy based on the trade of marketed commodities) and present circumstances under which city development is influenced to a significant degree by forces beyond its control. Today, investment decisions taken by managers in a transnational corporation with headquarters in one of the 'command cities' of the global economic system can have a direct effect on the well-being of a family living on a council estate in Britain. In order to confront such forces, cities in the modern world must seek to position themselves and, increasingly, compete in global society. The fact that cities vary greatly in their capacity to meet the challenge posed by globalization is reflected in the extent to which each can shape or simply react to global forces. This challenge is particularly acute for cities in advanced industrial states like the UK which have been destabilized by a process of industrial restructuring that has accelerated since the early 1970s as part of the transition to advanced capitalism.

It is important to recognize, however, that although the power of global forces to influence urban growth or decline is profound it is not omnipotent. Economic and political action undertaken at a national level or by regional groupings of nation states can modify the effects of the global economic system. Regulatory and tax policies shape the environments that attract or repel investors; decisions about public investment determine whether infrastructure will be rebuilt or allowed to deteriorate; government procurement policy stimulates the private economy, and intergovernmental transfer payments can prevent the collapse of a local economy. Furthermore, the impacts of global processes are manifested in particular local contexts, the nature of which varies between places. Thus while the UK as a whole may be influenced by the shockwaves of the global economy, the response of cities in different parts of the country, as well as of different areas within individual cities, varies according to local conditions. As we have indicated, to achieve a full understanding of the process of urban restructuring in Britain requires knowledge of both national and international structural forces as well as of local mediating factors underlying the process of urban change.

Accordingly, the five chapters that comprise Part I of this book examine the principal structural forces acting on Britain's cities and establish a contextual framework for the detailed consideration of local conditions presented in Part II. In this introductory chapter I provide an overview of the major factors relating to the process of urban restructuring in the United Kingdom. The chapter is organized into five main parts. In the first part the discussion establishes the *economic context* of urban restructuring by describing the nature of the capitalist system and examining the post-war restructuring of the UK space economy. In part two attention is focused on the *policy context*, with particular consideration directed to the nature of urban policy. In the third section the *planning context* of urban restructuring is explained, while in section four the *social context* with particular reference to the changing role of the welfare state, is discussed. Finally, section five examines the geography of the current urban crisis, focusing on the nature and incidence of poverty and related disadvantages, and the growth of social polarization and exclusion in urban Britain. In combination this set of contextual analyses provides a foundation and framework for the remainder of the book.

THE ECONOMIC CONTEXT

The nature of capitalism

Capitalism, as defined by Marx, is a specific mode of production: a set of institutionalized practices by which societies organize their productive activities, provide for their material needs and reproduce their socio-economic structure. Capitalism refers not only to economic mechanisms of resource allocation but views economic activity as a *social enterprise* governed by political and legal regulations that embody a particular set of cultural values. (The rise of the new conservatism in the UK and USA during the 1980s provides clear evidence of the effect of political change on economic organization and, through policies such as privatization, de-regulation and public spending restraint, on the spatial structure of society.) Capitalism attaches particular importance to individual freedom as a cultural value and capitalist societies are marked by a strong commitment to private property ownership and preference for limited government intervention in the decisions of individuals to allocate resources on the basis of the values set by the market. Capitalism is one of five major modes of production – the others being subsistence, slavery, feudalism and socialism. Capitalism is distinguished from other modes of production by its expansionary dynamic encapsulated in the profit-maximizing goals of the actors within the capitalist mode of production (and enshrined in Marx's dictum: production for production's sake, accumulation for accumulation's sake). The other distinguishing feature of capitalism is its *uneven results*.

The unevenness of capitalism

The unevenness of capitalist economic development is evident both *temporally* and *geographically*. The temporal unevenness of capitalist development is revealed in the growth cycles identified by economic historians. Of particular significance in interpreting the recent evolution of the global economic system are the Kondratieff 'long waves' that have appeared since the Industrial Revolution. These alternate phases of growth and stagnation show a high correlation with technological change, and the regenerative effects of *technological innovation* underlies the major non-Marxist explanations of long waves (Freeman 1987). The first Kondratieff expansion phase (*c.*1790–1815) was associated with the period of the original Industrial Revolution (with the introduction of mechanized textile production and improved iron production).

Subsequent periods also appear to fit the pattern with steam engines, railways and Bessemer steel in the second growth phase (*c.*1844–74); chemicals, electricity and automobiles in the third phase (*c.*1890–1920); and aerospace, electronics and nuclear technology in the fourth phase (*c.*1940–73). The fifth Kondratieff identified by Hall (1985) highlights the role of micro-electronics (in, for example, computing, industrial process control and telecommunications) and biotechnologies in transforming the economy.

In Marxist interpretations these regular fifty-year cycles of growth and recession in the modern world economy are linked to the inherent dynamics of capitalist economies (Mandel 1980). According to this analysis, over-investment by competing firms each seeking to capture larger market shares during periods of expansion, together with the success of labour in increasing their wages when business is buoyant, eventually lead to excess productive capacity and uncompetitive labour costs. Recession sets in with redundancies and bankruptcies marking a period of intense competition from which surviving firms emerge 'leaner and meaner' with a more profitable configuration of production facilities and labour costs to supply the prevailing level of market demand. The recovery of economic growth – which is by no means automatic – sets in motion a new round of profit-enhancing innovations embodied in industries whose *locational preferences* may well differ from those of more traditional forms of production (Wallace 1990).

As Mandel (1980) acknowledges, however, the emergence and development of a new technological system is dependent not only on the actions of capital but on the development of an appropriate *socio-institutional structure*. The socio-institutional environment comprises a complex of management strategies, skill requirements, firm structures, infrastructural investments, consumption norms and government policies – all of the institutions and institutional arrangements that impinge upon economic production, investment, consumption and employment. This environment, also described as the 'social structure of accumulation' (Gordon 1980), embraces the money and credit system, the nature and degree of labour organization, the structure and operating roles of labour and product markets, and the structures and conditions of the international institutions that influence the organization and operation of the national economy. This concept of a social structure of accumulation or *mode of regulation* is central to the French Regulationist interpretation of the current economic transition which is depicted as a shift from a post-war Fordist paradigm to a post-Fordist regime of 'flexible accumulation' (Tables 1.1 and 1.2). Significantly, post-Fordism is characterized not only by greater flexibility in the means of production but in the *location* of productive facilities. Places that cannot provide the necessary socio-institutional environment for flexible accumulation are likely to be disadvantaged in the competition for growth and investment. As we shall see later, many of Britain's cities and urban communities fall into this category.

Table 1.1 The post-war 'Fordist' expansionary regime of the late 1940s to early 1970s

Characteristic	*Key features*
	Accumulation regime: monopolistic
Industry	Monopolistic; increasing concentration of capital; steady growth of output and productivity, especially in new consumer durable goods sectors; secular expansion of private and especially public services.
Employment	Full employment: growth of manufacturing jobs up to mid-1960s; progressive expansion of service employment; growth of female work; marked skill divisions of labour.
Consumption	Rise and spread of mass consumption norms for standardized household durables (esp. electrical goods) and motor vehicles.
Production	Economies of scale; volume, mechanized (Fordist-type) production processes; functional decentralization and multinationalization of production.
	Socio-institutional structure: collectivistic
Labour market	Collectivistic; segmented by skill; increasingly institutionalized and unionized; spread of collective wage-bargaining; employment protection.
Social structure	Organized mainly by occupation, but tendency towards homogenization. Income distribution slowly convergent.
Politics	Closely aligned with occupation and organized labour; working-class politics important; regionalist.
State intervention	Keynesian-liberal collectivist; regulation of markets; maintenance of demand; expansion of welfare state; corporatist; nationalization of capital for the state.
Space-economy	Convergent; inherited regional sectoral specialization (both old and new industries) overlaid by new spatial division of labour based on functional decentralization and specialization: regional unemployment disparities relatively stable.

Source: Martin 1988

The nature of the world economy

As we have indicated, Britain's cities must increasingly compete within a global economic system. Wallerstein (1979) has identified two types of world system – 'world-empires' (characterized by a common political system) and 'world-economies' (the prime example of which is the contemporary capitalist economic system). A feature of the capitalist world economy is that because the economic factors operate within an area larger than that which

Table 1.2 The 'post-Fordist' regime of flexible accumulation, from the mid-1970s to the present

Characteristic	*Key features*
	Accumulation regime: flexible
Industry	Rationalization and modernization of established sectors to restore profitability and improve competitiveness: growth of high-tech and producer service activities, and small firm sector.
Employment	Persistent mass unemployment; generalized contraction of manufacturing employment, growth of private service sector jobs; partial de-feminization (in manufacturing); flexibilization of labour utilization; large part-time and temporary segment.
Consumption	Increasingly differentiated (customized) consumption patterns for new goods (esp. electronics) and household services.
Production	Growing importance of economies of scope; use of post-Fordist flexible automation; small batch specialization; organizational fragmentation combined with internationalization of production.
	Socio-institutional structure: competitive-individualist
Labour market	Competitive; de-unionization and de-rigidification; increasing dualism between core and peripheral workers; less collective, more localized wage determination.
Social structure	Trichotomous and increasingly hierarchical; income distribution divergent.
Politics	De-alignment from socio-economic class; marked decline of working-class politics; rise of conservative individualism; localist.
State intervention	Keynesianism replaced by free-market Conservatism; monetary and supply-side intervention rather than demand stabilization; de-regulation of markets; constraints on welfare; self-help ideology; privatizing the state for capital.
Space-economy	Divergent; decline of industrial areas (pre- and post-war); rise of new high-tech and producer services complexes; increasingly polarized spatial division of labour; widening of regional and local unemployment disparities.

Source: Martin 1988

any political entity can control, capitalists – financial institutions and multinational companies – have a freedom to manoeuvre that is structurally based. The exercise of this freedom by 'footloose' capital can result in tension between them and territorially bounded national governments and lead to adverse consequences (e.g. redundancies and economic decline) for less competitive places.

The main features of the modern capitalist world economy may be summarized as follows:

1 It comprises a single world market within which production is for exchange rather than use with prices fixed by the self-regulating market. This means that, for a period of time, more efficient producers can undercut others and increase their market share to the detriment of other producers and *places*.
2 Territorial divisions between states in the world economy lead to a competitive state system in which each seeks to insulate itself from the rigours of the world market (e.g. by trade barriers) while attempting to turn the world market to its advantage (e.g. by offering tax incentives to incoming industries).
3 The modern world economy comprises a core (characterized by relatively high incomes, advanced technology and diversified production) and periphery. The core needs the periphery to provide the surplus to fuel its growth. Core–periphery relationships exist at all geographic scales including, for example, differences in levels of prosperity between the 'sunbelt' towns of southern England and the older industrial cities of north Britain.
4 As we have seen, the world economy has followed a temporally cyclical pattern of growth and recession characterized by Kondratieff cycles.
5 Finally, it is important to reiterate that every part of the world has its own particular relationship to the world economy based on locally specific modes of socio-economic organization.

The world economy in the post-war era

The end of the post-war industrial boom in the early 1970s heralded a crisis for industrial capitalism. The system-shock precipitated by the quadrupling of petroleum prices in 1973 as a result of the OPEC cartel has been indicted as a major cause of the economic downturn. However, as Hamilton (1984) explained, a number of other trends originating in the previous decade or before were also responsible for the recession. These included:

1 A slow-down in economic growth and falling profits in the industrial core countries of the OECD associated with falling levels of demand for capital goods (particularly in transport, steel and building).
2 Rising levels of inflation, which reduced profits and hampered capital accumulation, led to greater dependency on the banking sector for investment funds. High interest rates, however, restricted technological investment and so hindered competitiveness. Inflation also raised labour costs. The net outcome was the widespread depression of *both* capital-intensive (e.g. steel, ship building, vehicles) and labour-intensive industries (e.g. textiles, clothing).

3 Increased international monetary instability which took two forms:
   (i) Under- or overvaluation of exchange rates as a result of the transition in the early 1970s from fixed exchange rates to floating exchange rates. In countries whose currencies were undervalued (e.g. Germany and Japan) industrial production was stimulated by increased demand for exports whereas in those countries with overvalued currencies (e.g. oil and gas producers such as Norway and the UK) the loss of international competitiveness led to import penetration and a consequent decline in domestic industrial capacity.
   (ii) Indebtedness among newly industrialized countries (NICs) and some underdeveloped countries as a result of massive borrowing from the 'petrodollar' surpluses of the OPEC states. This stimulated debtor countries to increase exports of cheap manufactured goods to the core regions in order to obtain necessary foreign exchange. This increased the competitive pressure on the labour-intensive sectors of the core economies.
4 The growth of new social values (discussed later) related to social welfare and environmental protection increased industrial costs and contributed to a higher tax burden for both producers and consumers.
5 The introduction of technological innovations in response to escalating energy and labour costs led to reduced demand in some traditional industrial sectors. For example, energy saving designs in transport cut demand for steel, while innovations in micro-electronics reduced demand for electro-mechanical products.
6 A resurgence of political volatility (e.g. in the Middle East and South East Asia) reduced the area of stable business and constrained world trade.
7 An increasing intensity of international competition arose from the post-war liberalization of trade, the spread of industrialization to the periphery, the aggressive role of governments in NICs, and the post-1970 stagnation of world markets.

The net effect of these trends was to initiate a process of industrial reorganization and structural change that signalled a transition from *industrial capitalism* to post-industrial or *advanced capitalism* (alternatively described as a move from the fourth to the fifth Kondratieff). These trends have had a particular effect on the UK space-economy.

The post-war restructuring of the UK space-economy

The unparalleled prosperity of the long post-war boom (*c.*1945–73) stimulated the Conservative Prime Minister Harold Macmillan to proclaim to the people of Britain that they 'never had it so good'. Between the early 1950s and early 1970s labour productivity and real wages at least doubled. The fastest growth rates were in the new industries of electrical engineering, vehicle manufacture and chemicals and petroleum products. The spread of mass consumption norms, rising incomes and expansion of home and export demand for new standardized consumer goods (such as cars, television and refrigerators) led to the adoption of mass assembly line production methods and the growth of large, often multiplant and multiregional, firms capable of exploiting the economies of scale afforded by expanding markets. The expansion of manufacturing was accompanied by growth in the range and output of personal, business and public services. At the high point in the early 1950s, Britain recorded a surplus on manufacturing trade equivalent to 10 per cent of GDP (Martin 1988). The buoyancy of the economy was such that rapid growth, rising prosperity and full employment combined to disguise the problems of the old industries of ship building, coal, iron and steel and heavy engineering. By 1960 Britain's share of total world exports had declined from 25.5 per cent in 1950 to 16.5 per cent – a trend that presaged further decline in Britain's international competitiveness to 9.7 per cent in 1979. Import penetration had become a significant drain on the balance of payments even in sectors that during the 1950s had been dominated by domestic producers (e.g. consumer durables). By the early 1970s it was evident that the post-war boom was coming to an end. This had fundamental consequences for the British economy and for Britain's cities.

One of the most striking features of the transition to advanced capitalism has been the rapid and intense de-industrialization of Britain's manufacturing base. Initially this took the form of relative decline (with manufacturing growth being less than that of the service sector) but with the advent of advanced capitalism there has been an *absolute* decline in manufacturing. Between 1966 and 1976 more than one million manufacturing jobs disappeared in net terms, a loss of 13 per cent. The decline affected most sectors of manufacturing – both traditional industries of ship building (–9.7 per cent), metal manufacturing (–21.3 per cent), mechanical engineering (–14.5 per cent) and textiles (–27.6 per cent) and the former growth sectors (and bases of the fourth Kondratieff) of motor vehicles (–10.1 per cent) and electrical engineering (–10.5 per cent). In the West Midlands the net loss of 151,117 manufacturing jobs between 1978–81 represented almost a quarter of all manufacturing employment in the region and helped redefine the once-prosperous area as part of Britain's 'rust belt' (Flynn and Taylor 1986). In Coventry alone, between 1978 and 1982, 39,286 manufacturing jobs (39 per cent of the total) were lost (Healey and Clark 1985). The severity of the job losses in the West Midlands were in part due to the highly integrated nature of a local economy centred on metal and car industries which, during the 1960s and 1970s, were subjected to intensive overseas competition. A further contributory factor was that national economic policies of the early 1980s (discussed later) based on a free market and anti-interventionist philosophy failed to mitigate (and arguably exacerbated) the effects of recession and uncompetitiveness in this and other sectors of British industry.

Plate 1.1 The silence of the cranes in London's Docklands bears testimony to the widespread process of de-industrialization that has affected large parts of urban Britain

The problems of de-industrialization in Britain have been felt most acutely in the peripheral regions of Clydeside, Tyneside and in Lancashire where the textile industry alone shed over half a million jobs (Martin and Rowthorn 1986). The impact of de-industrialization has had ripple effects beyond the manufacturing sector with, for example, the coal industry adversely affected by the fall in demand from manufacturing (as well as from the power generators due in part to the government decision to substitute cheaper oil and gas supplies in order to reduce energy costs for manufacturing industry).

The growth of the service sector represented the other side of the coin, with over 3 million jobs created between 1971 and 1988 in areas such as R & D, marketing, finance and insurance (collectively referred to as producer services); transport and communications (distributive services); leisure and personal services (consumer services); and central and local government administration (public services). Growth of service sector employment, however, has not been sufficient to cancel out the loss of employment opportunities in manufacturing. Evidence from individual cities provides a graphic illustration of the decline in non-service sector employment (Table 1.3). In the Clydeside conurbation manufacturing employment fell from 387,000 in 1961 to 187,000 in 1981. In the city of Glasgow employment in manufacturing virtually collapsed from 227,000 in 1961 to 87,651 in 1981 (–61.4 per cent) and 48,782 by 1991 (–44.3 per cent). In Glasgow, between 1981 and 1991, while manufacturing shed 38,869 jobs, service sector employment increased by only 2,795. Nor are the new service sector jobs necessarily suited to the skills of redundant manufacturing workers. The result has been rising unemployment. In Glasgow in 1993, 40,261 males and 10,775 females were unemployed; 46 per cent of males and 31 per cent of females had been unemployed for more than a year while 9 per cent of males and 5 per cent of females had been unemployed for over five years. Significantly, youth unemployment (i.e. among those aged 16–25 years) accounted for 28 per cent of male and 44 per cent of female unemployment (Pacione 1995a). These trends have been repeated in all of the large provincial cities of Britain since the onset of recession in the mid-1970s.

One of the major factors underlying urban employment loss over recent decades has been the internationalization of economic activity as private businesses have sought to respond to intense competition in stagnating markets by expanding the scope of their activities both functionally (through vertical and horizontal integration) and geographically (via the new international division of labour). This has produced a major paradox whereby the poor domestic record of British industrial capital in the post-war years is in contrast to the substantial growth that has taken place in the number of British multinational companies with extensive investment and trading interests overseas. By 1970 all of the largest 100 British manufacturing companies had become multinationals (Gamble 1981). While these investments abroad

Table 1.3 Employment change by sector in major British cities, 1981–91

	Glasgow	*Newcastle*	*Birmingham*	*Leeds*	*Sheffield*	*Manchester*	*Liverpool*	*Cardiff*
	(%)	*(%)*	*(%)*	*(%)*	*(%)*	*(%)*	*(%)*	*(%)*
Primary	–16.3	–38.3	–18.6	–39.3	–52.0	–17.3	–64.8	–10.0
Chemicals etc.	–67.4	–46.3	–48.3	–14.8	–73.0	–44.5	–41.9	–28.9
Engineering etc.	–45.4	–49.0	–37.8	–26.2	–31.7	–32.8	–49.4	–12.2
Other manufacturing	–40.8	–37.2	–19.8	–26.9	–18.7	–48.6	–60.4	–3.7
Construction	–15.0	–41.1	–10.0	–14.0	–13.5	–26.2	–24.7	–24.4
Producer services	22.0	46.2	33.0	54.1	32.2	16.2	10.3	53.4
Distributive services	–25.3	–20.0	–6.4	9.8	11.1	–16.0	–36.9	–16.3
Personal services	4.6	16.7	22.2	19.3	32.5	–3.6	3.3	13.0
Non-market services	11.1	26.3	5.3	2.6	7.5	–0.7	–11.2	–1.0
Total	–13.1	–4.5	–10.8	–0.6	–15.1	–12.9	–25.7	–0.7

Source: Pacione 1995a

facilitated the penetration of foreign markets, the comparative lack of *domestic* investment by the leading sector of British industrial capital was an important factor underlying the failure of British industry to compete in the international economy. The internationalization of Britain's economy has also occurred in part through inward investment. During the 1980s in both manufacturing and service industries there was an increased reliance on foreign-owned companies, and an increased dependence of domestic employment upon decisions taken by non-UK companies.

Government policy has also played a major role in the restructuring of Britain's economy. The impact of government policy was clearly illustrated during the 1950s and 1960s when a key policy objective was to preserve the stability of sterling to enable it to continue to function as a top international reserve currency and as a major medium of international trade (and thereby maintain the key international role of the City of London). This necessitated efforts to avoid deficits on the balance of payments in order to maintain confidence in sterling. Since this could not be achieved by imposing restrictions on the free flow of capital investment overseas, successive governments deflated the domestic economy in order to reduce demand for imports. At the same time, the need to maintain domestic levels of employment and to finance state expenditure limited the possibilities of such a policy, with the result that the UK economy experienced alternating periods of expansion and stagnation (stop–go). Not only did this fail to halt the long-term decline in the roles of sterling and Britain in the international system but, as the periods of 'go' got shorter and those of 'stop' longer, it became increasingly difficult for firms to improve levels of investment, productivity and output (Rees and Lambert 1985).

A further factor in the explanation of Britain's post-war economic problems was the defensive power of organized labour. For much of the period the better organized workers established a significant degree of control over the labour process which often led to restrictive practices concerning demarcation, manning levels, work-rates and overtime as well as shop-floor resistance to the reorganization of production. One consequence was that industrial capital was reluctant to re-equip and restructure, and often preferred to invest in lower cost, non-unionized labour markets overseas. The commitment of post-war governments to the maintenance of full employment through Keynesian demand management and to the provision of a substantial Welfare State further strengthened the power of organized labour (as well as being, in part, a reflection of that strength). During the 1950s and 1960s conditions of relative labour shortage aided labour to resist any reductions in real wages, thereby compounding the difficulties of industrial capital. In addition, financing of the state welfare system necessitated a rise in taxation levels. This impacted upon both industry (affecting profitability) and earned income and consumption (which fuelled trade union militancy and industrial disputes).

The slow growth of the economy meant that, by the late 1960s, the British state was beset by both a crisis of industrial relations and an emergent fiscal crisis (Jessop 1980), a combination that engendered descriptions of 'the British malaise' (Mackintosh 1977) and of the post-1973 era as a 'decade of discontent' (Gardner 1987). A measure of the relative decline of the British economy in world markets was the reduction in the nation's share of world trade in manufacturing goods from 10 per cent by value in 1972 to under 7 per cent in 1988. Much of this fall was the result of competition from the rapidly growing Far Eastern economies which challenged Britain in a variety of manufacturing industries including the vehicle and textile sectors. There was also intense competition from Europe and Japan. At the end of 1986 unemployment had reached 3.25 million. By the end of the decade Britain was running a serious balance of payments deficit with export prices above those of other West European economies. Recurrent fears were expressed, in both Labour and Conservative parties, about the uncompetitiveness of the British economy.

Much of the burden of recession was felt in Britain's cities which had retained a disproportionate share of those industries most vulnerable to the demands of advanced capitalism. Urban areas with old factories employing outmoded production techniques and with uncompetitively low levels of labour productivity were affected most severely by capital reorganization aimed at countering declining rates of profit. These measures included, in many urban areas, plant closures and transfer of production to other locations with reduced labour inputs. Thus between 1966 and 1974, 27 per cent of the job loss in manufacturing in Greater London was the result of firms relocating, 44 per cent was due to plant closures, and 23 per cent due to labour shedding by firms remaining in situ (Dennis 1978). Furthermore, many of the firms that survived in inner city areas imposed regimes of long working hours, low wages and poor conditions upon a workforce drawn from exploitable social groups, including women and ethnic minorities (Community Development Project 1977a).

The loss of industry and jobs from inner urban areas was accompanied by decentralization of population from overcrowded central areas to suburban locations, outer estates and New Towns. Between 1951 and 1981 the largest cities lost on average a third of their population (Robson 1988). The effects of these population shifts were heightened by the composition of migration flows with the more able, affluent and self-sufficient departing by choice, leaving the elderly, young adults and those with below average incomes to await the arrival of the redeveloper's bulldozer. In some of Britain's major cities the vacated spaces in the inner areas were occupied by immigrants. As we shall see later, these economic, demographic and social trends contributed to a growing social polarization within British cities and led to state recognition of an 'inner-city problem', the symptoms of which have appeared subsequently in other parts of Britain's cities.

THE POLICY CONTEXT

Urban policy is concerned with the management of urban change. It is a state activity that seeks to influence the distribution and operation of investment and consumption processes in cities. It is important to recognize, however, that urban policy is not confined to activity at the urban scale. National and international economic and social policies are as much urban policy, if defined by their urban impacts, as land use planning or urban redevelopment. In effect, urban policy is often made under another name. Urban policy is dynamic. Its formulation and implementation is a continuing process, not an event. Measures that are introduced cause changes that may resolve some problems but create others for which further policy is required. Furthermore, only rarely is there a simple optimum solution to an urban problem. More usually a range of policy options exists.

Urban policy is the product of the power relationships between the different interest groups that constitute a particular social formation. Foremost among these agents are the state, both local and national, and capital in its various fractions. Capital and state pursue specific goals which may be either complementary or contradictory. For capital, as we have seen, the prime directive is profit maximization. The state, on the other hand, in addition to facilitating the process of accumulation, must also satisfy the goal of legitimation. These political and economic imperatives have a direct influence on the nature of urban policy. Urban policy is also conditioned by external forces operating within the global system, as well as by locally specific factors and agents.

The form of urban policy employed depends on the problem to be addressed and, most fundamentally, on the ideological position of the state. Adherents to market capitalism view the production of unevenly developed cities as the inevitable outcome of technological change within an economic system that readily adapts to innovation. The negative socio-spatial effects of this restructuring that impinge on disadvantaged people and places are regarded as unavoidable consequences of a process that is of benefit to society as a whole. For those on the political Right, market forces are the most efficient allocators of capital and labour, and state intervention is considered unnecessary. Policies involving social welfare expenditure and government financial aid to declining cities are regarded as harmful because they anchor low-wage workers to sites of low employment opportunity, discourage labour force participation and inhibit labour mobility. Welfare state liberals, on the other hand, while accepting the central role of the market, acknowledge that the institutional and cyclical 'market imperfections' that have left certain people and places in prolonged economic distress must be rectified by compensatory government policies (Pacione 1992).

Urban policy in Britain

From the passage of the 1947 Town and Country Planning Act, which established a centralized system of planning in the UK, until the mid-1960s, urban problems were seen largely in physical terms. Issues of housing quality and supply, transport and industrial change were to be tackled by comprehensive redevelopment strategies and the dispersal of urban problems via regional policy, including construction of New Towns. Labour politicians of the time believed that they possessed the policy machinery to overcome unemployment and want. In 1964 Anthony Crossland wrote that 'primary poverty has been largely eliminated; the Beveridge revolution has been carried through; and Britain now boasts the widest range of social services in the world, and, as a result the appellation Welfare State'. At about the same time, empirical research was being carried out that led to the 'rediscovery of poverty' (Townsend 1962, Wedderburn 1962). Together with growing public concern over poverty and deprivation and evidence that living standards had declined since 1964 (Field 1969), this led to a refocusing of urban policy. The spark that fired government action to tackle the urban problem was Enoch Powell's 'rivers of blood' speech of 1968 in which he criticized the rate of immigration into Britain. This inflamed racial tensions in those urban areas where black and Asian immigrants had settled in large numbers and stimulated Harold Wilson's government to initiate the Urban Programme (Figure 1.1). This was intended to tackle 'needle points' of deprivation by offering local authorities a 75 per cent grant towards the cost of projects in the fields of education, housing, health and welfare in *areas of special need* (many of which would contain concentrations of coloured residents). Rees and Lambert (1985) have suggested that the strategy of targeting positive discrimination to pockets of residual social need and the specific concern afforded to social spheres related to children reflected contemporary thinking which located the origins of poverty in individual or family weaknesses (Table 1.4). The kind of projects fostered by the Urban Programme, however, did not address the fundamental issues of structural economic change that underlay the urban crisis. This was recognized by the Community Development Project which offered a starkly contrasting interpretation of the causes of urban poverty and in particular the role played by the state. The CDP rejected the 'blame the victim' perspective of the *social pathology* model and identified urban poverty as an inevitable by-product of uneven capitalist development (Community Development Project 1977b). Increasing tensions between the recommendations of the CDP teams (who encouraged residents in their areas to oppose the activities of the capitalist state) and the Home Office eventually led to withdrawal of funding for the project, so that by 1978 all were defunct.

Government awareness of the multifaceted and interconnected nature of urban problems was also aided by the Shelter Neighbourhood Action Project

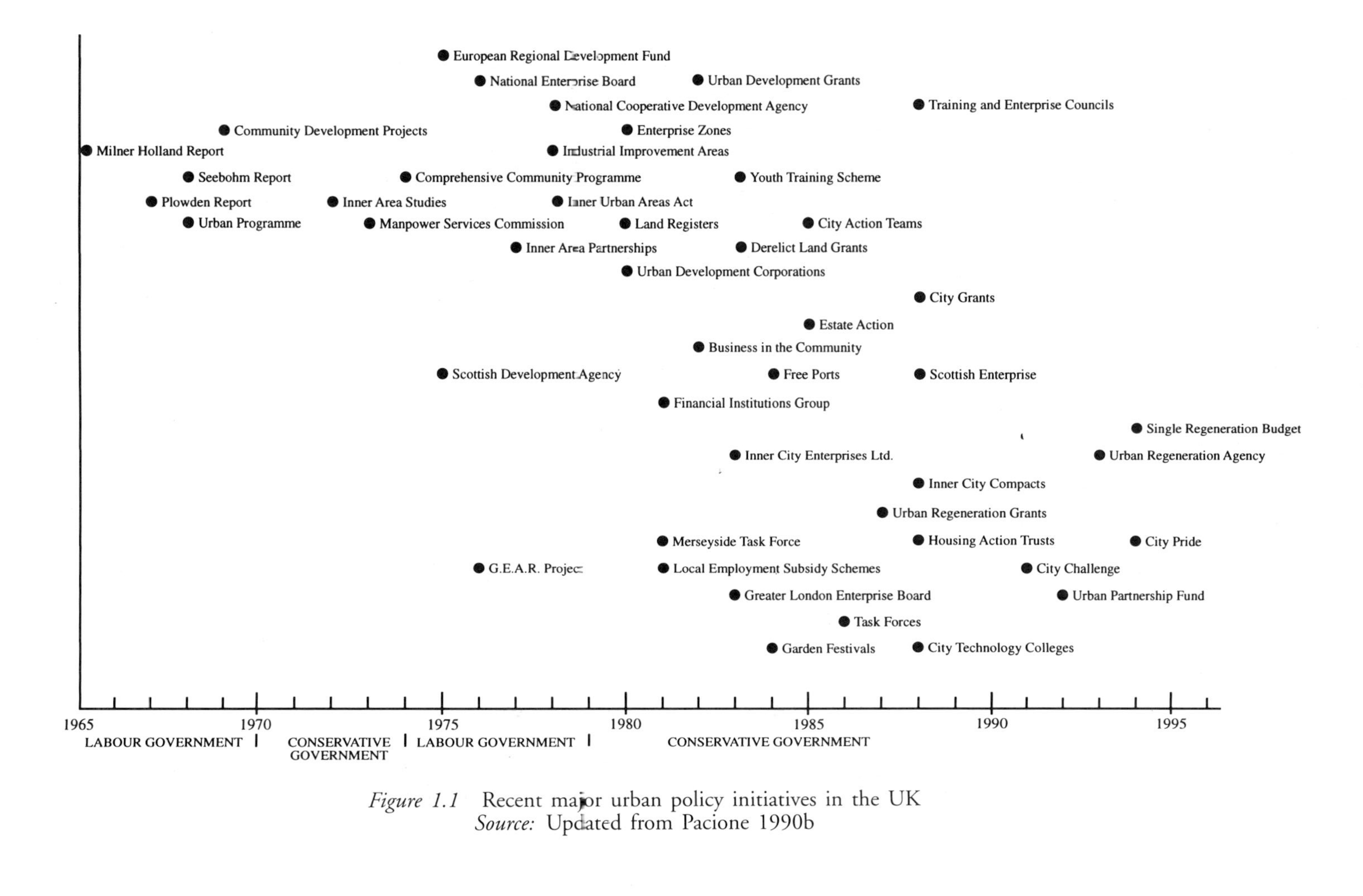

Figure 1.1 Recent major urban policy initiatives in the UK
Source: Updated from Pacione 1990b

Table 1.4 Principal theories of urban deprivation

Theoretical model	*Explanation*	*Location of problems*
1 Culture of poverty	Problems arising from the internal pathology of deviant groups	Internal dynamics of deviant behaviour
2 Transmitted deprivation (cycle of deprivation)	Problems arising from individual psychological handicaps and inadequacies transmitted from one generation to the next	Relationships between individuals, families and groups
3 Institutional malfunctioning	Problems arising from failures of planning, management or administration	Relationship between the 'disadvantaged' and the bureaucracy
4 Maldistribution of resources and opportunities	Problems arising from an inequitable distribution of resources	Relationship between the underprivileged and the formal political machine
5 Structural class conflict	Problems arising from the divisions necessary to maintain an economic system based on private profit	Relationship between the working class and the political and economic structure

Source: Community Development Project 1977b

(1972) in the Granby district of Liverpool. This concluded that a main policy requirement to tackle the problems of the inner city was better co-ordination of the key agencies of central and local government in a concerted programme of reinvestment. The need for a 'total approach' led to a series of Inner Area Studies (to research the causes of urban deprivation) and Comprehensive Community Programmes (to apply corporate management techniques to urban problems). Official recognition that tackling the root causes of urban deprivation would require more than marginal adjustments to existing social policies was also reflected in a major revision of urban policy marked by the 1977 White Paper on the inner cities (HMSO 1977). One of the most important elements of the White Paper was recognition that urban decline and poverty had structural causes located in economic, social and political relations that originated outside the areas concerned. This signalled more broadly based action on urban problems combining economic, social and environmental programmes and involving new organizational arrangements between central and local government.

The 1977 White Paper identified four basic components of the inner city problem.

1 Economic decline associated with the contracting industrial base due to recession in the UK economy; de-industrialization and the run-down of

traditional inner city services and industries (e.g. warehousing and dock-related activities); the closure of branch plants, often of multinational companies, and associated ripple effects on the local economy resulting in the failure of small dependent firms; failure to attract new industry (due to a series of disadvantages including the high cost of industrial land, local rates, shortage of suitable premises, problems of access and limited opportunities for expansion); and labour constraints, including a shortage of female workers.

2 The decaying condition of the physical environment, including housing, as a result of age, lack of investment, and planning blight (due to the stop–go nature of public expenditure which affected the continuity of local authority development programmes).

3 Social disadvantage, including high levels of unemployment and low wages due to the nature of available work; exclusion of sizeable proportions of the population from the labour force because of age or infirmity; and a pervasive sense of decay and neglect which diminishes community spirit assailed by crime and vandalism.

4 The concentration of ethnic minorities in parts of the inner city which may result in discrimination in job and housing markets and engender community and racial tensions, particularly in times of economic hardship.

The Labour government's response to the 1977 White Paper was the passage of the Inner Urban Areas Act of 1978, the main element in which was the creation of seven partnerships between central and local government in an attempt to harness private capital for urban economic revival. The emphasis on improving the economic environment of cities was also promoted by a shift of policy emphasis from the New Town programme to urban regeneration (with, for example, in Scotland the abandonment of Stonehouse New Town and the initiation of the Glasgow Eastern Areas Renewal project), and increased powers to enable local authorities to aid and attract industrial developments. The main vehicle for these measures was the expanded Urban Programme.

In practice, a wide-ranging national and local attack on urban problems failed to materialize. While the diagnosis implied long-term commitment of substantial financial resources to overcome the deep-rooted problems of the inner cities, doubts about the wisdom of 'excessive' public expenditure had taken root in the polity and the Treasury was wary of any significant long-term financial commitment. As a result, the rate support grant provided little additional resources to hard-pressed urban authorities at a time when the government's wider economic strategy meant that local authority spending was actually reduced. In this climate of retrenchment it proved difficult for urban local authorities to 'bend' their own main programmes towards the inner areas as they had been encouraged to do (Gibson and Langstaff 1982). A second difficulty for urban policy was represented by the growing strength of idea that state action was an inefficient, ineffective and uneconomic method of dealing with the urban problem, and that the private sector offered a better solution (Atkinson and Moon 1994). Thus in terms of the level of resources provided on the ground, the outcome of the 1977 White Paper was limited (Lawless 1986). The main value of the White Paper lay in its identification of the structural underpinnings of the inner city problem and its, albeit tentative, acknowledgement of the potential role of the private sector in urban renewal.

The shift to the Right

The election of a Conservative government under Margaret Thatcher in 1979 did not signal a clean break with past urban policy. Some of the initiatives begun by the Labour government in the late 1970s were retained (e.g. the partnership arrangements of the Inner Urban Areas Act), albeit with some streamlining and with a much greater role for the private sector within the partnerships (Stewart 1987). Nevertheless the change of government did represent a watershed in British urban policy. Since 1979 successive Conservative governments have sought to reduce central government involvement in urban regeneration and to shift the policy emphasis from the public to the private sector. The main role envisaged for the public sector is to attract and accommodate the requirements of private investors without unduly influencing their development decisions. This perspective underlay a number of new initiatives introduced by the Local Government Planning and Land Act of 1980, the most significant of which for urban policy were the concepts of enterprise zones and urban development corporations. The consistent belief of the government in the power of the private sector to initiate urban regeneration was also evident during the 1980s in a range of other schemes which included Derelict Land Grants, Urban Development Grants, Inner City Enterprise, Business in the Community, the Financial Institutions Group, City Action Teams, Urban Regeneration Grant, Task Forces, Simplified Planning Zones, City Technology Colleges, British Urban Development, Housing Action Trusts, Estate Action, Training and Enterprise Councils and Local Development Companies (Figure 1.1). The strength of the government's faith in the regenerative power of the private sector was voiced by the Prime Minister, who explained that 'one of the difficulties about inner cities is that some councils are positively hostile to the private sector which could solve their problems' (Sills *et al.* 1988: 63).

The reorientation of urban policy by the New Right Conservative government was part of a wider agenda to restructure Britain economically, socially, spatially and ideologically around a new consensus of free market individualism and an unequivocal rejection of the social democratic consensus of the post-war Keynesian Welfare State. As Martin (1988) observed, the thrust of state policy shifted from welfare to enterprise.

> The aim has been to reverse the post-war drift towards collectivism and creeping corporatism, to redefine the role and extent of state intervention in the economy, to curb the power of organised labour, and to release the natural, self-regenerative power of competitive market forces in order to revive private capitalism, economic growth and accumulation.
>
> (Martin 1988: 221)

The Keynesian commitment to the macro-economic goal of full employment was replaced by the objective of controlling inflation by means of restrictive monetary measures and supply-side flexibilization. From its inception, 'Thatcherism' has been a doctrine for remodernizing Britain's economy by exposing its industries, its cities and its people to the rigours of international competition in the belief that this would promote the shift of resources out of inefficient 'lame-duck' traditional industries and processes into new, more flexible and competitive high-technology sectors, production methods and work practices (Martin 1986). The principal mechanisms for achieving this transformation centred on tax cuts and deficit spending, deregulation and privatization, all of which had *geographically uneven* impacts. At the urban level these three macro-economic strategies are combined most strikingly in the concept of the enterprise zone.

Enterprise zones have formed a major plank in the Conservative government's strategy to remove state regulations on private capital investment and entrepreneurial activity. The underlying assumption is that by encouraging companies to develop derelict urban sites 'a boost will be given to the entire local economy, leading to jobs and opportunities for the nearby residents' (Butler 1982: 108). In practice the efficiency of the 'trickle down' mechanism has been less than predicted (Robson 1987) and despite advantages for capital, in particular the tax exemptions, EZs have not succeeded in engineering a fundamental revival of local economic activity. There is, in addition, evidence that the greatest gains have been made by landowners and property developers within the EZ with no guarantee that these are passed on to employers leasing property. A more fundamental criticism is that the powers granted to EZs are not sufficient to achieve either economic growth or employment creation (Lawless 1986). It can be argued that if economic growth was the main objective this would require the selection of more favourable areas, the expansion of newer industrial sectors, additional financial support and a more thorough dismantling of controls. Alternatively, pursuing a principal goal of employment creation would necessitate control over incoming companies to ensure they were leading to a net increase in employment, were job intensive, and were assisting the less skilled who form the majority of the urban employed. Paradoxically, the most successful EZs (such as Corby and Clydebank) have proved not to be those where the public sector has withdrawn (as envisaged in the initial EZ model) but those where

Plate 1.2 Welcome to the Metro Centre. Opened in 1986, 'Europe's largest shopping and leisure city' is a symbol of the market-led development promoted by the relaxation of planning controls over large-scale decentralization and the financial benefits of location within the Gateshead Enterprise Zone

a single public agency controlled much of the land and could effect an integrated development strategy (Lawless 1988).

In a review of urban policy measures introduced during the 1980s the Audit Commission (1989: 1) found that local authorities and business regarded inner city policy as 'a patchwork quilt of complexity and idiosyncrasy' and described urban initiatives as lacking co-ordination and having no sense of strategic direction, with programmes frequently operating in isolation and often in competition with one another. This supported Spencer's conclusion that there is 'no national urban policy in any meaningful sense – there are individual programmes, but no overarching policy framework' (Spencer 1987: 9). The government's response was to establish ten regional offices and combine the resources from twenty different programmes into a Single Regeneration Budget from 1994. The introduction of the Single Regeneration Budget, however, masked a reduction of almost £300 million in the funds available from £1.6 billion for SRB equivalent programmes in 1992–3 to £1.3 billion in 1996–7. This policy decision is in keeping with the government's goal of reducing public expenditure and privatization of urban redevelopment. As a result 'the amount of money going into urban policy is minuscule compared to the size of the problems which are being tackled, and the loss of mainstream money in many authorities has more than countered any increase in urban funds' (Bradford and Robson 1995: 53). In view of the uneven geography of public expenditure cuts, cynics might observe that electorally the Conservative government had little to gain in the run-down areas of Britain's cities. Government attention has also appeared to peak at times of possible legitimation crisis, as in the urban riots of the 1980s (Benyon and Solomos 1987). More positive assessments, however, see the Conservative's approach to the urban crisis as 'one component of a comprehensive strategy to regenerate Britain's ailing economy by creating an enterprise culture' (Parkinson and Judd 1988).

Justification for the market-led approach to urban regeneration centres on the concept of 'trickle down' which argues that, in the longer term, an expanded city revenue base created by central area revitalization provides funds that can be used to address social needs. In practice, however, these funds are generally recycled into further development activity. Nowhere has a substantial 'trickle down' effect been demonstrated. The available evidence suggests that urban revitalization through 'capitalism with a social conscience' is a chimera. It is naive to expect a 'morally aware' private sector to effect the revitalization of run-down estates. Private sector investment decisions are founded largely upon self-interest and not philanthropy. The privatization of urban development inevitably means accepting a policy of triage and concentrating on areas of greatest economic potential – with adverse consequences for other areas. As Pacione (1993) concludes, to address the problems of poverty and deprivation, urban policy must possess both a social and an economic perspective. It must be concerned as much about the *distribution* of wealth as about wealth *creation*. Such social criteria fit awkwardly into the conventional balance sheet of economic progress.

Local economic strategies

The divergence between economic and social goals is most striking where central and local authorities embrace different political ideologies. For most of the 1980s the Conservative central government sought to by-pass local authorities, particularly those under Labour control that did not share the government's views on how to tackle the urban crisis. Many local authorities did not accept this situation passively but instead attempted to formulate alternative development strategies more attuned to the needs of local communities and less compliant with the demands of capital (Cochrane 1986). These radical local economic initiatives sought to moderate the impact of uneven development on depressed urban regions by focusing attention on the social costs associated with the unfettered ability of corporate capital institutions to move investments between global locations in search of maximum profit. Local strategies have included equity financing of local firms, negotiation of planning agreements between local authorities and companies, the identification and promotion of 'socially-useful' production, as well as a direct role in the economy for the local authority which is often a major employer and purchaser.

Particular attention has been given to the creation of local enterprise boards to generate an investment programme that produces long-term local benefits. Although their impact in relation to the overall scale of the urban problem has been marginal, in terms of cost per job to the public exchequer they rate as one of the most cost-effective agencies of urban regeneration (Hasluck 1987). The most wide-ranging strategy of local economic intervention was formulated by the former Greater London Council (Boddy and Fudge 1984). However, the scale of the problems affecting London's deprived areas and tension between the economic and social objectives of the strategy meant that even this comprehensive initiative was incapable of combating the adverse structural trends affecting the urban economy. Following abolition of the metropolitan counties in 1985, some enterprise boards succeeded in attracting institutional funding to replace the Section 137 funds lost; but the strategy has not received the support of central government and faces an uncertain future (Cochrane and Clarke 1990). The community business is another form of local economic initiative which stems from recognition of the fact that market forces are largely impotent in the most deprived areas. This is essentially an organization owned and controlled by the local community which aims to create ultimately self-supporting jobs for local people. Again, however, the economic impact of community businesses has been marginal to the scale of the problem facing depressed communities. In addition, the jobs created tend to be low level and semi-skilled, although there are multiplier effects including enhanced local spending power and some reduction of the pool of long-term unemployed. Other basic difficulties for local economic strategies include conflict between the aim of ensuring that jobs created are 'good jobs' (with proper rates of pay and conditions) and the exigencies of the capitalist market place; doubts over the possibility of building an alternative economic strategy based on the rescue of failing capitalist firms; and the fact that strategies based on the funding of co-operatives and small community businesses inevitably encounter problems that affect the small firm sector overall. Although local economic initiatives are innovative and wide-ranging they alone cannot resolve the problems confronting disadvantaged city dwellers in the face of contrary tendencies originating from the central state and from capital itself. As presently constituted, radical local economic policies are of most value as ideological and political instruments, demonstrating that there are alternatives to the economic strategies of the New Right.

From the bottom up

The failure of 'top-down' strategies to service the needs of the disadvantaged within the urban arena has provoked interest in an alternative grassroots perspective. A large number of 'bottom-up' or locality-based initiatives have been applied, with varying degrees of success, in the attempt to resolve the economic, social and environmental problems confronting disadvantaged populations and places within the capitalist city. Examples include class-based and populist urban social movements (Fainstein 1987), pragmatic radicalism (Kraushaar 1979), greenlining, i.e. positive discrimination as opposed to redlining of areas (Smith 1980) and popular-democratic political opposition to business coalitions (Jacobs 1982), as well as local church involvement in social issues (Pacione 1990a).

As state responsibilities for public well-being are redefined through the mechanism of public sector retrenchment and privatization of welfare services, the *voluntary sector* has assumed increased importance (Wolch 1989). This trend is open to different interpretations. Those on the political Right claim that voluntary groups are cornerstones of democracy that counteract the expansion of an unresponsive state increasingly beyond popular control. Some explicitly stress their importance in promoting self-sufficiency and individual initiative and thus the capitalist system. More pragmatic reasons for supporting the voluntary sector include the economic argument that it can provide services more cheaply than government, and the political expedient that funding a variety of voluntary groups helps secure the support of affected groups. These sentiments underlie the social and economic policies of the New Right (Webb and Wistow 1987). Somewhat paradoxically, support for voluntarism also emanates from the political Left, who highlight the potential advantages of grassroots participation, local democracy, alternative economic strategies and greater self-determination, especially for the disadvantaged (Donnison 1984). The paradox is explained by different perceptions of the level of power to be devolved to the local level. The growth of the voluntary sector may be interpreted as a deliberate strategy of social control on the part of the state whereby limited access to the decision-making process is used to defuse dissent or co-opt opposition. On the other hand, there is ample evidence of voluntary groups playing a leading role in the struggle for progressive social change in the city (Lowe 1986).

In the UK there is a long history of grassroots protest by the poor in response to the inequities of the capitalist system. Examples of protest movements include tenants associations opposed to council house rent increases (Baldock 1982) and squatter movements seeking to effect a more equitable use of housing (Wates and Wolmar 1980). Two principal conclusions may be drawn from analysis of successful grassroots social movements. The first refers to the 'insider–outsider' debate over the most effective means for disadvantaged groups to acquire a voice in the political process. As Pacione (1988) has shown, most formal attempts at municipal democratization have redistributed a level of influence rather than power (i.e. control over decision-making). The limited ability of urban interest groups to exert a meaningful influence on decision-making through formal channels of participation (i.e.

representative democracy and government-mandated citizen participation) and the dangers of co-optation have persuaded some analysts of the benefits of *adversarial participation.* Among the advantages of operating outside the formal system are the strengthening of group solidarity and an ability to gain concessions from governments through fear, sympathy or successful mobilization of public groups to which elites are normally attentive. Direct action to obtain decisions favourable to disadvantaged groups and neighbourhoods is often necessary because 'no one gives up power to others unless he no longer needs it, can no longer sustain it for personal reasons, or is forced to do so' (Reidel 1972: 219). The second conclusion is that in the unequal struggle between spatially mobile capital or remote governments and locally rooted communities, the probability of success is highest for amalgamations of single-issue interest groups such as trade unions, senior citizens, environmental and women's organizations. Coalition can overcome many of the restrictions that are a function of the small size of most neighbourhood organizations. These include narrowness of issues, part-time leadership, cross-cutting cleavages, individual geographic immobility and limited financial resources. Neighbourhood social groups must avoid the pitfalls of parochialism and must be constantly aware of the regional and national economic forces that shape their opportunity space. Forming alliances without surrendering identity and penetrating the political system while reserving the right of direct action are among the major requirements for a shift from grassroots pressure to grassroots power in the shaping of urban policy.

THE PLANNING CONTEXT

A nationally organized planning system emerged in the UK as part of the post-war consensus between the major political parties. There was general agreement that an increased role for the state was necessary for the reconstruction of society, to avoid a return to the Depression conditions of the inter-war years, to resolve conflicts between competing land uses and to provide for urban and regional redevelopment. The co-existence of physical and social goals helped the proposals achieve popular acceptance for a system of planning 'in the public interest'. In practice, until the late 1960s land use planning dominated, with social planning relegated to a subsidiary role (Cullingworth and Nadin 1994). In Britain's major cities war damage and the results of the virtually unconstrained operation of nineteenth-century industrial capitalism had generated a host of problems that included an insufficient and sub-standard housing stock, concentrations of decaying industry, and inadequate and congested transport infrastructure. The main physical planning solutions centred on the dispersal of 'surplus' population to suburban estates and New Towns, comprehensive redevelopment and major road-building programmes.

The social inadequacies of land use planning were reflected in growing popular reaction against the insensitivity of slum clearance, the disruptive impact of new urban motorways on local communities, and the 'rediscovery of poverty'. New urban problems, including the growing numbers and spatial concentration of black and Asian peoples in Britain's cities, posed a major challenge for planning. By the late 1960s it was apparent that urban problems could not be resolved by physical planning alone. The credibility of planning was also undermined by the economic difficulties of the early 1970s when British capitalism entered a period of severe structural crisis or 'stagflation' in which *both* unemployment and inflation were rising. As we have seen, government efforts to manage the situation compounded the problem and led to a situation of over one million unemployed (high at that time) and inflation of almost 25 per cent. In this context the difficulty of tackling pressing social problems was compounded by a need to cut public expenditure and adjudicate between the claims of different groups and areas. The government's failure to defeat the crisis not only eroded the economic bedrock on which the post-war consensus rested but led to a loss of faith in established modes of decision-making, including the planning system. Planning was also criticized for its technocratic elitist approach and insensitivity to public participation.

The principles and practice of planning have come under attack from both Left and Right of the political spectrum. The value of planning has been dismissed by the far Left who regard it as a state apparatus attuned to the needs of capital and designed to maintain the unequal distribution of power in society. For some critics on the Right, planning is seen as a major *cause* of inner city decline and social unrest through its policies of clearance and decentralization, rigid land use zoning and imposition of standards. According to Steen (1981), because of planning, resources are fruitlessly channelled into deprived areas and wasted rather than encouraging wealth creation. For others the chief problems of planning lie in its interference with the market and in the fact that, contrary to the goals of planning, a dynamic and prosperous urban economy requires *inefficiency* in its structure and land use in order to permit innovation and experimentation (Jacobs 1970).

The perceived failings of planning provided ammunition for critics of the corporatist state and fuelled a sustained attack on the post-war consensus. This was articulated most forcibly by the Conservatives under Margaret Thatcher. Thatcherism emphasizes the virtues of individual liberty, the free market, entrepreneurial spirit and a minimalist state (although, paradoxically, this called for a centralization of power in the short term in order to reduce government in the long term). In line with an enhanced role for the private sector in urban regeneration, the right-wing Adam Smith Institute (1983) advocated the dismantling of the planning system, retention of a limited amount of centrally administered legislation relating to conservation issues, and the switching of many planning controls into the legal arena by using property law to resolve land use conflicts. In practice, the changes introduced since 1979 have stopped short of abolishing the planning system but rather

Plate 1.3 The 1.5-mile long Byker wall forms the northern boundary of a slum clearance scheme in inner Newcastle upon Tyne. The five-storey high density development is designed to shelter the interior layout of terraced housing from the noise of traffic

have restructured the scope of planning. Adhering to the principles of economic liberalism, the importance of plans and policies has been downgraded to become only one 'material consideration' alongside market pressure and demand in determining urban change. Decision-making procedures have also been streamlined, at the expense of public participation, primarily through a process of 'authoritarian decentralism' whereby decision-making power has been centralized from local authorities then redeployed to the market-place. This has been manifested in a more interventionist role for the Secretary of State, increased use of Circulars, financial controls over local government, inclusion of more reserve powers in legislation and curtailment of bureaucratic discretion.

The changes to planning can be seen as a reorientation of the *purpose* of planning towards market criteria, selective application of environmental criteria and removal of social criteria. This has been accompanied by reorientation of the *procedures* of planning away from community-based local democracy towards centralized government supervision. For Thornley (1991) the principle of a national planning system has been replaced by a disaggregated system comprising areas in which environment is important and planning controls are strong, areas where economic criteria dominate but decisions are still made with the albeit much modified procedures of local democracy, and areas where economic criteria dominate and decisions are made outside the framework of local democracy. In general the shift to the Right in British politics marked a move from market-critical planning (aimed at redressing imbalances and inequalities created by the market) to market-led planning (aimed at correcting inefficiencies while supporting market processes).

THE SOCIAL CONTEXT

A 'welfare state' is a state in which organized power is used deliberately through politics and administration in an effort to modify the operation of market forces in at least three ways. First, by guaranteeing individuals and families a minimum income irrespective of the market value of their work or property; second, by narrowing the extent of insecurity by enabling individuals and families to meet certain social obligations; and third, by ensuring that all citizens without distinction of status or class are offered the best standards available in relation to an agreed range of social services (Briggs 1961).

The British welfare state was the product of a long movement for social reform that began in the last quarter of the nineteenth century, was stimulated by the ideas of Keynes and wartime consensus, and culminated in the institutional structure established by the Labour government elected in 1945. The concepts of a mixed economy and a welfare state received general support from both Labour and Conservative governments in a 'Butskellite' post-war settlement of major ideological differences. Thus when Labour was defeated in 1951, the incoming Conservative government retained the institutional basis of the post-war settlement and the commitment to the welfare state and full employment. This 'end of ideology' era lasted until the 1970s when it was destroyed by the emergence of the New Right. A key factor underlying cross-party support for the welfare state was that the social policies of both Labour and Conservative governments were linked to the health of the national economy. Social expenditure was seen as necessary to engender the stability needed for the growth of the economy while, simultaneously, economic growth was required to support the additional expenditure. The corollary is that economic recession would pose difficulties for maintenance of high levels of welfare spending. As we have seen, during the 1960s the post-war boom began to slacken and the structural weaknesses of the British economy reasserted themselves. The emerging fiscal crisis necessitated reductions in state expenditure that undermined the rationale of the Keynesian welfare state and the social-democratic consensus on which it was based.

The limitations of the welfare state were also exposed by the rediscovery of poverty in the late 1960s which revealed that, despite the general affluence of the post-war years, significant groups were severely disadvantaged. Growing popular dissatisfaction with the ability of the state to combat poverty and social inequality contributed to the defeat of Harold Wilson's Labour government in 1970. By the time of the next Labour government in 1974 the effects of the 1973 oil price shock and the fundamental weaknesses of the British economy continued to impede economic growth. In response, the government sought to capitalize on its 'special relationship' with the trades unions to negotiate a voluntary wage restraint in return for policy concessions (such as repeal of the Heath government's industrial relations act). In the arena of social welfare policies – designed to form part of the 'social wage' of workers – the fiscal crisis posed major difficulties in responding to union demands, given that levels of expenditure on social services were reduced consistently after 1975. By 1978/9 non-defence spending was 8.5 per cent below its level of 1975/6. These major cuts in social welfare spending indicated the government's acceptance of a form of monetarism, seen in restriction of the Public Sector Borrowing Requirement, and a move away from the orthodoxy of Keynesian demand management. This contributed to the TUC and individual union leadership's problems of delivering their side of the bargain. The collapse of the Social Contract and widespread strike action, particularly in the public sector, during the 'winter of discontent' of 1978/9 contributed to the fall of the Labour government and the return of the Conservatives under Margaret Thatcher. Under Thatcherism the bulwarks of government economic policy were *monetarism* and *supply-side* economics. The former was based on the belief that control of the money supply, including bank credit and government borrowing and spending, can check inflation. The latter required a retreat from the corporatist state and reduction in 'unnecessary' bureaucratic intervention in the free market. In terms of social welfare policy, this philosophy saw little role for government in redistributing resources. Indeed Nozick (1974) considered welfare transfers as essentially immoral impositions on the individual's right to distribute the rewards of their own labour. Taxation as a means of financing state activities should be kept to a minimum. In the view of the New Right the removal of state interference with the market mechanism allows full pursuit of such natural instincts as individual initiative, the acceptance of inequality and adoption of self-help (Joseph and Sumption 1979).

As well as job losses in local government, which had been responsible for provision of many services, the stringent public expenditure regime under Thatcherism imposed severe constraints on local autonomy. These were seen in the practice of 'rate-capping', in the use of enterprise zones and urban development corporations to marginalize local authorities and, most directly, in the abolition of the metropolitan counties in 1985. Between 1978/9 and 1982/3, whereas central government expenditure increased by 10.7 per cent in real terms, local authority spending fell by 3.6 per cent. In order to legitimize public expenditure reductions the government sought to diminish public support for the apparatus of the post-war social-democratic welfare state. Particular attention was focused on the escalating costs of social expenditure, its effect on the tax burden, and on welfare fraud by the 'undeserving poor'. This brought the interests of taxpayers into direct conflict with the needs of the poor. Since the early 1980s welfare benefit levels have been linked to prices rather than real incomes, thereby increasing the gap between the incomes of those dependent on benefits and those of the working population. Together with the shift of the tax burden from direct to indirect taxation, this represented an attempt to produce a new benefits system that 'did not undermine the incentive to work' (Gaffikin and Morrissey 1994: 541). For those unable to compete in the arena of market capitalism the holes in the safety net grew wider (Sills *et al.* 1988, Pacione 1989, Keith and Rogers 1991). Support for the ethics and policies of the New Right carried Mrs Thatcher to three successive election victories, although significantly, in geographical terms, the Conservative vote was much weaker in northern Britain and in the inner cities. In view of the particular goals of government economic and social policy since 1979 it is not surprising that 'the gap between conditions and opportunities in deprived areas and other kinds of place . . . remains as wide as it was a decade and a half ago (and) in some respects the gap has widened' (Willmott and Hutchinson 1992: 82). In the next section, we will examine several key dimensions of this socio-spatial divide in more detail.

THE GEOGRAPHY OF URBAN CRISIS

The problems of poverty and deprivation experienced by those people and places marginal to the capitalist development process have intensified over the last decade and a half (Lansley and Mack 1991, Institute for Fiscal Studies 1994). During the 1980s poverty increased faster in the UK than in any other member state of the EC so that by the end of the decade one in four of all poor families in the Community lived in Britain (O'Higgens and Jenkins 1990). Of equal concern is the fact that since 1979 the gap between rich and poor has widened (Figure 1.2) and currently over 90 per cent of the nation's wealth is owned by the richest half of the population. One in three of the poorest group is unemployed, 70 per cent of the income of poor households comes from social security payments and nearly one in five are single parents.

A substantial proportion of the disadvantaged live in Britain's cities, large areas of which have been economically and socially devastated by the effects of global economic restructuring, the de-industrialization of the UK economy and ineffective urban policies (Robson 1988, Pacione 1990b). Older industrial cities have been affected most severely by the transition to advanced capitalism while market-led urban policies have only indirectly addressed the difficulties experienced by those locations which, although remaining meaningful places to their inhabitants, are not considered profitable spaces by capital (Hudson 1988).

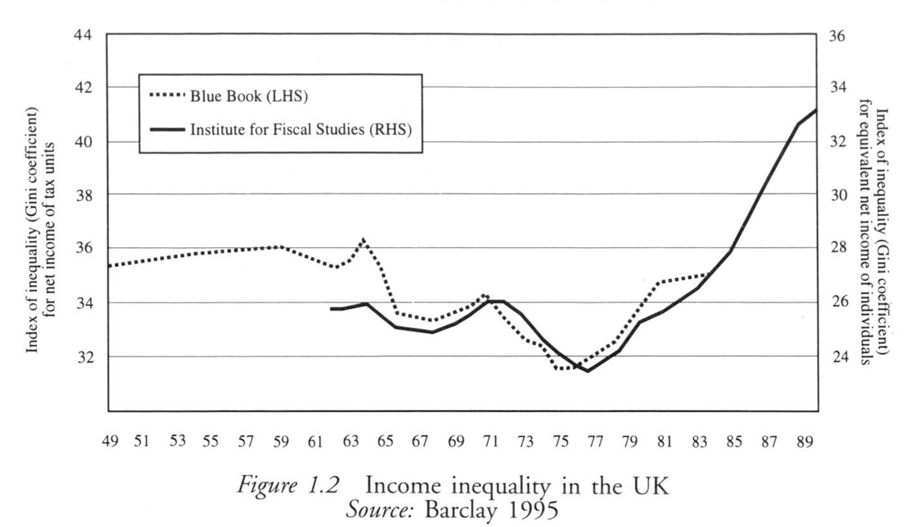

Figure 1.2 Income inequality in the UK
Source: Barclay 1995

The urban crisis is a multidimensional concept that involves a host of interrelated forms of disadvantage (Figure 1.3). The term *multiple deprivation* describes this situation in which a range of social, economic and other problems reinforce one another to create an environment of compound disadvantage for those affected (Figure 1.4).

Poverty and deprivation

A basic issue underlying the debate over the extent of poverty in Britain is the distinction between absolute and relative poverty. The absolutist definition of poverty derived from that formulated by Rowntree (1901: 186) contends that a family would be living in poverty if its 'total earnings are insufficient to obtain the minimum necessaries for the maintenance of merely physical efficiency'. This notion of a minimum level of subsistence, and the related concept of a poverty line, strongly influenced the development of social welfare legislation in post-war Britain. The system of national assistance benefits introduced following the Beveridge Report (1942) was based on calculations of the amount required to satisfy the basic needs of food, clothing and housing plus a small amount for other expenses. If, however, we accept that needs are culturally or socially determined rather than biologically fixed, then poverty is more accurately viewed as a relative phenomenon. The broader definition of needs inherent in the concept of relative poverty includes job security, work satisfaction, fringe benefits such as pension rights plus various components of the 'social wage' including use of and access to public services,

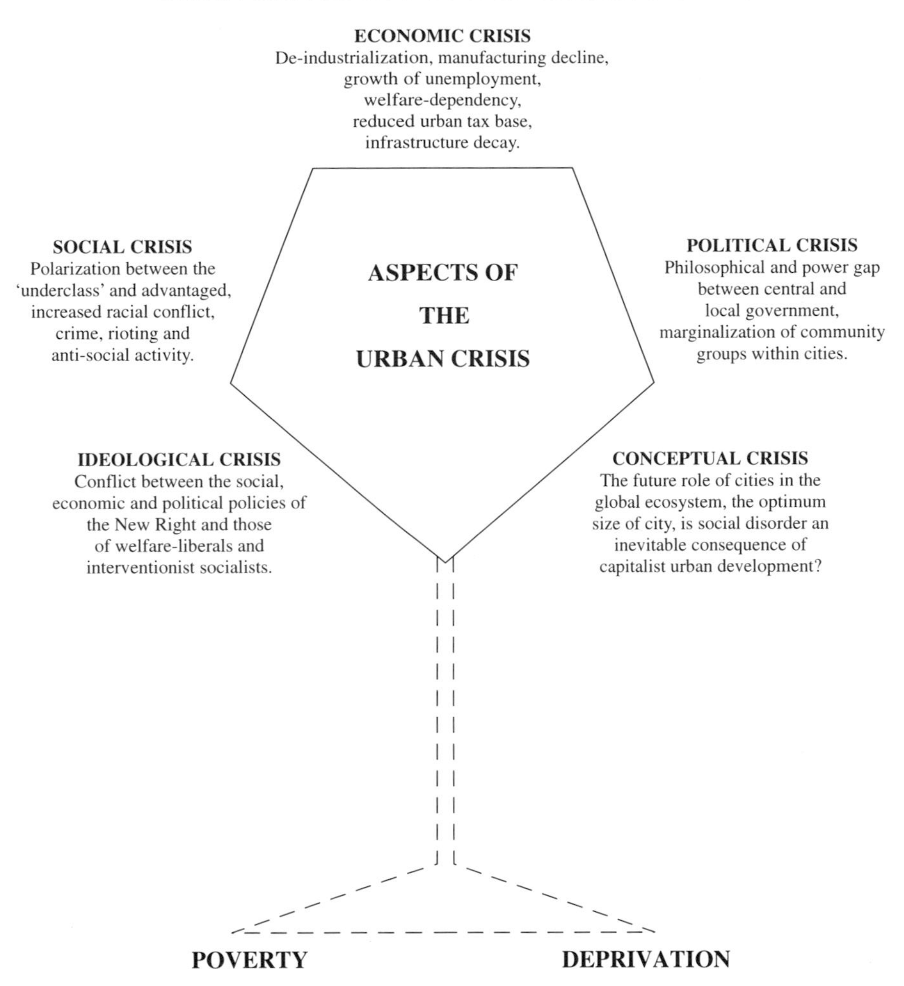

Figure 1.3 Aspects of the urban crisis

as well as satisfaction of higher-order needs such as status, power and self-esteem. This view of poverty is summarized by Townsend who states that

> individuals, families and groups in the population can be said to be in poverty when they lack the resources to obtain the types of diet, participate in the activities and have the living conditions and amenities which are customary, or at least widely encouraged or approved, in societies to which they belong. Their resources are so seriously below those commanded by the average individual or family that they are, in effect, excluded from ordinary living patterns, customs and activities.
> (Townsend 1979: 31)

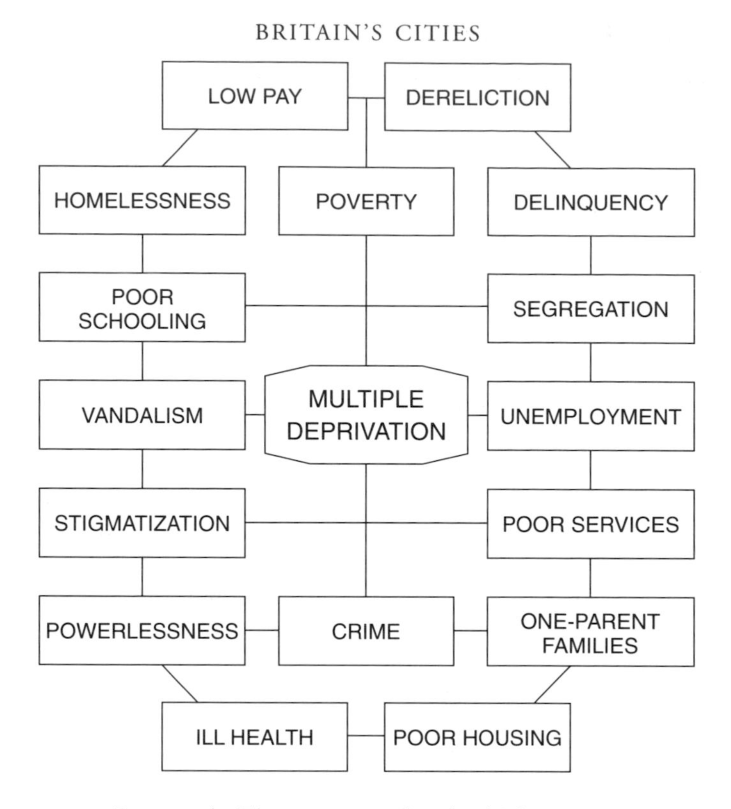

Figure 1.4 The anatomy of multiple deprivation
Source: Pacione 1995b

The absolutist perspective carries with it the implication that poverty can be eliminated in an economically advanced society, while the relativist view accepts that the poor are always with us.

Poverty is a key factor underlying multiple deprivation. The root cause of deprivation is economic and stems from two main sources (Thake and Staubach 1993). The first arises due to the low wages earned by those employed in declining traditional industries or engaged, often on a part-time basis, in newer service-based industries. The second cause is the unemployment experienced by those marginal to the job market such as single parents, the elderly, disabled and, increasingly, never-employed school-leavers. Since the 1960s, when poverty was largely age-related, the increasing number of unemployed and growing pool of economically inactive families (such as lone parents and long-term sick) have displaced pensioners as the poorest in society. There is also evidence of a growing 'feminization of poverty' since, in addition to an increase in the numbers of female-headed single families, women have in general continued to occupy the less advantaged positions within the workforce, with less security, lower pay and poorer levels of social protection.

The spatial concentration of poverty-related problems, such as crime, delinquency, poor housing, unemployment and increased mortality and morbidity, serves to accentuate the effects of poverty and deprivation for particular localities. Neighbourhood unemployment levels of three times the national average are common in economically deprived communities, with male unemployment rates frequently in excess of 40 per cent (Donnison and Middleton 1987). Lack of job opportunities leads to a dependence on public support systems. The shift from heavy industrial employment to service-oriented activities in major urban regions, and the consequent demand for a different kind of labour force, has also served to undermine long-standing social structures built around full-time male employment and has contributed to social stress within families. Dependence upon social welfare and lack of disposable income lowers self-esteem and can lead to clinical depression. Poverty also restricts diet and accentuates poor health. Infant mortality rates are often higher in deprived areas and children brought up in such environments are more likely to be exposed to criminal sub-cultures and to suffer educational disadvantage. The physical environment in deprived areas is typically bleak, with extensive areas of dereliction, and shopping and leisure facilities that reflect the poverty of the area. Residents are often the victims of stigmatization which operates as an additional obstacle to obtaining employment or credit facilities. Many deprived areas are also socially and physically isolated and those who are able to move away do so, leaving behind a residual population with limited control over their quality of life. Although support groups, community organizations and pressure groups can engender social cohesion and make tangible gains, the scale and structural underpinnings of the problem of multiple deprivation generally precludes a community-based resolution of the difficulties facing such localities.

Examination of a number of deprivation-related indicators (e.g. unemployment, inactivity rate and car-less households) from the 1991 census provides a striking indication of the contrasting geographies of poverty and affluence among local authority districts in Britain. Compare, for example, the positions of Glasgow district and the spatially contiguous suburban district of Bearsden and Milngavie (Tables 1.5 and 1.6). Green (1994) has also shown that in terms of unemployment and inactivity rates the gap between the best and worst wards in the UK increased between 1981 and 1991 to accentuate an already significant degree of polarization. Significantly, poverty was found to exhibit a consistent spatial correlation over the decade. This conclusion was also demonstrated at the conurbation level by Pacione's (1995b) analysis of multiple deprivation in Strathclyde which revealed the spatial and temporal consistency of deprivation in specific parts of the main urban areas. There is also ample evidence of the differential incidence of poverty and deprivation within cities. Evidence from Scottish cities has

Table 1.5 Most deprived local authority districts in Britain, 1991

Unemployment (%)		*Inactivity rate (20–59)(%)*		*Households owning no car (%)*	
1 Hackney	22.5	1 Rhondda	32.3	1 Glasgow City	65.6
2 Knowsley	22.1	2 Easington	31.2	2 Hackney	61.7
3 Tower Hamlets	21.8	3 Merthyr Tydfil	30.4	3 Tower Hamlets	61.6
4 Liverpool	21.1	4 Afan	30.2	4 Islington	59.9
5 Newham	19.3	5 Rhymney Valley	29.3	5 Clydebank	58.7
6 Glasgow City	19.1	6 Blaenau Gwent	29.3	6 Southwark	58.0
7 Manchester	18.7	7 Cynon Valley	29.2	7 Westminster	57.7
8 Southwark	18.2	8 Knowsley	28.2	8 Liverpool	57.0
9 Haringey	17.7	9 Cumnock and Doon Valley	28.0	9 Manchester	56.6
10 Lambeth	17.1	10 Glasgow City	27.8	10 Camden	55.8

Source: Hills 1995

Table 1.6 Most affluent local authority districts in Britain, 1991

Households from social classes 1 and 2 (%)		*Adults with high level qualifications (%)*		*Households with two or more cars (%)*	
1 City of London	71.4	1 City of London	35.7	1 Surrey Heath	51.0
2 Richmond upon Thames	63.3	2 Bearsden and Milngavie	31.4	2 Hart	50.5
3 Bearsden and Milngavie	63.0	3 Richmond upon Thames	29.2	3 Wokingham	48.5
4 Eastwood	62.3	4 Cambridge	27.9	4 Chiltern	47.7
5 Elmbridge	60.5	5 Kensington and Chelsea	27.6	5 South Buckinghamshire	47.3
6 Kensington and Chelsea	60.3	6 Eastwood	26.5	6 Uttlesford	44.4
7 Wokingham	59.8	7 Oxford	26.2	7 Tandridge	43.1
8 Chiltern	59.7	8 Camden	25.8	8 Wycombe	42.8
9 St Albans	59.5	9 St Albans	25.7	9 Mole Valley	42.7
10 Mole Valley	57.3	10 Chiltern	24.0	10 East Hampshire	42.3

Source: Hills 1995

revealed that while the relative intensity of multiple deprivation has shown some spatial change between 1971 and 1991 with, in general, a centrifugal shift from the inner city to the outer city, in social terms the problems have remained concentrated in the private-rented and, in particular, council-rented housing areas (Pacione 1986, 1987, 1989, 1993). The findings of Noble *et al.* (1994) in Oldham also identify the importance of housing tenure as an indicator of multiple deprivation. Nationally, since 1970 council housing has become occupied predominantly by those with low incomes. Whereas in 1961 and 1971 fewer than half of those in council housing were in the poorest 40 per cent by 1981 this had increased to 57 per cent and by 1991 to 75 per cent (Hills 1995).

There is also increasing evidence of a strong relationship between poverty and deprivation and other facets of the urban crisis. In the following sections we illustrate this with particular reference to recent findings related to crime, health and ethnic status.

Crime

The emergence of mass unemployment in Britain since the mid-1970s has coincided with an increase in recorded crime. The British Crime Survey confirmed that crime grew by 49 per cent, or 4.1 per cent per annum, between 1981 and 1991. This rise was dominated by property crime which increased by 95 per cent over the period. The nature of the relationship between property crime and fluctuations in the business cycle is shown graphically in Figures 1.5 and 1.6. Figure 1.5 reveals several clear trends for England and Wales. The first is a strong upward trend in property crime as Britain leaves full employment behind and enters an era of mass unemployment. Second, and most striking, property crime, though rising strongly during the period, fell during each of the upturns in economic activity – during the 'Barber boom' (1972–3), the partial economic recovery of 1978–9 at the end of the last Labour government, the most intense phase of the 'Lawson boom' (1988–9) and during 1993 in line with another economic upswing. Conversely, recorded property crime rose during each of the downturns in economic activity – during the recession in the mid-1970s following the OPEC price shock, during the early 1980s Thatcher–Howe recession (due to the twin economic shocks of monetary and fiscal deflation and the real sterling exchange rate appreciation), and during the early 1990s Major recession. As Figure 1.6 shows, time-series data for housebreaking in Scotland also exhibit the same sensitivity to the business cycle. The association between property crime and unemployment is supported by records from police force areas in England and Wales which show that 'the unemployment black spots in the country (Cleveland, Merseyside, Northumbria, Greater Manchester, South Yorkshire, West Midlands, Greater London and South Wales) are also crime black spots' (Wells 1994: 33), and by findings that the unemployed are responsible for a disproportionate volume of crime (Tarling 1982). In an attempt to identify factors conditioning criminal behaviour, Farrington *et al.* (1988), in a longitudinal study of 400 predominantly white, inner city males born in 1953, found several factors at age 8–10 that significantly predicted chronic convicted offenders. There were:

- economic/material deprivation, including low income, poor housing and unemployment periods experienced by parents;
- family criminality, including convicted parents and delinquent siblings;

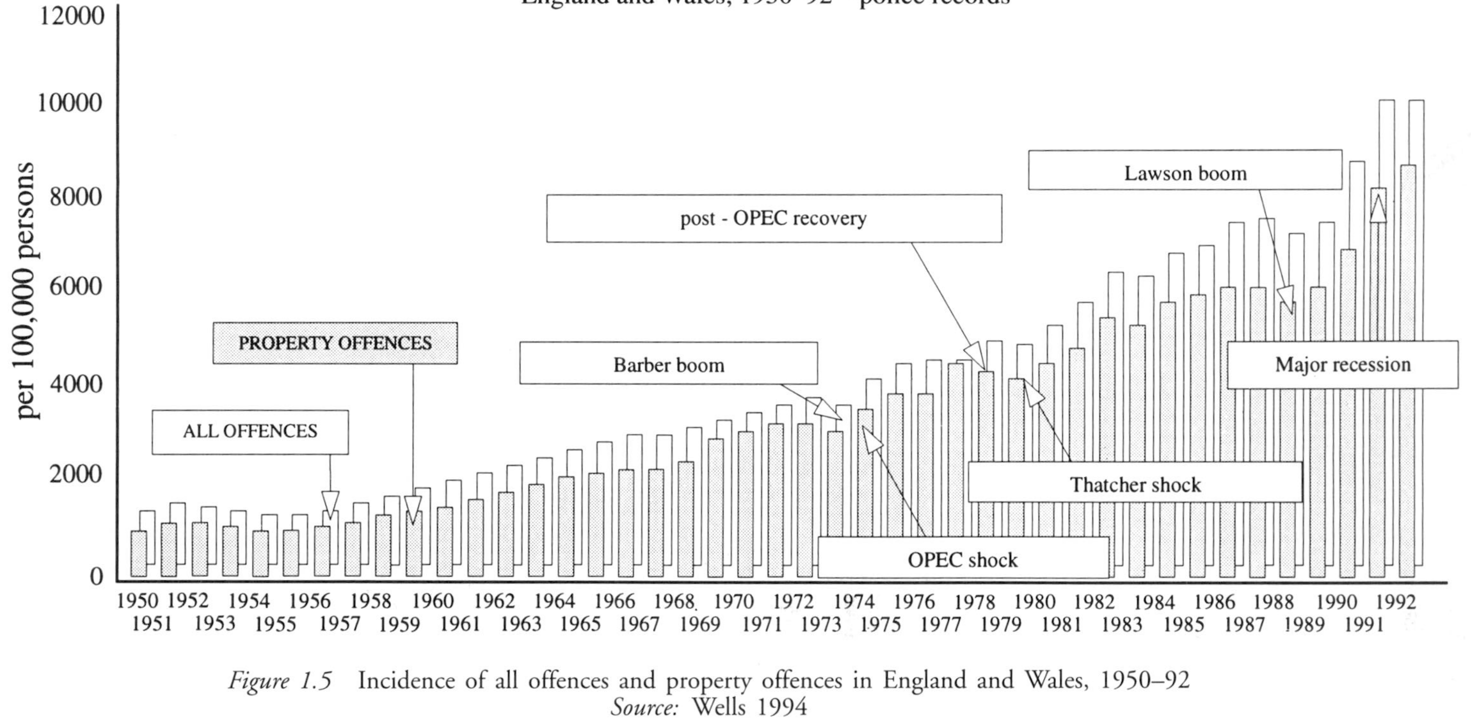

Figure 1.5 Incidence of all offences and property offences in England and Wales, 1950–92
Source: Wells 1994

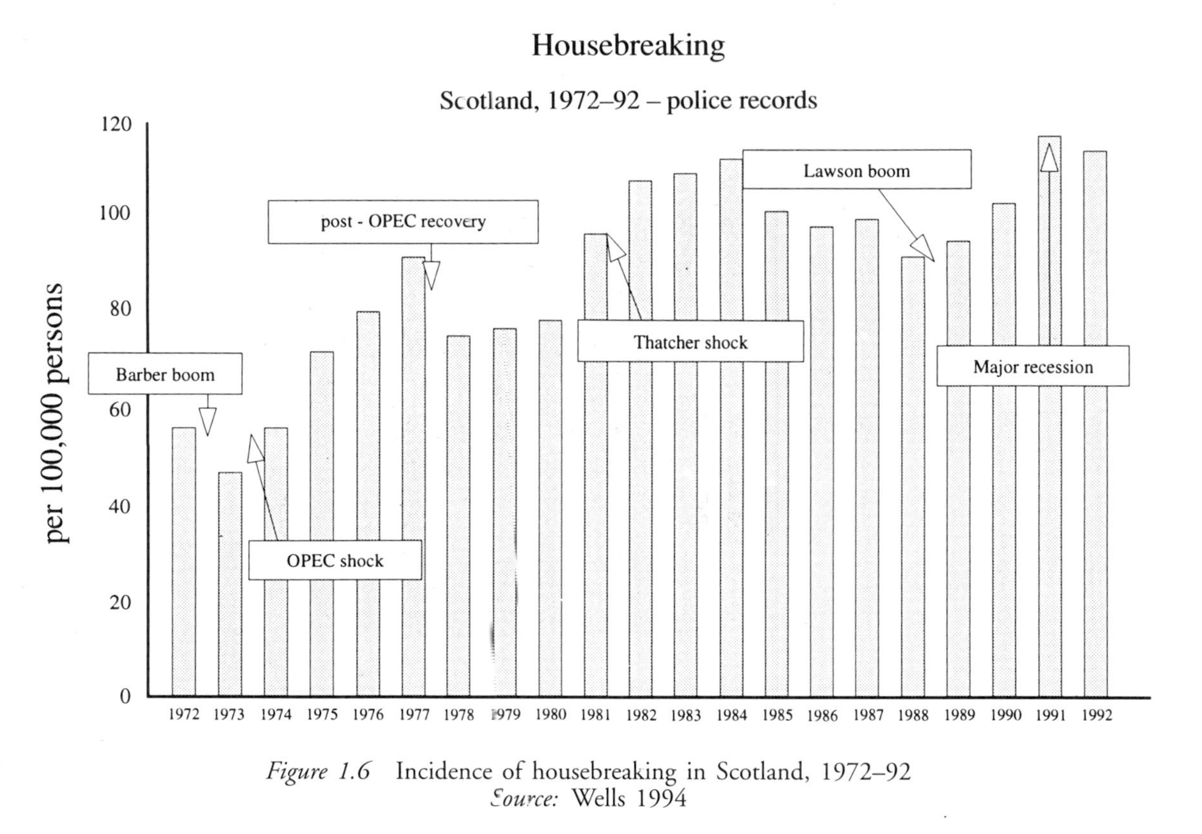

Figure 1.6 Incidence of housebreaking in Scotland, 1972–92
Source: Wells 1994

Plate 1.4 The CCTV camera installed in the Benwell district of Newcastle upon Tyne has been provided with a set of defence mechanisms for its own protection

- unsatisfactory parenting, being either too authoritarian or too unbounded;
- school failure.

A key predictor of persistent offending at age 14–18 was being unemployed at age 16, while for latecomers to crime in their twenties the only significant predictor was unemployment at age 18.

While the association between unemployment and criminal activity is unequivocal, the question of the extent to which there is a *causal* relationship is a contentious issue. The Conservative government rejects the existence of any causal relationship between unemployment and crime (Field 1990). Clearly, one must avoid the suggestion that crime is predetermined by poverty or unemployment not least since this slanders the majority of the unemployed who lead respectable law-abiding lives. Unemployment provides the motivation in the form of material deprivation, frustrated aspirations, boredom and anger (particularly in a society that extols material success) whilst the generally high level of material possessions enjoyed by the majority in employment provides the opportunity. The catalyst lies in the loosening of the individual's moral or ethical constraint on unlawful behaviour. This may be fostered by resentment on the part of the disadvantaged which questions the legitimacy of a social order which countenances mass unemployment and increasing income inequality. Kempson *et al.* (1994) have shown how the poor, lacking legal means to obtain a reasonable level of living, may as a last resort turn to criminal activities such as welfare fraud and burglary. In urban areas where poverty and deprivation are spatially concentrated, the weakening of individual moral restraint on a more general scale can provide a fertile breeding ground for anti-social behaviour and, as discussed below, may foster development of a marginalized criminal subculture. Wells (1994) is convinced of the causal relationship between unemployment and criminality and concludes that if we are going to be hard on the causes of crime we must come down very hard on unemployment and poverty.

Health

In 1980 the Working Group on Inequalities in Health concluded that, despite more than thirty years of a National Health Service committed to offering equal care to all, there remained a marked class gradient in standards of health in Britain. The report found that over the two decades up to the early 1970s the mortality rates for men and women aged 35 and over in occupational classes I and II had steadily diminished while those in classes IV and V changed little or deteriorated. The report was dismissed by the then Secretary of State, Patrick Jenkin, on the grounds that 'additional expenditure on the scale which could result from the report's recommendations – the amount involved could be upwards of £2 billion a year – is quite unrealistic in present or any foreseeable economic circumstances, quite apart from any judgement that may be formed of the effectiveness of such expenditure in dealing with the problems identified' (Townsend and Davidson 1982: 16).

Clearly, while some of the factors that affect health, such as age, gender and genetic make-up cannot be changed by public policy or individual choice, a number of 'external' factors are recognized to be of significance for health status. These include the physical environment (e.g. adequacy of housing, working conditions and air quality), social and economic factors (e.g. income and wealth, levels of unemployment) and access to appropriate and effective health and social services.

There is a strong association between quality of housing and health. Lack of shelter or homelessness can have a direct impact on the individual's health. Bronchitis, tuberculosis, arthritis, skin diseases and infections as well as alcohol- and drug-related problems and psychiatric difficulties are all more prevalent among single people who are homeless (Barry *et al.* 1991). Even families occupying temporary bed and breakfast accommodation often find it difficult to maintain hygiene while washing, eating and sleeping in one overcrowded room. One of the clearest indications that 'bad housing damages your health' (Ineichen 1993) is the effects of inadequate heating and dampness. Insufficient warmth (due to poor housing design and excessive fuel costs), leading to hypothermia, is reflected in higher proportions of deaths among older people in winter than in summer. Dampness encourages the spread of dust mites and fungal spores which lead to respiratory illnesses (Platt *et al.* 1989) while cockroaches thrive in the warm, damp conditions characteristic of many of the system-built tower blocks erected in the 1960s and 1970s. Other health problems related to inadequate housing include stress-related illness arising from poor sound insulation between neighbouring homes, lack of privacy and overcrowding. As women generally spend more time in the home, the effects of bad housing often impact on them to a greater degree (Gake and Williams 1993). It is difficult, however, to separate out the impact of poor housing because of the interrelated nature of the effects of social, economic and environmental factors on individual well-being. There is, however, ample evidence of the fundamental importance of income and wealth status for health (Phillimore *et al.* 1994, Eames *et al.* 1993, Carstairs and Morris 1991).

In the UK death rates at most ages are two to three times higher among the growing numbers of disadvantaged people than they are for their more affluent counterparts (Figure 1.7). Individuals experiencing poor socio-economic circumstances also have higher levels of illness and disability. Low income can impact upon family health in at least three ways: physiologically, with income providing the means of obtaining the fundamental prerequisites for health; psychologically, in that living with inadequate resources creates stress and reduces the individual's coping ability; and behaviourally, with poverty leading to health-damaging actions such as smoking and recourse to a low nutrient diet. In the light of the body of evidence it is difficult to refute the conclusion that 'it is one of the greatest of contemporary social injustices that people who live in the most disadvantaged circumstances have more illness, more disability and shorter lives than those who are more affluent' (Benzeval *et al.* 1995: 1).

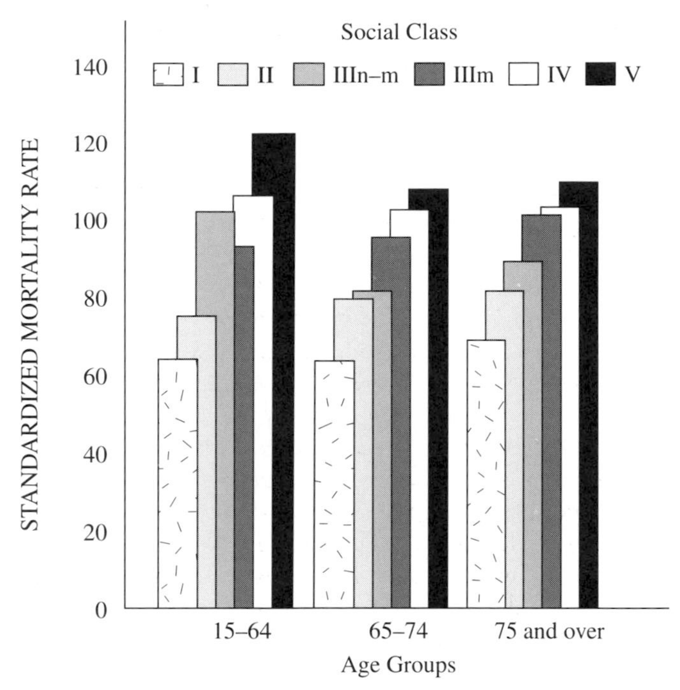

Figure 1.7 Male mortality rates by social class and age group in England and Wales, 1976–81
Source: Benzeval *et al.* 1995

Ethnic status

A recent Joseph Rowntree Foundation (Barclay 1995) investigation into income and wealth in the UK found that ethnic minority incomes tend to be lower than those of the rest of the population. Disaggregation by ethnic origin revealed that the Indian population was disproportionately concentrated in above-average income groups, whereas 40 per cent of the West Indian population and over half of the Pakistani and Bangladeshi population were in the lowest income group. This is supported by Green's (1994) findings that over 60 per cent of the ethnic minority population were living in wards that were ranked in the worst 20 per cent nationally for unemployment. Over half of the Bangladeshi population were living in wards that were in the most deprived 10 per cent nationally.

The main reasons why people from ethnic minorities are more likely to be living in poverty are summarized in Figure 1.8. In the early post-war era migrants were recruited to meet labour shortages in manufacturing industries (such as iron foundries and textile mills) and public services (such as transport

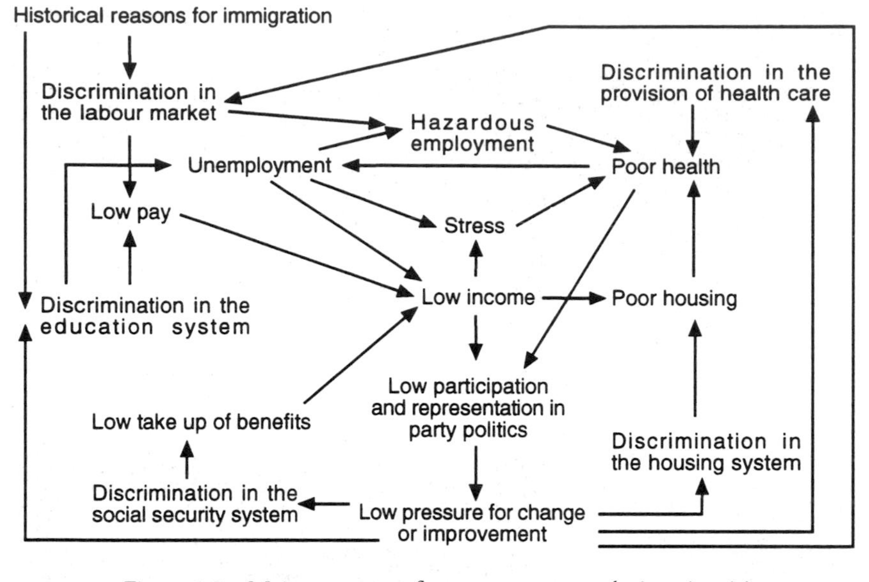

Figure 1.8 Major sources of poverty among ethnic minorities
Source: Amin 1992

and health) largely in manual occupations and often with low pay and minimum employment rights. The restructuring of the UK economy, and in particular the reduction in manufacturing employment and rationalization of public sector services, has resulted in large-scale unemployment among ethnic minorities. Government policies to restrict employment rights and reductions in welfare support have also had a heavy impact on the ethnic minority workforce because of its concentration in low-paid sectors of the economy. Evidence that ethnic minority groups have suffered disproportionately from the recession and effects of urban restructuring is provided by Amin (1992) who found that unemployment rates among ethnic minority groups increased more rapidly than those of the general population after 1979, reaching 22 per cent or twice the national average by 1984. By 1990, while nationally 8 per cent of all men were unemployed, the figure for ethnic minority groups was 14 per cent, and for Pakistani men aged 16–24 years it was 31 per cent. With 48 per cent of Pakistanis aged 16–24 and 54 per cent of Bangladeshis in the same age group having no formal qualifications, the culture of advanced capitalism represents a formidable challenge for many ethnic minorities (Jones 1993). The social and geographical segregation of the ethnic minority population was confirmed by Owen's analysis of the 1991 census which showed 'a clear pattern of the relative exclusion of minority ethnic groups from the successful parts of the British space-economy and a tendency for them to concentrate in areas of decline' (Owen 1995: 32).

Plate 1.5 Evidence of ethnic conflict is clearly visible on the door of an Asian welfare centre in the inner west end district of Elswick in Newcastle upon Tyne

Despite legislation, discrimination can compound the social and economic difficulties of ethnic minorities. Research into housing allocation practices in the London borough of Hackney found that black applicants received lower-quality housing than their white counterparts (Commission for Racial Equality 1984), while in Tower Hamlets local authority housing staff displayed 'suspicion and resentment' and sometimes 'overt racism' against Bengalis (Morris and Winn 1990). There is also evidence to suggest that the National Health Service fails to meet the particular needs of ethnic minorities who exhibit higher than average rates of certain illnesses (such as ischaemic heart disease and diabetes) as well as higher levels of peri-natal and infant mortality (Cox and Bostock 1989).

The excluded minority

As we have seen, the gap between those on low incomes and those who benefited from the economic growth of the 1980s has increased. The principal causes underlying this social polarization are the process of de-industrialization and increasing levels of unemployment; the bifurcation of a rapidly expanding service sector into high-paid and low-paid jobs; the increase in lone-parent, lone-elderly and dual-income families; the continuing economic and social marginalization of ethnic minorities; and the residualization of state welfare services (Pinch 1993).

At the disadvantaged pole of the quality of life spectrum there is evidence of the emergence of an alienated and excluded *urban underclass* comprising those forced to the margins or out of the labour market by the advent of advanced capitalism. Myrdal viewed the underclass as 'an unprivileged class of unemployed, unemployables and underemployed who are more or less hopelessly set apart from the nation at large and also do not share in its life, its ambitions and its achievements' (Myrdal 1962: 10). More specifically, Room (1993) identifies four principal characteristics of the underclass. First, its members are suffering or face the prospect of experiencing persistent poverty. Second, they act in ways that are contrary to those of mainstream society – they prefer to depend on welfare benefits or illicit earnings rather than seeking employment; males eschew stable family relationships; and there are high rates of school dropout, teenage pregnancy and criminality. Third, the members of the underclass tend to be spatially concentrated. Over a long term this can increase social exclusion and intensify anti-social behaviour. Finally, it is suggested that such behaviour is shaped by a deviant sub-culture through which 'dysfunctional' behaviour is transmitted to the next generation.

Explanations for the rise of the underclass either emphasize the causal importance of the behavioural characteristics of members of the group or stress the structural underpinnings of the phenomenon. For Murray (1984) the underclass are identified most clearly by their behavioural traits of unemployment, criminality and in particular by the high incidence of unmarried mothers (which is related to the disintegration of family values and absence of an appropriate male role model for children). This interpretation shares an affinity with the culture of poverty thesis of Lewis (1968), according to which children born into poverty have usually absorbed the basic attitudes and values of their sub-culture and are not psychologically geared to take full advantage of changing conditions or increased opportunities that may occur in their lifetime. This view identifies a lack of a will to work as a feature of the underclass. Murray and others on the political Right also indict an overgenerous welfare state as promoting a 'dependency culture' among the unemployed. Others focus causal attention on structural factors (Wilson 1987). In his definition of the underclass Field (1989) includes the frail elderly pensioner, the long-term unemployed and the single-parent *with no chance of escaping welfare* (emphasis added). This perspective emphasizes the failure of the economy to generate sufficient jobs to absorb an expanding labour supply, which leads to long-term unemployment, and identifies inadequate welfare provision as contributing to poverty. Thus while one viewpoint tends to assign blame to personal failings, the other indicts failures of the prevailing political economy in explaining the re-emergence of an urban underclass in contemporary society. The continued existence of a disadvan-

Plate 1.6 Breakfast consists of a cigarette and can of beer for one of the 'urban underclass' starting a new day on the steps of a hostel for homeless men in the east end of Glasgow

taged social stratum – literally redundant and alienated from the mores of society – represents the future for large parts of urban Britain unless the underlying difficulties can be resolved (Audit Commission 1987).

Socio-spatial divisions between rich and poor and the presence of excluded groups in Britain's cities is a long-standing phenomenon and may even be regarded as an unavoidable consequence of capitalist urban development. For some the rationale for seeking to narrow the gap between haves and have-nots rests on a moral obligation to fellow human beings, but the reduction of social division is to the advantage not only of the poor. All citizens share a common (self-) interest in social cohesion. The existence of a significant marginalized group increasingly excluded from mainstream society is likely to impinge upon the lifestyles of the privileged majority. While in the cities of Victorian Britain it was in the interests of all to support public health measures, so today it is in the common interest to eradicate the factors that foster the social ills of crime and anomie.

Since the late 1970s an increasing number of people in Britain's cities have come to rely on state welfare benefits to maintain a minimum standard of living. However, the rising costs of such provision and the ideological convictions of the Conservative government have resulted in major changes in social welfare policy, and have called into question the ability and willingness of the state to maintain the welfare system in its traditional form. The problems of socio-spatial division within Britain's cities seem set to continue for the foreseeable future. It is highly appropriate, therefore, that the principal focus of this book is an analysis of the geographies of division that continue to characterize much of urban Britain as we approach the third millennium.

GUIDE TO FURTHER READING

An account of the nature of the global economic system is provided by I. Wallace (1990) *The Global Economic System* London: Unwin Hyman, while the post-war restructuring of the UK space-economy is described in R. Martin (1988) 'Industrial capitalism in transition' in D. Massey and J. Allen *Uneven Development* London: Hodder and Stoughton: 202–31. R. Atkinson and G. Moon (1994) *Urban Policy in Britain* London: Macmillan offer a readable account of urban policy in the UK, M. Campbell (1990) *Local Economic Policy* London: Cassell presents a collection of papers on local economic strategies, while S. Lowe (1986) *Urban Social Movements* London: Macmillan offers a general discussion of grassroots social movements. The changing nature of the UK welfare state is discussed in N. Deakin (1987) *The Politics of Welfare* London: Methuen, and an overview of planning in the UK is afforded by J. Cullingworth and V. Nadin (1994) *Town and Country Planning in Britain* London: Routledge. In an examination of 'The geography of the urban crisis' *Scottish Geographical Magazine* 109(2): 87–95, M. Pacione (1993) provides a detailed insight into the nature and distribution of poverty and deprivation in the contemporary British city.

REFERENCES

Adam Smith Institute (1983) *Omega Report: Local Government Policy*, London: ASI.

Amin, K. (1992) *Poverty in Black and White: Deprivation and Ethnic Minorities*, London: Child Poverty Action Group.

Atkinson, R. and Moon, G. (1994) *Urban Policy in Britain*, London: Macmillan.

Audit Commission (1987) *The Management of London's Authorities*, London: Audit Commission.

——(1989) *Urban Regeneration and Economic Development*, London: HMSO.

Baldock, P. (1982) 'The Sheffield rent strike of 1967–68' in P. Henderson *et al. Successes and Struggles on Council Estates*, London: Association of Community Workers.

Barclay, P. (1995) *Inquiry into Income and Wealth* vol. 1, York: Joseph Rowntree Foundation.

Barry, A. *et al.* (1991) 'Homelessness and Health', *Discussion Paper 84*, Centre for Health Economics, University of York.

Benyon, J. and Solomos, J. (1987) *The Roots of Urban Unrest*, Oxford: Pergamon.

Benzeval, M. *et al.* (1995) *Tackling Inequalities in Health*, London: Kings Fund.

Beveridge, W. (1942) 'Social insurance and allied services', *Cmnd 6404*, London: HMSO.

Boddy, M. and Fudge, C. (1984) *Local Socialism*, London: Macmillan.

Bradford, M. and Robson, B. (1995) 'An evaluation of urban policy' in R. Hambleton and H. Thomas *Urban Policy Evaluation*, London: Paul Chapman: 37–54.

Briggs, A. (1961) 'The welfare state in historical perspective', *European Journal of Sociology* 2(2): 221–58.

Butler, S. (1982) *Enterprise Zones*, London: Heinemann.

Carstairs, C. and Morris, R. (1991) *Deprivation and Health in Scotland*, Aberdeen: Aberdeen University Press.

Cochrane, A. (1986) 'Local Employment Initiatives' in P. Lawless and C. Rabin *The Contemporary British City*, London: Harper & Row: 144–62.

Cochrane, A. and Clarke, A. (1990) 'Local enterprise boards: the short history of a radical initiative', *Public Administration* 68: 315–36.

Commission for Racial Equality (1984) *Race and Council Housing in Hackney*, London: CRE.

Community Development Project (1977a) *The Cost of Industrial Change*, London: CDP Editorial Team.

——(1977b) *Gilding the Ghetto*, London: CDP Editorial Team.

Couclelis, H. (1982) 'Philosophy in the construction of geographic reality' in P. Gould and G. Olsson *A Search For Common Ground*, London: Pion: 105–40.

Cox, J. and Bostock, S. (1989) *Housing and Social Equality*, London: Penrhos.

Crossland, A. (1964) *The Future of Socialism*, London: Jonathan Cape.

Cullingworth, J. and Nadin, V. (1994) *Town and Country Planning in Britain*, London: Routledge.

Dear, M. and Scott, A. (1981) *Urbanization and Urban Planning in Capitalist Society*, London: Methuen.

Dennis, R. (1978) 'The decline of manufacturing employment in Greater London', *Urban Studies* 15: 69–73.

Donnison, D. (1984) 'The progressive potential of privatization' in J. Le Grand and R. Robinson *Privatization and the Welfare State*, London: Allen & Unwin.

Donnison, D. and Middleton, A. (1987) *Regenerating the Inner City*, London: Routledge & Kegan Paul.

Eames, M. *et al.* (1993), 'Social deprivation and premature mortality: regional comparisons across England', *British Medical Journal* 307: 1097–102.

Fainstein, S. (1987) 'Local mobilization and economic discontent' in M. Smith and J. Feagin *The Capitalist City*, Oxford: Basil Blackwell: 323–42.

Farrington, D. *et al.* (1988) 'Cambridge study in delinquent development: long term follow up', *Final Report to the Home Office*, Cambridge Institute of Criminology.

Field, F. (1969) *Poverty and the Labour Government*, London: Child Poverty Action Group.

——(1989) *Losing Out*, Oxford: Basil Blackwell.

Field, S. (1990) 'Trends in crime and their interpretation', *Home Office Research Study 119*, London: Home Office.

Flynn, N. and Taylor, A. (1986) 'Inside the rust belt: an analysis of the decline of the West Midlands economy', *Environment and Planning A* 18: 865–900.

Freeman, C. (1987) 'Technical innovation, long cycles and regional policy' in K. Chapman and G. Humphreys *Technical Change and Industrial Policy*, Oxford: Basil Blackwell: 10–25.

Gaffikin, F. and Morrissey, M. (1994) 'Poverty in the 1980s: a comparison of the United States and the United Kingdom', *International Journal of Urban and Regional Research* 22(1): 43–58.

Gake, J. and Williams, P. (1993) 'Women, crowding and mental health' in R. Burridge and D. Ormandy *Unhealthy Housing*, London: E. & F.N. Spon: 191–208.

Gamble, A. (1981) *Britain in Decline: Economic Policy, Political Strategy and the British State*, London: Macmillan.

Gardner, N. (1987) *Decade of Discontent: The Changing British Economy Since 1973*, Oxford: Basil Blackwell.
Gibson, M. and Langstaff, M. (1982) *An Introduction to Urban Renewal*, London: Macmillan.
Gordon, D. (1980) 'Stages of accumulation and long economic cycles' in T. Hopkins and I. Wallerstein *Processes of the World System*, London: Sage.
Green, A. (1994) *The Geography of Poverty and Wealth*, Coventry: University of Warwick.
Hall, P. (1985) 'The geography of the fifth Kondratieff' in P. Hall and A. Markusen *Silicon Landscapes*, London: Allen & Unwin: 1–19.
Hamilton, F. (1984) 'Industrial restructuring: an international problem', *Geoforum* 15: 349–64.
Harvey, D. (1985) *The Urbanization of Capital*, Oxford: Basil Blackwell.
Hasluck, C. (1987) *Urban Unemployment*, London: Longman.
Healey, M. and Clark, D. (1985) 'Industrial decline in a local economy: the case of Coventry 1974–1982', *Environment and Planning A* 17: 1351–67.
Hills, J. (1995) *Inquiry into Income and Wealth* vol. 2, York: Joseph Rowntree Foundation.
HMSO (1977) *Policy for the Inner Cities, Cmnd 6845*, London: HMSO.
Hudson, R. (1983) 'The question of theory in political geography' in N. Kliot and S. Waterman *Pluralism and Political Geography*, London: Croom Helm.
——(1988) 'Uneven development in capitalist societies', *Transactions of the Institute of British Geographers* 13: 484–96.
Ineichen, B. (1993) *Homes and Health*, London: E. & F.N. Spon.
Institute for Fiscal Studies (1994) *Update* Autumn/Winter, London: IFS.
Jacobs, J. (1970) *The Economy of Cities*, London: Jonathan Cape.
——(1982) 'DARE to Struggle', *Socialist Review* 63/64: 85–104.
Jessop, B. (1980) 'The transformation of the State in post-war Britain' in R. Scase *The State in Western Europe*, London: Routledge: 23–93.
Johnston, R. (1980) 'On the nature of explanation in human geography', *Transactions of the Institute of British Geographers* 5: 402–12.
Jones, T. (1993) *Britain's Ethnic Minorities*, London: Policy Studies Institute.
Joseph, K. and Sumption, J. (1979) *Equality*, London: John Murray.
Keith, M. and Rogers, A. (1991) *Hollow Promises? Rhetoric and Reality in the Inner City*, London: Mansell.
Kempson, E. *et al.* (1994) *Hard Times?*, London: Policy Studies Institute.
Kraushaar, R. (1979) 'Pragmatic radicalism', *International Journal of Urban and Regional Research* 3: 61–80.
Lansley, S. and Mack, J. (1991) *Breadline Britain in the 1990s*, London: Harper Collins.
Lawless, P. (1986) *The Evolution of Spatial Policy*, London: Pion.
——(1988) 'British inner urban policy', *Regional Studies* 22(6): 531–42.
Lewis, O. (1968) 'The culture of poverty' in D. Moynihan *Understanding Poverty*, New York: Basic Books.
Lowe, S. (1986) *Urban Social Movements*, London: Macmillan.
Mackintosh, J. (1977) *The British Malaise: Political or Economic?*, Southampton: University of Southampton Press.
Mandel, E. (1980) *Long Waves of Capitalist Development: The Marxist Interpretation*, Cambridge: Cambridge University Press.
Martin, R. (1986) 'Thatcherism and Britain's industrial landscape' in R. Martin and B. Rowthorn *The Geography of Deindustrialization*, London: Macmillan.
——(1988) 'Industrial capitalism in transition: the contemporary reorganization of the British space-economy' in D. Massey and J. Allen *Uneven Re-Development*, London: Hodder & Stoughton: 202–31.

Martin, R. and Rowthorn, B. (1986) *The Geography of Deindustrialization*, London: Macmillan.
Morris, J. and Winn, M. (1990) *Housing and Social Inequality*, London: Hilary Shipman.
Murray, C. (1984) *Losing Ground*, New York: Basic Books.
Myrdal, G. (1962) *Challenge to Affluence*, New York: Pantheon.
Noble, M. *et al.* (1994) *Changing Patterns of Income and Wealth in Oxford and Oldham*, Department of Applied Social Studies and Social Research, Oxford University.
Nozick, R. (1974) *Anarchy, State and Utopia*, Oxford: Basil Blackwell.
O'Higgens, M. and Jenkins, S. (1990) 'Poverty in the EC' in R. Teehens and B. van Praag *Analysing Poverty in the European Community*, Luxembourg: Eurostat.
Owen, D. (1995) 'The spatial and socio-economic patterns of minority ethnic groups in Great Britain', *Scottish Geographical Magazine* 111(1): 27–35.
Pacione, M. (1986) 'Quality of life in Glasgow – an applied geographical analysis', *Environment and Planning A* 18: 1499–520.
——(1988) 'Public participation in neighbourhood change', *Applied Geography* 8: 229–47.
——(1989) 'The urban crisis: poverty and deprivation in the Scottish city', *Scottish Geographical Magazine* 105: 101–15.
——(1990a) 'The ecclesiastical community of interest as a response to urban poverty and deprivation', *Transactions of the Institute of British Geographers* 15(2): 193–204.
——(1990b) 'What about people? A critical analysis of urban policy in the United Kingdom', *Geography* 75: 193–202.
——(1992) 'Urban policy' in M. Hawkesworth and M. Kogan *Encyclopaedia of Government and Politics*, London: Routledge: 777–91.
——(1993) 'The geography of the urban crisis: some evidence from Glasgow', *Scottish Geographical Magazine* 109(2): 87–95.
——(1995a) *Glasgow: The Socio-spatial Development of the City*. Chichester: John Wiley & Sons.
——(1995b) 'The geography of multiple deprivation in the Clydeside conurbation', *Tijdschrift voor Economische En Sociale Geografie* 86(5): 407–25.
Parkinson, M. and Judd, D. (1988) 'Urban revitalisation in America and the United Kingdom' in M. Parkinson, B. Foley and D. Judd *Regenerating the Cities,* Manchester: Manchester University Press: 1–8.
Phillimore, P. *et al.* (1994) 'Widening inequality of health in northern England 1981–1991', *British Medical Journal* 308: 1125–8.
Pinch, S. (1993) 'Social polarisation: a comparison of evidence from Britain and the United States', *Environment and Planning A* 25: 779–95.
Platt, S. *et al.* (1989) 'Damp housing, mould growth and symptomatic health state', *British Medical Journal* 298: 1673–8.
Rees, G. and Lambert, J. (1985) *Cities in Crisis*, London: Edward Arnold.
Reidel, J. (1972) 'Citizen participation: myths and realities', *Public Administration Review* 32: 211–20.
Robson, B. (1987) *Managing The City*, London: Croom Helm.
——(1988) *These Inner Cities*, Oxford: Clarendon Press.
Room, G. (1993) *Anti-Poverty Action Research in Europe*, School for Advanced Urban Studies, Bristol University.
Rowntree, S. (1901) *Poverty: A Study of Town Life*, London: Macmillan.
Shelter Neighbourhood Action Project (1972) *Another Chance for Cities*, London: Shelter.
Sills, A., Taylor, G. and Golding, P. (1988) *The Politics of the Urban Crisis*, London: Hutchinson.

Smith, M. (1980) *The City and Social Theory*, Oxford: Basil Blackwell.
Smith, N. (1984) *Urban Development: Nature, Capital and the Production of Space*, Oxford: Basil Blackwell.
Spencer, K. (1987) 'Developing an urban policy focus in the 1990s', *Local Government Policy Making* 14: 9–11.
Steen, A. (1981) *New Life for Old Cities*, London: Aims of Industry.
Stewart, M. (1987) 'Ten years of inner cities policy', *Town Planning Review* 58(2): 129–45.
Tarling, R. (1982) 'Unemployment and crime', *Home Office Research Bulletin* 14: 28–33.
Thake, S. and Staubach, R. (1993) *Investing in People*, York: Joseph Rowntree Foundation.
Thornley, A. (1991) *Urban Planning Under Thatcherism*, London: Routledge.
Townsend, P. (1962) 'The meaning of poverty', *British Journal of Sociology* 18: 210–17.
——(1979) *Poverty in the United Kingdom*, Harmondsworth: Penguin.
Townsend, P. and Davidson, N. (1982) *Inequalities in Health: The Black Report*, Harmondsworth: Penguin.
Wallace, I. (1990) *The Global Economic System*, London: Unwin Hyman.
Wallerstein, I. (1979) *The Capitalist World Economy*, Cambridge: Cambridge University Press.
Wates, N. and Wolmar, C. (1980) *Squatting*, London: Bay Leaf Books.
Webb, A. and Wistow, G. (1987) *Social Work, Social Care, and Social Services since Seebohm*, London: Longman.
Wedderburn, D. (1962) 'Poverty in Britain today', *Sociological Review* 10: 257–82.
Wells, J. (1994) *Mitigating the Social Effects of Unemployment: Crime and Unemployment*, Report for the House of Commons Select Committee on Employment, University of Cambridge.
Willmott, P. and Hutchinson, R. (1992) *Urban Trends 1*, London: Policy Studies Institute.
Wilson, W. (1987) *The Truly Disadvantaged*, Chicago IL: University of Chicago Press.
Wolch, H. (1989) 'The Shadow State' in J. Wolch and M. Dear *The Power of Geography*, London: Unwin Hyman: 197–221.

Part I

THE STRUCTURAL CONTEXT OF URBAN DIVISION

2

GLOBAL RESTRUCTURING AND LOCAL IMPACT

John Lovering

INTRODUCTION

Since the early 1980s British city centres have been visibly transformed by gleaming new shopping malls, entertainment complexes and office buildings. But over the same period, beggars have become common, homelessness has become visible in every city, and parts of many towns and cities have become no-go areas for women, the middle-aged and elderly. The social and economic gap between the urban haves and have-nots has widened conspicuously.

Most people would agree that any attempt to explain disadvantage in Britain's cities must take into account wider changes in the international, or 'global', economy. However, quite what such a 'taking into account' involves is open to debate. This chapter looks at this debate. The first part sets the scene by outlining the economic environment in Britain's cities. The second part explores interpretations of the causal relationships linking global restructuring and local economic change, and the connections to urban disadvantage. It argues that some of the most influential explanations, and associated policy agendas, are inadequate.

THE CHANGING URBAN ECONOMY

Between the 1940s and mid-1970s working life in the advanced capitalist countries was shaped by a combination of a long economic boom and the creation of the Welfare State. Incomes rose steadily, not least in working-class communities. Inequalities in income and wealth were reduced, and the worse extremes of urban poverty were virtually eradicated (Hobsbawm 1994, Dahrendorf 1995). But the 'Long Boom' is long over. European economic growth rates are now half those of the mid-1960s while unemployment is three times higher,[1] and the Welfare State is being unravelled.

The stagnation in job opportunities

The British labour market has stopped growing. In terms of aggregate employment the British economy stopped growing in the 1980s. Manufacturing

Table 2.1 Changes in workforce in employment (thousands)

	Employees			Full-time	Part-time	Self-employed
	All	*Male*	*Female*	*Male & female*	*Male & female*	*Male & female*
1979–83	−2,132	−1,456	−676	76	457	261
1983–90	1,798	24	1,774	586	1,213	1,143
1990–93	−1,221	−1,050	−171	−1,384	162	−313
1993–95	156	56	100	−48	204	176

Source: Employment Gazette 1995

employment has been declining since the late 1960s, accelerating sharply with the politically induced recessions of 1979–83 and 1990–3. During the 1980s this was largely offset by the growth of jobs in the service sector. But this too has now faltered, and technological changes are shrinking the employment potential of the industries that soaked up workers in the Thatcher era. Despite the return to what official statisticians coyly describe as 'subdued growth' (*Economic Trends*, Sept 95: 21), the total number of full-time jobs appears to be either static or falling. Most of the new jobs created since the 1970s have been part-time, and even the flow of these appears to be drying up. The number of part-time jobs created during 1990–3 was under half that created during the similar stage of the economic cycle a decade earlier (Table 2.1).

Aggregate employment statistics give a misleading impression of the number of people in work (as opposed to the number of jobs). At least 1.3 million of those in employment (in 1995) have more than one job whilst 330,000 of the self-employed also have other jobs.

The much proclaimed growth of an 'Entrepreneurial Culture' in the 1980s has done little to compensate for the stagnation of employment opportunities. Self-employment in Britain almost doubled in the decade to 1991 to just under 3.3 million, 14 per cent of the labour force. But the increase during the early 1980s was outstripped by the decline in the early 1990s (see Table 2.1). Self-employment in Britain seems to reflect a shortage of the kind of jobs that people want, rather than an ascendant entrepreneurialism that augurs well for future opportunities.

Increasing polarization in the labour market

In terms of pay, conditions and prospects for advancement, the set of job opportunities in urban Britain has become much more polarized. Insecure employment is growing, especially in the cities (Gordon and Sassen 1993: 126, Morris 1995). A quarter of men are self-employed, in part-time work, or temporary work (Taylor 1995). Some 1.5 million jobs are temporary (*Employment Gazette* Sept 1995). Workers in insecure jobs are especially unlikely to gain transferable skills or credentials. Workers in poorly paid or unhealthy jobs are more likely to suffer poor health, making them even less attractive to employers and less able to engage in effective job search. The unemployed tend to become isolated from the informal networks through which information about vacancies is transmitted, and credibility with a prospective employer is established (Morris 1994, 1995). They are also prone to exclusion from the formal banking system, sliding into the hazardous world of semi-illicit finance (Leyshon and Thrift 1995). They suffer disproportionately from physical and psychological distress and thereby tend to become even less attractive to employers. The slippery slope is getting steeper; the chances of the unemployed or partners of the unemployed getting a new job, compared to those already in employment, have deteriorated since the 1970s (Barclay 1995: 26).

In effect, the labour market sorts job-seekers into a queue. The most public of the criteria used to do this are educational qualifications. The British workforce is becoming more highly qualified. In part this reflects rising skill levels (Cabinet Office 1993). But 'credential inflation' is also an important, and possibly the dominant, influence (Brown 1995, Gordon and Sassen 1993: 110). Young women appear to be much more astute than young men in recognizing that without qualifications they stand little chance of getting a job at all, let alone one with career prospects.

At the exhausting end of the queue, a growing number of people have given up looking for formal work or bothering to register as unemployed. Between 1975 and 1993 the proportion of working-age men in employment in Britain fell by a quarter, to 77 per cent. Over one in five British men of working age, and even one in seven in the 'prime age group' of 22–54, are jobless (Borooah and Hart 1995, Schmitt and Wadsworth 1994). In de-industrialized northern industrial towns, Scotland and the Welsh valleys, the male participation rate is even lower. But some deprived inner city areas in London have higher activity rates than the national average, alongside higher unemployment rates. The economic activity rate amongst women rose over the same period from 65.4 per cent to 71.7 per cent, and is forecast to continue to rise to nearly 75 per cent in 2006, while male rates continue to fall (Ellison 1995). But while married women have been increasingly drawn into the labour market, lone mothers have been driven out. Employment is increasingly concentrated in dual-earner households in the 25–49 age group (Harrop and Moss 1995).

The new Victorian economy

From the 1940s to the 1970s the real income of even the worst-off in Britain steadily improved. Since 1979 real hourly wages for the lowest paid men have hardly changed, the real disposable income of the poorest tenth of the population has fallen by a sixth, and the poorest third of the population have stopped benefiting from economic growth (Hills 1995: 29). Between 1979 and 1995 the number of 'rich' people and 'poor' people in Britain grew by 2.1 million and 800,000 respectively (to 11.1 million and 17.3 million) (Grant 1995: 3). Hutton (1995) suggests that Britain is now a '30–30–40' society: 30 per cent of the working-age population are 'disadvantaged', 30 per cent 'insecure' and 40 per cent 'advantaged', but even they face new insecurities. According to Dahrendorf (1995), 40 per cent can expect to see their earnings decline. The replacement of Unemployment Benefit by the Job Seekers Allowance in 1996 will reduce the annual income of the jobless by an estimated £210 million (Finn 1995).

Deprivation kills. The most deprived tenth of men aged 45–65 have a mortality rate one-and-a-half times that of the least deprived tenth (Drewer and Whitehead 1995: 24).

Labour market disadvantage is a largely unintended outcome of the independent actions of numerous employers and job-seekers. Yet the effects are remarkably systematic, dividing the labour force in effect into a set of distinct 'groups' (Morris 1995). Joblessness, for example, is exaggerated amongst non-whites. Ethnic minority activity rates are 9 per cent lower than whites for men, and 20 per cent lower for women (falling to 49 per cent and 26 per cent respectively for Pakistani-Bangladeshis). The proportion of the 1.3 million economically active ethnic minorities who are self-employed is the same as for whites, but the proportion unemployed is twice as high (Sly 1995). The proportion of the unemployed who are long-term unemployed is also higher, ranging from 52 per cent for Indians to 60 per cent for black people, compared to 44 per cent for whites.

The urban dimension

Although there are many poor people in rural areas in Britain (Cloke *et al.* 1994), poverty is becoming more concentrated, and more prominent, in the cities. Across the advanced capitalist world, the larger cities are, the more severe the deprivation within them (Robson 1992). The big British cities have not yet reached the nightmarish depths of their US counterparts, but many are moving in that direction. Deprivation in many smaller towns, such as Holyhead or Plymouth, is also intensifying, but often less visibly.

The collapse of full-time employment in deprived urban areas dramatically surpassed the fall in population (Willmott 1994: 32–5). These areas also missed out on part-time work, which grew in the areas examined by Willmott (1994) by only 1 per cent in the 1980s, less than a tenth of the national rate, and on the growth in self-employment (with the exception of some London boroughs where there are unusual opportunities to capitalize on affluent consumer markets and specialist business demands).

Most of the wards showing multiple indicators of 'concentrated poverty' are in de-industrialized urban areas (Hills 1995: 86, Forrest and Gordon

1993). Analysis at a closer scale would show deprivation concentrated on council estates in the inner and outer cities. The greatest increase in deprivation indicators is in Inner London boroughs (Willmott 1994). The high mortality districts in Britain are overwhelmingly urban (Drewer and Whitehead 1995: 24).

The non-white British population is concentrated in the cities. While the population of deprived urban areas fell 16 times faster than that of the country as a whole, the ethnic minority population in them grew by two-thirds, although it was static in the UK as a whole (Willmott 1994: 13). Over the last decade, community, business, leisure and residential developments, new parks and trees, have improved the appearance of St Paul's in Bristol, Handsworth in Birmingham, Chapeltown in Leeds, Toxteth in Liverpool, Brixton in London, and so on. But employment opportunities have also collapsed, especially for the non-white population living there.

Meanwhile, the semi-rural, and overwhelmingly white, suburban fringes surrounding those cities have expanded. Liverpool is close to some of the most affluent populations and smartest properties in Britain in the Wirral; the upwardly mobile of Leeds can live in affluent Harrogate or dozens of pretty villages; Bristol decants its professionals to the 'Middle England' villages of south Gloucestershire and Wiltshire, used as sets in BBC classic serials; whilst the Vale of Glamorgan is a laboratory of post-industrial gentrification a million miles away, socially, from the blighted ex-mining valleys just 15 minutes drive away.

However, many of the well-off still live in all but the most blighted parts of the cities. In Inner London (uniquely) the number of resident professional men actually rose in the 1980s (Willmott 1994: 10). Even in the Rhondda about a fifth of men are in professional, managerial or technical occupations. Some analysts expect the British urban pattern increasingly to emulate that of the US, as suburbanization joins up towns and cities to form a seamless, semi-urban sprawl across the lowlands (Robson 1992). If this turns out to be correct, the cities and large towns are likely to be distinguishable as islands of intense deprivation, adjacent to zones of consumption and offices staffed by commuters.

Who and where are the disadvantaged?

However, not *all* women, young people, single parents, the low-skilled working class or even 'ethnic minorities' can be regarded as systematically disadvantaged. Men's and women's income distributions have become more alike so that the greatest variations are within the sexes rather than between them. Similarly, income polarization is greatest amongst the young,[2] where the income gap between graduates and the unqualified in the 16–30 age range grew by over a fifth in the 1980s (Hills 1995). Different ethnic minority groups and households within them have very different incomes and employment experiences.

The truly disadvantaged form sub-sets of these broad social categories and are most likely to be found where they overlap (Hills 1995, Morris 1994).

By the same token, disadvantage cannot be identified very closely with broad spatial units such as local authority areas (Drewer and Whitehead 1995). Areas greater than a street or neighbourhood are likely to contain well-off people as well as the poor. But the urban affluent are increasingly insulated from the disadvantaged, living, travelling and socializing in separate worlds, increasingly surrounded by an apparatus of surveillance and security (thereby creating at least one small source of urban employment).

UNDERSTANDING THE GLOBAL–LOCAL INTERACTION AND ITS IMPLICATIONS FOR THE URBAN DISADVANTAGED

The connections between global economic trends and local economic problems have been a subject of frequent discussion in Britain since at least the late 1960s. Versions of the idea that urban decline and poverty are the local end result of inexorable global processes have appeared not only in numerous academic books and articles and official reports, but also in television and radio programmes, press articles, films and pop songs.

The core ideas can be summarized in a Simple Story that has become almost as much a part of popular political culture as it has become an academic and policy orthodoxy.[3] According to this story, the problems of the great Victorian industrial cities (plus large parts of London) are rooted in Britain's century-long decline from being the 'workshop of the world'. Industries have disappeared into oblivion under the impact of technological change (whither the steam locomotive industry?), or moved to cheaper workforces elsewhere (ship building, cars, radios, toys, etc.). Newly industrializing countries, with cheaper labour and new factories, will continue to take over the production of consumer goods, moving up-market from low- to high-tech products. The global mobility of capital, including the growth of multinational corporations, has increased this likelihood, making it easier to transfer work to cheaper labour areas abroad. Britain's cities have therefore ceased to be centres of production. The more fortunate amongst them have become instead centres of consumption and administration. Out with factories, in with offices, shopping malls and clubland.

But this, in its turn, creates different kinds of jobs for which different kinds of people are favoured: women rather than men, graduates rather than unqualified school-leavers, commuters rather than local residents. There is an increasing 'mis-match' between urban job opportunities and the skills and capacities of a significant part of the urban population, especially the residents of formerly industrial working-class areas (Begg *et al.* 1986). Ethnic minorities are especially likely to miss out because they also suffer from racial discrimination.

Like all narratives, this story does more than describe; it also suggests how the reality portrayed is to be understood. The Simple Story conjures up the image of a queue of people looking for work. In every queue some unfortunates will be at the wrong end. The story suggests that these will naturally tend to be people whose skills and personal characteristics, or personal misfortune, make them less productive.

Moreover, joblessness tends to become pathological. Unemployed people tend to lose what skills they had and become less motivated. At the extreme this can lead to the formation of a virtually unemployable 'underclass'.[4] This group includes some older workers who have been unable to adapt, but it is increasingly dominated by young people who have never learnt how to get, or keep, a job. Social and cultural changes tend to reinforce these developments. The decline in male employment in deprived areas puts strains on family structures leading to single motherhood (a markedly urban phenomenon – Forrest and Gordon 1993). This increases the number of children growing up in poverty, and this, together with the lack of appropriate male role models, tends to exacerbate the unemployability of the next generation (Dennis 1993). Alienation amongst jobless young men (Campbell 1991) in its turn leads to more crime, further blighting urban life and driving employers away.

The underlying assumptions of the Simple Story: economic and cultural mechanisms

The Simply Story explains deepening urban social problems as the effect of a combination of economic and cultural mechanisms that interact through time to create a vicious circle.[5] On the one hand, international economic change is raising the stakes in what has become a global competition between cities for investment and jobs. On the other hand, social-psychological processes rooted in human nature and social interaction tend to create dysfunctional 'sub-cultures of poverty' amongst the casualties, so that insiders and outsiders move further apart through time.

It is increasingly argued that the 'cultural dimension' has taken over as the most important factor underlying urban social deterioration. Some also argue that this has political roots. Many of the developments that have exacerbated the unemployability of the urban disadvantaged can allegedly be traced to state intervention and the misguided cultural leadership of elites, which have together fomented a short-sighted 'dependency culture' and a failure to acquire appropriate values (Murray 1994). The Left used to argue that these cultural dimensions are less important than the failings of British industrialists and investors (Coates 1994). In the Labour Party in the 1990s this view seems to be giving way to the 'cultural failure' doctrine (as expressed, for example, following the disturbance in Brixton in late 1995).

The pervasive influence of the Simple Story is reflected in the consensus that little can be done to rescue the deprived parts of the cities from further decline, short of a major social-economic upheaval. This will require economic flexibility and geographical mobility (a return to the early 1930s view that the problem of the Welsh valleys would be solved only by out-migration). It will also require cultural and personal change, a return to prudent Victorian values (Dennis 1993, Murray 1994).

Images of the global–local relationship and the disadvantaged

The Simple Story evokes powerful images that are seductively familiar. It is reflected in many academic studies and official policy documents (Solesbury 1993, Cmnd 2867 1995), and dominates the political 'centre-ground'. But it is severely oversimplified. It is also ideological in the sense that it encourages a way of looking at things that is advantageous for particular interests. It is no accident that it is often explicitly invoked by property speculators and developers, who argue that the only economic solution for the cities is to bow to world market pressures and seek to attract international firms to new office complexes (Fitch 1994), or by politicians, business people and cultural commentators, who seek to demonstrate that little can be done to improve the lot of the urban poor. Dahrendorf's argument that Western governments seeking to raise competitiveness have to choose between social cohesion and political freedom (1995: 38) is essentially an extrapolation of the Simple Story.

However, the underlying narrative is based on a set of assumptions that have the effect of obscuring the political character of both the global–local relationship and urban exclusion. This is a result of its economistic conception of the way economic forces operate, and its dualistic spatial conception of the relationships between the 'global' and the 'local'. The former is the space of autonomous economic change; the latter the place where its effects are felt. The global is the domain of world-historical economic forces, symbolized by the transnational corporation and international financial flows; the local is the domain of responsivity, adaptation or resistance (the latter more often than not resulting in failure). The global is conceived as the space of irresistible change, decentring, anonymity; the local as a space of stasis, identity, community (*Gemeinschaft*) and the past (Smith 1995).

This vision echoes a nineteenth-century conception of Modernity (Robertson 1995: 29). A global, universal 'civilization' stands against a local, particular 'culture'. The fatalism inherent in this view is reproduced in the conceptualization of the disadvantaged as losers at the end of the queue, casualties of an unavoidable sorting process.

The 'Locality debate': the politics of the capitalist spatial division of labour

It was precisely this sort of loaded, compartmentalized thinking that provoked the body of research and discussion in the 1970s and 1980s sometimes referred to as the 'Locality debate' or the 'Restructuring debate' (Massey 1994, Cooke 1989, Lovering 1989). The contributors wished to refute the argument that economic deterioration in the cities was largely the fault of the people who lived there (Massey 1994). Many participants were influenced by the Marxist idea that capitalism is *always* restlessly restructuring, and this jars with the need to invest fixed capital in particular places. The result is an inherent tendency to uneven development (Harvey 1985).

The Locality debate highlighted the significance of the political character of economic decisions – a dimension missing in the Simple Story. Massey's (1994) notion of a 'spatial division of labour' pointed to the active construction of industrial geographies through the investment decisions of firms. Firms perceive different localities as offering different opportunities and problems, influenced by the locally unique legacies of previous 'rounds of investment' embedded in local traditions of industrial attitudes, expectations and militancy. These perceptions influence their location decisions, leading to a tendency to relocate work to different groups of workers in different locations, and generally to move work from high-wage cities to areas with lower labour costs and more amenable workforces – notably semi-rural areas with no tradition of trades unionism (Massey and Meegan 1982). At the same time, high-level management and R & D work is often separated off from production work and located in more 'middle-class' areas.

The state plays a modest role in the creation of these spatial divisions of labour. Regional aid can influence the choice of one location over another, but not the wider trend towards relocation out of established industrial towns. The Locality debate was greatly influenced by the idea that the state, especially at the local level, could potentially have a much greater and more productive impact than this. It is no accident that in Britain the Locality debate flourished alongside experiments by Labour-controlled local authorities to introduce new economic development strategies. These were driven by the idea that the local state might become a forum in which different social interests could be brought to the fore, repoliticizing economic change, and thereby possibly leading to strategies that might bias restructuring 'for labour' rather than simply 'for capital'. The slogan 'Think Globally, Act Locally', now a cliché of multinational business, was at that time the slogan of an optimistic urban Left (Boddy and Fudge 1984).

From the Locality debate to the New Localism

Since the early 1980s, the academic and real-world context of the debate on global–local issues has changed remarkably.[6] First, local economic development strategies have become mainstream statutory activities for all local authorities and are no longer associated with 'radical' local governments. The emphasis is firmly on productivist issues, seeking to attract and promote businesses, rather than distributional issues, although these usually still appear lower down in the mission statement. The 'New Localism' in theory and policy (Lovering 1995) has created a major market for economic development ideas, and the journalistic, practitioner and academic literatures have expanded rapidly.

Second, the world economy is now, allegedly, exhibiting new spatial tendencies. It is widely asserted that we are in the midst of a historic change in the way economies work, a claim often presented in the idea of a transition from Fordism to post-Fordism, or 'Flexibility' (Scott 1988). Increasing economic flows across national boundaries, alongside a technological revolution based in particular on the spread of communications and information technologies, are creating a new form of global economic integration (Freeman and Soete 1994, Best 1990). The core process here is 'globalization' (Dahrendorf 1995). According to Castells (1994), this means that transnational firms operate 'as global units in real-time'. It is possible as never before for firms to pick and choose between labour forces (and markets) around the world. As a result, regions and cities now compete with each other in a global economic space.

This is creating, paradoxically, a new world economy 'which is essentially local or at least sub-national' (Lash and Urry 1994: 283, see also Piore and Sabel 1984, Scott 1988). The motors of world economic forces are not countries but key regions, associated with particular concentrations of industries (Cooke 1995, Fukuyama 1995, Ohmae 1995). Some cities and city-regions have been able to exploit the changing economic environment to the advantage of their citizens (Piore and Sabel 1984, Castells and Hall 1994). Moreover, these 'competitive' regions are not only places where successful clusters of industries have developed; they are also places where poverty and disadvantage is much less severe (Castells and Hall 1994).

Successful regions in older economies are able to attract and retain investment because they have become sources of innovative new products and processes. The most important group of workers from the local/regional economic development point of view are therefore 'symbolic analysts' who work with the ideas, symbols and discourses that are entailed in developing new products and services (Reich 1991). If they flourish, they will be able to sustain the shops, restaurants and other services that provide employment for a further group of 'personal service' workers. For cities and regions in the old industrial countries there is no choice but to 'Get Smarter or Get Poorer' (Freeman and Soete 1994: 67).

Competitive regions not only have localized clusters of interrelating industries (Porter 1990), they also have 'smart' labour forces. These two features are interconnected, both depending on a localization of business relationships and a fruitful local business culture (Castells and Hall 1994, Storper 1993, Ostrey and Nelson 1994). Recent research has focused on the ways in which economic activity is 'embedded' in local cultural and social institutions (Granovetter 1990, Cooke and Morgan 1993). It suggests that a critical role is played by institutions of co-ordination which help build a distinctive public sphere, nurturing a business culture of 'trust' and co-operation adapted to long-term growth, systematic innovation and mutual benefit (Fukuyama 1995). It is precisely because they are geographically compact that it has been possible for some regions, rather than countries, to develop beneficial networks and business cultures (Castells and Hall 1994).

Cities in the 'Innovative Spaces' perspective

This debate suggests that cities face new challenges, but they also face new possibilities unrecognized in the Locality debate. The Innovative Spaces debate has shifted the focus to the potential for change arising from the adoption of local economic strategies (Castells and Hall 1994). Something *can* be done, this side of a radical social upheaval, to alter the way in which localities such as cities are inserted into global economic flows. The global–local relationship is not necessarily one of cause and effect (Smith 1995).

A developing branch of the innovative regions literature is looking at Regional Systems of Innovation (Nauwelaers and Reid 1994). The reality in Britain is that there aren't any. This debate is really about ways to *develop* new regional or sub-regional networks that might nurture more innovation. Some analysts hope this sort of neo-Schumpeterian 'boosterism' can form the basis for wider strategies and a greater participation of various social interests in the formation of local economic policy (Cooke and Morgan 1993, Hutton 1994a).

Many agree that a precondition for urban improvement is civic leadership, networking and 'concentration' between local economic actors (Judd and Parkinson 1990, Zeitlin 1989). Some also believe that forms of spatial governance are changing in ways that are likely to make this more possible (Healey *et al.* 1995). Economic policy-making in the US and Europe is shifting downwards from the state to the city-region, although Britain is behind the trend. If local economic policy communities can develop and work together to nurture the social preconditions for greater innovation, building a long-term developmental partnership between private firms and public bodies, they may be able to rebuild the city economies, replacing declining Victorian activities with twenty-first-century ones (Robson 1992).

The new focus on competitiveness and innovation is enormously influential, and has opened new research questions and policy debates. However, it all too easily slips into becoming a theoretical justification for changes that are already underway (Lovering 1995, Byrne 1995). And the Innovative Spaces literature, dominated by a 'can-do' rather than an analytical spirit, smuggles in many dubious claims. It has narrowed the analytical and policy focus on urban problems onto issues that may not be as universally important as it implies. In fact, employment patterns and change in the major urban areas cannot be adequately explained simply in terms of local 'competitiveness'. The

Plate 2.1 A world of difference exists between (a) the small-scale antiquated metals recycling business located in the inner city of Glasgow and (b) the mix of private and public capital engaged in the regeneration of the quayside in Newcastle upon Tyne. *Source:* Michael Pacione

structure of the labour market, and the construction of disadvantage, are shaped by many other influences. Similarly, on the policy front, the set of measures suggested by the new orthodoxy may not be appropriate to all cities, and especially the disadvantaged.

Getting internationalization/globalization in perspective

Although the notion of globalization has become common currency, it is often used imprecisely, and many misleading claims are made about its extent and significance (Gordon 1988, Glyn 1995, Piven 1995). Concerning the former, the evidence that globalization is increasing rapidly is not entirely clear. If globalization is measured by the share of economic activity controlled by multinational companies, the British economy has not become much more globalized over the past decade and a half (Hay 1994).[7] And few of the world's multinational corporations are genuinely global, most being firmly identified with, and based in, a home country and enjoying close relations to home governments (Archibugi and Michie 1995). Even fewer operate as Castell's 'global units in real-time' other than at the level of flows of finance. The details of the organization of production and employment are not often determined from central offices.

The globalization of the labour market is minimal. The world's jobless and low-paid workers simply cannot move around the globe to compete for jobs (except through long-term migration, which is an important factor influencing some world cities). Some employees in British cities do operate in what are in effect international labour markets, but they are relatively few in number, and mainly at elite levels. Most, especially the low paid, work in sectors in which output is not internationally traded (a corollary of the predominance of service activities, especially consumer services, and public employment, in the cities). Even those employed in manufacturing, a minority urban activity, do not all work on tasks in which low-wage countries are competitors (Krugman 1994: 48, Singh and Zammit 1995). For the majority of the urban workforce, the problem is not that they are directly competing with low-paid labour abroad,[8] but that many British employers offer low-wage, low-skill, jobs. This is not simply an inevitable result of a world-historical process of 'globalization'.

Prominent users of the term 'globalization' often elevate to the status of social scientific fact what is really only a partisan view from the boardroom of a corporation with global reach. This is disarmingly explicit in the puzzlingly influential work of Ohmae (1995), but only slightly less so in that of the 'marxisant' sociologist Manuel Castells (1989, 1994). Perhaps globalization makes a particularly strong impression on elite academics and policy makers for whom 'the world' is accessed everyday by e-mail and every month by a flight to a conference. But for many urban workers and their dependants, that world is as alien as the moon, and is likely to remain so.

'Competitiveness', 'innovation' and the urban policy agenda

The emphasis on international competitiveness and the imperative of innovation directs attention to some influences on the local (urban) economic system and its possibilities, but only some. This is obscured by the broad-brush use of the term 'competitiveness' as applied to areas such as countries, regions or cities (see Cmnd 2867, 1995). This usage derives from the highly publicized work of a group of Strategic Trade policy advocates in the US (Cmnd 2867, 1995); but it is misleading.[9]

A locality may become more 'competitive' in the sense that firms based in it increase their rate of product- or process-innovation and improve their position in the market, but this will not necessarily have a beneficial impact on the urban economy as a whole, and on workers and job-seekers generally. In fact innovation is currently destroying jobs in Europe and the US (CEC 1993, Fitch 1994: 19). In many cases new 'informational' or high-technology jobs are disappearing as fast as older manufacturing jobs. The finance sectors of London and New York contain many globally successful firms, but are shedding workers. The same is true of the 'successful' German car industry in Baden-Württemberg, and many companies in the much-admired 'Third Italy' (Harrison 1994a and b). Successful Innovative Spaces may be characterized by energetic innovation, but that has not stopped beggars emerging in their streets too. Significantly, the European Commission's Regional Innovations Systems initiative recognizes that innovation *per se* will not even begin to address labour market disadvantage (Landabaso 1995, CEC 1993).

The emphasis on competitiveness and innovation invites the creation of a new proactive local policy. Most British cities are now playing this game. But the only certain benefits are those for the service class involved (amongst whom the advocates and practitioners of this model are, not surprisingly, to be found). Whether others will benefit in the longer term remains an act of faith, and how long a term this might be is rarely discussed. The new policy approach will not address the causes of and remedies for urban disadvantage in the foreseeable future.

'Smartness' and the structure of the urban labour market

This is made abundantly clear in the ubiquitous emphasis on 'flexibility' and the closely related imperative of 'smartness'. The labour market is central to the Innovative Spaces paradigm, but only insofar as it relates to flexibility, 'learning', the production and retention of graduates, and the interfirm mobility of highly valued labour. Policies targeting these might benefit a majority of workers in a 'post-Fordist' labour market, but unfortunately this creature has yet to be spotted in Britain. As we saw in Part I, the bulk of new jobs are not learning intensive, but are low-wage, part-time and routine. There are few signs that this is changing. In Britain there is not enough innovation-driven employment to utilize the existing supply of suitable people. The employment of R & D workers is declining (Cabinet Office 1994) and graduate unemployment is almost twice the average.

Policies, especially local policies, that aim to promote innovation have a long haul ahead, and even if they are successful, they are unlikely to result in a major increase in employment, and even less likely to benefit those most in need. Thurow (1993: 52) highlights the elitist implications: 'if the route to success is inventing new products the education of the smartest 25 per cent of the labour force is critical'. Those unfortunates at the bottom of the remaining three-quarters remain 'a drain on the economy' (Friedmann 1995: 41, Lash and Urry 1994, Robson 1992). The affinities between this idea and the ideological underclass thesis are indicated by Castells' (1994: 25) reference to the 'black holes' of our society, those social conditions from which 'there is no return'.

RETHINKING THE GLOBAL–LOCAL RELATIONSHIP: FOUR SUGGESTIONS

Neither the mainstream Simple Story, nor the mildly alternative Innovative Spaces account, provide an adequate framework for understanding urban economic development and the construction of disadvantage in urban labour markets. The Innovative Spaces literature opens a fracture in the wall of gloom, but it says little about job creation and disadvantage. The prevailing theories of local economic development do not offer an appropriate framework for analysing and considering policies for urban labour markets. With this in mind I would like to conclude by offering four suggestions.

Abandon the dualistic conception of the global–local relationship

We should avoid thinking of urban economies as unitary entities. They are not actors who 'respond' to global restructuring but are made up of networks within which 'the global is present in the local' in various ways (Massey 1994, Robertson 1995: 40–1), and vice versa. Cities, like any 'locality', are no more than bits of territory carved out by local authority boundaries, travel-to-work areas, business communities or some other spatial invention which is bound to be somewhat arbitrary in terms of real social, economic, cultural and political relations (Massey 1994). The creation and distribution of incomes within this territorial space is the outcome of a number of social processes, a range of social relationships (economic, political, juridical, etc.). All this is obscured by the idea of the local (over here) versus the global (over there).

The political economy of a city is shaped by the way in which diverse discourses as to the implications of wider economic change and the purposes of business are debated, sorted and turned into agendas for action (see for example Thomas *et al.* 1996). The actors in this process, in both the public sector and in private companies, look to and intervene on various spatial levels. It is helpful to think of the peoples of a city as engaged in social and economic interactions on different spatial scales, some being more immediately local, others more international or even global, but few entirely one or the other (Lash and Urry 1994, Beauregard 1995). Participation in some networks confers much more power than that in others.

Pay more attention to the state

These asymmetries of power draw attention to politics and the state.[10] The state plays a critical role in the local balance of economic power and the social structuring of the urban labour market. Through education and training, recognition of employee rights, legitimization of professionalism, anti-discrimination policies, etc. it underwrites (or sometimes moderates) the processes of inclusion/exclusion through which distinct labour market groups are formed and lived-out (Sengenberger and Wilkinson 1995, Morris 1995). This is not often a solely 'local' matter: policies on immigration and the wider construction of national identities affect social differentiation on the international level (Featherstone *et al.* 1995).

Through housing and welfare policies, the national and local states also impact on the spatial clustering of social groups (Forrest and Gordon 1993). Disadvantage in British cities has been directly affected by the deliberate creation of very high levels of unemployment (Buxton *et al.* 1995, Glyn 1995), at the same time as encouragement for suburbanization (through both credit liberalization and road building programmes), the dismantling of welfare, privatization of housing and lack of action against discrimination. Despite the new interest in local economic policy-making, few urban and regional researchers in the UK are studying the ways in which the *de facto* local states actually work (but see Meegan 1989, Byrne 1995, Thomas *et al.* 1996).

The significance of the state is far wider than this, however. National and regional policies are also deeply implicated in the character and strength of the international economic pressures that the Simple Story and the Innovative Spaces approach perceive as 'external' influences. Although there has been a growth of literature on the internationalization of finance and its urban significance (Corbridge *et al.* 1994, Knox and Taylor 1995), it is not often noted that national policies are implicated in both, not just the latter. The ideological emphasis on 'globalization' portrays 'Global Money' as an untethered force unleashed by technological change (communication technologies) through which Chaos is erupting into the world economy (e.g. Dahrendorf 1995). Such fables obscure the fact that the new global mobility of finance capital is a recent historical construction anchored in political decisions (Michie and Smith 1995).

Since the 1970s, a core of major Western governments have applied variants of neo-liberal economic strategy while subscribing to the deregulation of international finance and exchange rates. Domestic economic and social policies have brought about an enormous shift of income towards capital (markedly so in Britain), along with a suppression of demand. The combined effect has been to enhance the relative attractiveness of speculative investments and to encourage a short-termist bias in corporations. The cross-border short-term mobility of capital has increased vastly: 90 per cent of international exchange transactions are now speculative, compared to 10 per cent in 1971 (Eatwell 1995). This has had both local and global effects. In the Golden Age of capitalism (1950–80) the rich countries invested in developing poor countries. Now they spend their surplus on speculative holdings on each other's territory, claims on the products of others, rather than investment which puts labour to work. Despite the profit boom, fixed investment in Britain in the 1980s was lower than in the 1960s and 1970s, and dominated by commercial property rather than new industrial plant and equipment (Buxton *et al.* 1995: 116). This shift has influenced the visible physical and architectural restructuring of the cities, as well as the industrial structure (Fainstein *et al.* 1993, Fitch 1994).

National policies are causally involved in creating, as well as responding to, the vicious circle of competitive deflation, rising joblessness and place-competition around 'innovation'. Querying the desirability of innovation has always been regarded as a form of Luddite simple-mindedness, on the theoretical ground that innovation will cut employment only if demand rises more slowly than productivity (OECD 1994b: 27). For every new product, there ought to be new demand. But this is necessarily true only in the tidy world of economic textbooks. In the real world, demand *is* radically restricted, and this is a result of the same government economic strategies and policy paradigms that promote innovation (Singh and Zammit 1995: 107, Glyn 1995).

Job loss in the older industrial countries such as Britain has actually been *less* severe in the past two decades than it would have been because the recession slowed productivity growth (Eatwell 1995: 272). Strategies to promote innovation, without corresponding increases in aggregate demand, are bound to remove this prop to employment, leading not to increased prosperity for all but to beggar-my-neighbour competition. Rather than being merely local adaptations to global economic change, national, regional and local economic policies are also very much a part of the global restructuring, and a pernicious part.

Recognize that the construction of disadvantage is generic to labour markets

The development of a 'casino' model of world capital flows has given financial speculators enormous influence, and their decisions are necessarily based on highly subjective assessments: 'the markets are driven by average opinion about what average opinion will be' (Eatwell 1995: 277). It is not only in international exchange markets that the 'subjective' character of the decisions of those with economic power are becoming more significant. All markets are structured by political and cultural influences. Labour markets are thereby not only an economic resource, they are also social ranking devices. Since they involve social interactions that mobilize cultural 'markers', they thereby tend to act as systems of institutional discrimination affecting identifiable social groups unevenly (Brown 1995).

But this discrimination is socially and in most cases also economically irrational. This is not an abstract theoretical point. A Department of Employment survey (Meadows *et al.* 1988) found that the alleged 'skills mismatch' of the long-term unemployed in London was largely mythical. Nor did they have unrealistic pay aspirations. They lose out because employers '*assume* them to be inadequate' (Philpott 1994: 140). Similar self-fulfilling misjudgements are likely to be rife throughout labour markets characterized by excess supply, for employers can continue to make 'mistakes' indefinitely when it is others who have to bear the costs (Boddy *et al.* 1986, chapter 5).[11] The stagnation in the aggregate demand for labour, a corollary of the international economic policies to which national and regional economic development policies generally subscribe, feeds the socially divisive character of the labour market.

Recognize that capitalist vitality has another side

The choice of a theoretical approach, and policy emphasis, is always bound up with wider world views. It is impossible to escape the influence of some conception of 'the present as history'. Recent debates over economic change in British cities seem to have been heavily influenced by the idea that the current turbulence in the world economy will lead to a new era of economic expansion and long-term social improvement.

This notion is encouraged by a number of influential writings from different traditions. For liberal thinkers, the inherent potentials of free markets, long suppressed by excessive government and collectivism, are now at last being realized (Fukuyama 1992). Marxist-influenced thought offers its own tribute to capitalist rebirth in the form of the claim that we are in the midst of a transition to a post-Fordist regime of flexible specialization (Storper and Scott 1992). Another more mechanistic variant anticipates an upsurge of activity with the long-awaited arrival of the fifth Kondratieff long-wave (Hall 1981, Freeman and Soete 1994, Tylecote 1992).

However, if capitalism is revitalizing itself, this is only part of the picture (Callinicos 1995). The basic facts of recent economic change in the UK (and the US), for example, hardly suggest renewed vitality. There is little evidence of a general radical improvement in the productivity of industrial investment

(Buxton *et al.* 1995, Glyn 1995). The profit revival in the UK since the 1970s, and the improved UK trade position since the mid-1980s, is due mainly to a series of policy-driven constraints on wages, especially the rise in unemployment and withdrawal of employee rights and unemployment benefits, and *de facto* devaluations.

World-wide, the pace of industrialization has slowed sharply since the Golden Age. Much of the south is now experiencing both de-industrialization and falling per capita incomes (Singh and Zammit 1995: 98). Total unemployment in the OECD has risen to 35 million and world unemployment is around 120 million (World Bank 1995). In Europe, economic growth is unlikely to be sufficient to prevent the further rise of unemployment by the turn of the century, let alone reduce it (OECD 1994a).

CONCLUSION

From within the increasingly glossy centres of redeveloped cities, the offices of innovation agencies, or the new market-oriented universities, it is easy to be seduced by the idea that we are in the midst of changes that express the vitality of our socio-economic system. Disadvantage appears as a regrettable, but unavoidable, pain of transition. It can be soothed by charity, and its disruptive effects contained by stronger coercion, until a new economic order arrives to benefit us all. But in the forseeable future, the world economy will be characterized by 'persistent instability, or the stability of permanent recession' well into the new century (Eatwell 1995: 281). In Britain the demand for work will continue to outstrip the supply.[12] The cities will become harsher places to live in, and more threatening ones to visit.

In this context, as Darhendorf (1995: 40) notes, 'Asian values have become the new temptation, and political authoritarianism with them.' The urban policy debate is increasingly marshalled under the banners of 'competitiveness' and 'tough love'. This lamentable reworking of the ruling ideas of the last century is evident in the policy postures of both major British political parties. It ignores, and thereby underwrites, the political character of both global and local change.

If we are to understand economic and social transformation in the cities we need a wider intellectual agenda than that which currently absorbs the energies of many analysts. If we are to deal with it, we need a wider policy agenda than that of innovation agencies and local boosters. Above all, we need more democratic participation in the decisions that are remaking our cities and countryside.

NOTES

1. Britain differs from its European partners in that an unusually high proportion of the unemployed are long-term unemployed, and registered female unem-

ployment is lower than male (OECD 1994a: 11). On most employment measures, British women are relatively more disadvantaged than elsewhere in Europe (Rubery 1994: 46).

2. The transition from school to work has been disrupted, consigning many youngsters to periods in what Rees *et al.* evocatively describe as 'Status Zero'. This augurs ill not only for their individual futures but also for that of the communities in which they live (Rees *et al.* 1996).
3. In the first draft of this chapter the Simple Story had to be constructed from a range of sources. It can now be found in unified form thanks to Dahrendorf (1995).
4. In the US the underclass debate is largely a gloss on a discourse of racial difference. In Britain it is largely a variant of a discourse of 'deviance', associated with overly fertile young women, errant young women, and economically dysfunctional sub-cultures. The term is ideological rather than scientific (see Morris 1994), but is entering into common use.
5. Curiously, the more theoretically sophisticated versions of both of these draw on Schumpeter. Innovation theorists have turned to Schumpeter's suggestions concerning the bunching of inventions (Freeman and Soete 1994). Meanwhile, writers on the 'sociology of social affairs' (Dennis 1993) have turned to his writings on the class and cultural contradictions of capitalism. According to Dennis, since the 1960s a policy-influencing intellectual elite, out of touch with economic realities, has helped to construct a new cultural orthodoxy. This has encouraged the 'dismemberment of the family' and the formation of the underclass.
6. The academic Locality debate went through an unproductively vituperative phase before losing much of its audience to the Innovative Spaces debate. The arrival of post-modernism in academic geography helped to confuse the issues, turning some away from 'economic' analyses altogether. See the special issue of *Environment and Planning A*, vol. 23 (1991) and the reprinted papers and later comments in Massey (1994). For interesting recent perspectives see, for example, Beauregard (1995) and Thomas *et al.* (1996).
7. As Michie points out (Michie and Smith 1995: xxv), the argument that globalization has integrated and transformed the world economy is hardly new: Marx highlighted it 150 years ago. See also Piven (1995). This is not to suggest that globalization is not important, but to draw attention to the tendentious ways in which the concept is now being used.
8. The increase in imports from developing countries between the 1970s and 1990s reduced demand for unskilled workers in the northern countries by only 3–9 million (less than 3 per cent of the increase in unemployment) (Glyn 1995).
9. The 'strategic trade' concept of competitiveness is strictly intellectually incoherent. A term that has a precise and clear meaning in the carefully specified abstract theories of economics (referring to individual profit-seeking capitalist firms operating in competitive markets) has been plucked out of context and blown up into an imperative for social collectivities (Krugman 1993: 260). Economic theory shows that the social benefits of participation in international trade depend not on the *absolute* advantages of individual firms or industries, but on the *comparative* advantages of the regional or national economy as a whole.
10. The lack of attention to the state in recent global–local debates is remarkable. It is due in large part to the post-structuralist strand in the 'culturalist turn' in urban and regional studies in the late 1980s. The state is 'almost absent' in Foucault and other post-structuralist gurus, who were interested in establishing a different conception of politics (Giddens 1982: 223; see also Pringle and Watson 1992).
11. Job-seekers from deprived areas are usually well aware that they are stigmatized, and they are not wrong (Lovering 1991, Morris 1995). They know only too

well that the labour market tends to implement the principle of 'to them that hath it shall be given'.

12. Even the very few economists who believe economic growth in Britain might return to something like that of the 'Golden Age' concede that 'a substantial proportion of the population' will not feel any benefit (Sentance 1995: 21).

GUIDE TO FURTHER READING

The following books provide useful and provocative overviews of aspects of the economic and social problems and prospects in the cities associated with their changing economic functions and prospects. Other important texts are identified in the references that follow.

Campbell, B. (1991) *Goliath: Britain's Dangerous Places*, London: Methuen.
A controversial and stimulating analysis of the gender dimension of urban social problems and unrest in Britain's cities, focusing on the riots of the early 1990s. Needs to be counterbalanced by an awareness of the dangers of imagining an underclass, of whatever gender (see Morris below).

Fainstein, S., Gordon, I. and Harloe, M. (eds) (1993) *Divided Cities: New York and London in the Contemporary World*, Oxford: Basil Blackwell.
A detailed, and depressing, examination of recent trends in these two untypical, but important 'world cities'.

Hall, P. (1987) *Cities of Tomorrow*, Oxford: Basil Blackwell.
A highly readable and well-illustrated account of thinking about and planning for cities since capitalist urbanization, with a pessimistic conclusion.

Healey, P., Cameron, S., Davoudi, S., Graham, S. and Madani-Pour, A. (eds) (1995) *Managing Cities: The New Urban Context*, London: Wiley.
An analytical and international collection reviewing aspects of economic and governance issues in modern cities.

Morris, L. (1994) *Dangerous Classes: The Underclass and Social Citizenship*, London: Routledge.
An excellent empirical and theoretical review of the concept of the 'underclass' and the complex realities that it obscures. Essential reading for those who are tempted to demonologize the victims of economic change.

Savage, M. and Warde, A. (1993) *Urban Sociology, Capitalism and Modernity*, Macmillan/British Sociological Association.
A good review of a hundred years of social scientific thought concerning urban problems, which helps put today's ideas into perspective.

REFERENCES

Archibugi, G. and Michie, J. (1995) 'Introduction: the economics of innovation', *Cambridge Journal of Economics* 32: 1–4.

Barclay, P. (1995) *Inquiry into Income and Wealth*, York: Joseph Rowntree Foundation.

Beauregard, R.A. (1995) 'Theorizing the global–local connection' in P.L. Knox and P.J. Taylor (eds) *World Cities in a World-System*, Cambridge: Cambridge University Press: 232–48.

Begg, I., Moore, B. and Rhodes, J. (1986) 'Economic and social change in urban Britain and the inner cities' in Victor Hausner (ed.) *Critical Issues in Urban Economic Development*, Oxford: Oxford University Press: 10–49.

Best, M. (1990) *The New Competition*, Oxford: Polity Press.
Boddy, M. and Fudge, C. (1984) *Local Socialism*, London: Macmillan.
Boddy, M., Lovering, J. and Bassett, K. (1986) *Sunbelt City?*, Oxford: Oxford University Press.
Borooah, V. and Hart, M. (1995) 'Labour market outcomes and economic exclusion', *Regional Studies* 29: 433–8.
Brown, P. (1995) 'Cultural capital and social exclusion: some observations on recent trends in education, employment and the labour market', *Work, Employment and Society* 9: 29–51.
Buxton, T., Chapman, P. and Temple, P. (eds) (1995) *Britain's Economic Performance*, London: Routledge.
Byrne, D. (1995) 'Radical geography as "mere political economy": The local politics of space', *Capital and Class* 56: 117–38.
Cabinet Office (1993) *Realising our Potential*, London: HMSO.
——(1994) *Annual Review of Government-funded Research and Development*, London: HMSO.
Callinicos, A. (1995) *Theories and Narratives*, Oxford: Polity Press.
Campbell, B. (1991) *Goliath: Britain's Dangerous Places*, London: Methuen.
Castells, M. (1989) *The Informational City*, Oxford: Basil Blackwell.
——(1994) 'European cities, the informational society, and the global economy', *New Left Review* 204: 18–32.
Castells, M. and Hall, P. (1994) *Technopoles of the World: The Making of 21st Century Industrial Complexes*, London: Routledge.
CEC (1993) *Competition, Growth, Employment*, Brussels: Commission of the European Communities.
Cloke, P., Milbourne, P. and Thomas, C. (1994) *Lifestyles in Rural England*, London: Rural Development Commission.
Cmnd 2867 (1995) *Competitiveness: Forging Ahead*, London: HMSO.
Coates, D. (1994) *The Question of UK Decline: The Economy, State and Society*, Hemel Hempstead: Harvester Wheatsheaf.
Cooke, P. (ed.) (1989) *Localities: The Changing Face of Urban Britain*, London: Unwin Hyman.
——(ed.) (1995) *The Rise of the Rustbelt*, London: UCL Press.
Cooke, P. and Morgan, K. (1993) 'The network paradigm: new departures in corporate and regional development', *Environment and Planning D: Society and Space* 11: 543–64.
Corbridge, S., Martin, S. and Thrift, N. (eds) (1994) *Money, Power and Space*, Oxford: Basil Blackwell.
Dahrendorf, R. (1995) 'Preserving prosperity', *New Statesman* 15/29 December 1995: 36–41.
Dennis, N. (1993) *Rising Crime and the Dismembered Family*, Institute of Economic Affairs: 2 Lord North St, London.
Drewer, F. and Whitehead, M. (1995) 'Mortality in regions and local authority districts in the 1990s: exploring the relationship with deprivation', *Population Trends* 8: 219–26.
Eatwell, J. (1995) 'The international origins of unemployment' in J. Michie and J.G. Smith (eds) (1995) *Managing the Global Economy*, Oxford: Oxford University Press.
Ellison, R. (1995) 'Labour force projections for countries and regions in the United Kingdom', *Employment Gazette* August: 303–58.
Fainstein, S., Gordon, I. and Harloe, M. (1993) *Divided Cities: New York and London in the Contemporary World*, Oxford: Basil Blackwell.
Featherstone, M., Lash, S. and Robertson, R. (eds) (1995) *Global Modernities*, London: Sage.

Finn, D. (1995) 'The Job Seeker's Allowance – workfare and the stricter benefit regime', *Capital and Class* 57: 1–11.
Fitch, R. (1994) 'Explaining New York City's aberrant economy', *New Left Review* 207: 17–48.
Forrest, R. and Gordon, D. (1993) *People and Places*, School for Advanced Urban Studies, University of Bristol.
Freeman, C. and Soete, L. (1994) *Work for All or Mass Unemployment? Computerized Technical Change into the 21st Century*, London: Pinter Publishers.
Friedmann, J. (1995) 'Where we stand: a decade of world city research' in Paul L. Knox and Peter J. Taylor (eds) *World Cities in a World-System*, Cambridge University Press: 21–48.
Fukuyama, F. (1992) *The End of History and the Last Man*, Harmondsworth: Penguin Books.
——(1995) *Trust*, London: Hamish Hamilton.
Giddens, A. (1982) *Profiles and Critiques in Social Theory*, London: Macmillan.
Glyn, A. (1995) 'Social democracy and full employment', *New Left Review* 211 May/June: 33–55.
Glyn, A. and Miliband, D. (eds) *Paying for Inequality: The Economic Cost of Social Injustice*, London: IPPR/Rivers Oram Press.
Gordon, D. (1988) 'New edifice or crumbling foundations?', *New Left Review* March/April: 112–40.
Gordon, I. and Sassen, S. (1993) 'Restructuring the urban labour markets' in S. Fainstein, I. Gordon and M. Harloe (eds) *Divided Cities: New York and London in the Contemporary World*, Oxford: Basil Blackwell: 105–28.
Granovetter, M. (1990) *Entrepreneurship, Development and the Emergence of Firms*, WZB, Berlin, discussion paper FSI 90–2.
Grant, S. (1995) 'Poverty: definitions, causes and suggested remedies', *British Economic Survey* 25: 1–4.
Hall, P. (1981) 'The geography of the fifth Kondratieff cycle', *New Society* 26 March: 535–7.
Harrison, B. (1994a) 'The Italian industrial districts and the crisis of the cooperative form Part I', *European Planning Studies* 2: 3–22.
——(1994b) 'The Italian industrial districts and the crisis of the cooperative form Part II', *European Planning Studies* 2: 159–74.
Harrop, A. and Moss, P. (1995) 'Trends in parental employment', *Work, Employment and Society* 9: 421–44.
Harvey, D. (1985) *The Urbanisation of Capital*, Oxford: Basil Blackwell.
Hay, R. (1994) 'Recent trends in Overseas Direct Investment', *Economic Trends* 491 September: 19–23.
Healey, P., Cameron, S., Davoudi, S., Graham, S. and Madani-Pour A. (eds) (1995) *Managing Cities: The New Urban Context*, London: Wiley.
Hills, J. (1995) *Inquiry into Income and Wealth*, York: Joseph Rowntree Foundation.
Hobsbawm, E. (1994) *The Age of Extremes: The Short Twentieth Century 1914–1991*, London: Michael Joseph.
Hutton, W. (1994a) *The State We're In*, London: Jonathan Cape.
——(1994b) *Global Markets, Regional Clusters, and Liberal Democracy*, Occasional Paper No. 2, Centre for Advanced Studies in the Social Sciences, University of Wales College of Cardiff.
——(1995) 'High risk strategy', *The Guardian* October 30.
Judd, D. and Parkinson, M. (eds) (1990) *Leadership and Urban Regeneration*, Sage Urban Affairs Annual Reviews 37: 29–52.
Knox, P.L. and Taylor, P.J. (eds) (1995) *World Cities in a World System*, Cambridge: Cambridge University Press.

Krugman, P. (1994) *Peddling Prosperity: Economic Sense and Nonsense in the Age of Diminished Expectations*, New York and London: W.W. Norton and Co.

Landabaso, M. (1995) *The Promotion of Innovation in Regional Community Policy: Lessons and Proposals for a Regional Innovation Strategy*, Paper to Regional Science and Technology Policy workshop organized by the Japanese Institute of Science and Technology Policy, Himejl.

Lash, S. and Urry, J. (1994) *Economies of Signs and Space*, London: Sage.

Leyshon, A. and Thrift, N. (1995) 'Geographies of financial exclusion: financial abandonment in Britain and the United States', *Transactions of the Institute of British Geographers* 20: 312–41.

Lovering, J. (1988) 'The local economy and local economic strategies', *Policy and Politics* 16: 145–57.

——(1989) 'The restructuring debate' in R. Peet and N. Thrift (eds) *New Models in Geography Vol. 2*, London: Unwin Hyman: 198–223.

——(1990) 'Fordism's unknown successor: a comment on Scott's theory of flexible accumulation', *International Journal of Urban and Regional Research* 14: 159–74.

——(1991) *Bridging the Gap: Skills Training and Barriers to Employment in Bristol*, Inner City Task Force, Bristol City Council, Avon County Council.

——(1995) 'Creating discourses rather than jobs: the crisis in the cities and the transition fantasies of intellectuals and policy makers' in P. Healey *et al.* (eds) *Managing Cities: The New Urban Context*, London: Wiley: 109–26.

Malecki, E. (1991) *Technology and Economic Development: The Dynamics of Local, Regional and National Change*, London: Longmans.

Massey, D. (1994) *Space, Place and Gender*, Oxford: Polity Press.

Massey, D. and Meegan, R. (1982) *The Anatomy of Job Loss*, London: Methuen.

Meadows, P., Cooper, H. and Bartholomew, R. (1988) *The London Labour Market*, Employment Department, London: HMSO.

Meegan, R. (1989) 'Paradise postponed, the growth and decline of Merseyside's outer estates' in P. Cooke (ed.) *Localities: The Changing Face of Urban Britain*, London: Unwin Hyman: 198–234.

Michie, J. and Smith, J.G. (eds) (1995) *Managing the Global Economy*, Oxford: Oxford University Press.

Morris, L. (1994) *Dangerous Classes: The Underclass and Social Citizenship*, London: Routledge.

——(1995) *Social Divisions: Economic Decline and Social Structural Change*, Cambridge Studies in Work and Social Inequality 2, London: UCL Press.

Murray, C. (1994) *The Underclass: The Crisis Deepens*, Institute of Economic Affairs, Choice in Welfare Series No. 20, in association with *The Sunday Times*, London.

Nauwelaers, C. and Reid, A. (1994) *Innovative Regions? A Comparative Review of Methods of Evaluating Regional Innovation Potential*, RIDER and DG XIII-D, Luxemburg: Commission of the European Communities.

OECD (1994a) *The OECD Jobs Study: Facts, Analysis, Strategies*, Paris: OECD.

——(1994b) *The OECD Jobs Study: Implementing the Strategy*, Paris: OECD.

Ohmae, K. (1995) *The End of the Nation State: The Rise of Regional Economies*, London: McKinsey & Co.

Ostrey, S. and Nelson, R.E. (1995) *Techno-nationalism and Techno-globalism*, Washington: Brookings Institution.

Philpott, J. (1994) 'The incidence and cost of unemployment' in A. Glyn and D. Miliband (eds) *Paying for Inequality: The Economic Cost of Social Injustice*, London: IPPR/Rivers Oram Press: 130–44.

Piore, M.J. and Sabel, C.F. (1984) *The Second Industrial Divide: Possibilities for Prosperity*, New York: Basic Books.

Piven, F.F. (1995) 'Is it global economics or neo-*laissez-faire*?', *New Left Review* 213: 107–14.

Porter, M. (1990) *The Competitive Advantage of Nations*, New York: Free Press.
Pringle, R. and Watson, S. (1992) 'Women's interests and the post-structuralist state' in M. Barrett and A. Phillips (ed.) *Destabilising Theory: Contemporary Feminist Debates*, Oxford: Polity Press: 121–40.
Rees, G., Williamson, H. and Istance, D. (1996) '"Status Zero": a study of jobless school-leavers in South Wales', *Research Papers in Education* forthcoming.
Reich, R. (1991) *The Work of Nations: A Blueprint for the Future*, New York: Simon & Schuster.
Robertson, R. (1995) 'Glocalisation: Time–space and Homogeneity–Heterogeneity' in M. Featherstone, S. Lash and R. Robertson (eds) *Global Modernities*, London: Sage: 24–44.
Robson, B. (1992) 'The twenty-first century city: A British perspective' in A.P. Cohen and K. Fukui (eds) *Humanising the City: Social Contexts of Urban Life at the Turn of the Millennium*, Edinburgh: Edinburgh University Press: 36–51.
Rubery, J. (1994) 'Women and men in the European labour market' in F. Brouwer, V. Linter and M. Newman *Economic Policy Making and the European Union*, London: Federal Trust for Education and Research/PSI Publishing: 43–50.
Schmitt, J. and Wadsworth, J. (1994) 'The rise in economic inactivity' in A. Glyn and D. Miliband (eds) *Paying for Inequality: The Economic Cost of Social Injustice*, London: IPPR/Rivers Oram Press: 114–30.
Scott, A.J. (1988) 'Flexible production systems and regional development', *International Journal of Urban and Regional Research* 12: 71–185.
Sengenberger, W. and Wilkinson, F. (1995) 'Globalisation and labour standards' in J. Michie and J.G. Smith (eds) (1995) *Managing the Global Economy*, Oxford: Oxford University Press: 111–34.
Sentance, A. (1995) 'Are we entering a new golden age of economic growth?', *Economic Outlook* 20: 12–21.
Singh, A. and Zammitt, A. (1995) 'Employment and unemployment, north and south' in J. Michie and J.G. Smith (eds) (1995) *Managing the Global Economy*, Oxford: Oxford University Press: 93–110.
Sly, F. (1995) 'Ethnic groups and the labour market', *Employment Gazette* June 251–62.
Smith, P.M. (1995) 'The disappearance of world cities' in P.L. Knox and P.J. Taylor (eds) *World Cities in a World-System*, Cambridge: Cambridge University Press: 249–66.
Solesbury, B. (1993) 'Reframing urban policy', *Policy and Politics* 21: 31–8.
Storper, M. (1993) 'Regional "worlds" of production: Learning and innovation in the technology districts of France, Italy and the USA', *Regional Studies* 27: 433–55.
Storper, M. and Scott, A.J. (eds) (1992) *Pathways to Industrialization and Regional Development*, London: Routledge.
Taylor, R. (1995) *Flexible Employment in Britain*, Equal Opportunities Commission: Manchester.
Thomas, H., Stirling, T., Brownhill, S. and Razzaque, K. (1996) 'Locality, urban governance and contested meanings of place', *Area* forthcoming.
Thurow, L. (1993) *Head to Head: The Coming Economic Battle among Japan, Europe and America*, London: Nicholas Brealey Publishing.
Tylecote, A. (1992) *The Long Wave in the World Economy: The Current Crisis in Historical Perspective*, London: Routledge.
Willmott, P. (ed.) (1994) *Urban Trends 2: A Decade in Britain's Deprived Urban Areas*, London: Policy Studies Institute.
World Bank (1995) *Workers in an Integrating World; World Development Report 1995*, Oxford: Oxford University Press for the World Bank.
Zeitlin, J. (1989) 'Local industrial strategies', *Economy and Society* 18: 367–73.

3

NATIONAL ECONOMIC POLICY IN THE UNITED KINGDOM

Rob Imrie

INTRODUCTION

For the last eighty years or so the city, as a theoretical and empirical referent, has been a staple and significant component of social scientific analysis. From Simmel's (1964) seminal accounts, in seeking to specify the city as the locus of modernity, to Wirth's (1938) contention that cities are definable in terms of a generic urban culture, scholars have been preoccupied with the idea of the city as the kernel of society (Robins 1995, Savage and Warde 1993). In particular, researchers have come to conceptualize the city as the locus of industrial production and consumption, indicating that its socio-spatial structures are both a reflection and expression of the dynamic flux of the wider national and international economies (Healey *et al.* 1995). Indeed, much of urban sociology and geography have evolved from their concerns with the interrelationships between economic processes and the changing social structures of the cities and, as Savage and Warde (1993) have commented, the inequalities constantly generated by the mechanisms of capitalist accumulation are a persistent and constant feature (of the cities).

As Kleinman (1994) argues, the debate on the interrelationships between economic policy and the social structure of the cities has shifted over time (see also Robson 1988). In the 1960s and 1970s, in a context of de-industrialization, the focus was related to the absolute losses of both employment and population from the cities or with managing urban decline (Davies *et al.* 1982). Indeed, such was the concern with the economic decline of the British cities in the 1970s that the then Secretary of State for the Environment, Peter Shore, announced that 'if cities fail, so to a large extent does our society; that is the urgency of tackling the problems' (quoted in Davies *et al.* 1982: 2). Since the 1980s, in a context of aggregate economic growth in the UK, academic and policy analysis has been more concerned with what appears to be the opening up of significant economic differences between the poor and the rich in the cities and with what Kleinman (1994) terms the classic paradox, that while the cities appear to be the engines of economic growth they are, simultaneously, the locus of poverty and social disadvantage.

This locus has been a perennial concern of academics and policy makers alike and a targeted inner cities policy has existed in the UK for nearly thirty years as one attempt to address its contours. Yet, all of the evidence suggests that policy attempts to redress what many have termed 'the crisis' of the inner cities have systematically failed. Indeed, authors like Lawless (1991) and Robson (1988) have castigated policy for the cities as incoherent in lacking institutional co-ordination and political support, while floundering on the paltry level of resources committed to the main policy programmes. In particular, inner city policy has become increasingly aligned with national economic policy inasmuch as its shifting programmes have reflected the broader manoeuvre of policy from what Fainstein *et al.* (1993) characterize as social service and welfare provision to economic development initiatives (see also Keating 1993, Mingione 1995).

In turn, a broader economic rationality has taken centre stage, forcing a disjuncture between what Keating (1993) characterizes as the economics of the 'real world' and the notion of the social economy or providing for groups that require some subsidization or protection from economic policy. Indeed, national economic policy over the last ten years or so has been implicated in extending socio-spatial divisions in the cities by simultaneously pursuing policies of monetary restraint, the diminution of controls on corporate behaviour, and the withdrawal of state support from a range of public sector expenditures. Policy has also become partial and selective, especially in cultivating strategies of financial deregulation while starving manufacturing industry of investment capital. As Amin and Tomaney note, while the former strategy has favoured the City and the south-east of England, the latter has tended to hit hardest the older industrial cities so actively fostering 'social and economic inequality and segmentation' (Amin and Tomaney 1995: 174). Fainstein *et al.* are also unequivocal in claiming that the effects of such policies in New York and London 'worsened the condition of a large proportion of their populations and may have contributed to long term instability' (Fainstein *et al.* 1993: 3).

There is, then, little doubt that national economic policies have the capacity to influence the socio-economic structures of the cities; yet, in expressing this, a number of crucial qualifications need to be made. Foremost, it is important to avoid a reification of national economic policy, as though policy, in and of itself, exists independently of wider, determinate, socio-political forces. It is also important to avoid an essentialist perspective and, in this sense, policy does not just simply reflect the imperatives of global economic forces but is interconnected with them in quite complex, often contradictory ways. There is also a question of interlocking scales and of the possibilities that what is termed 'national' economic policy is little more than a series of stratagems linked into scale divisions of governance that transcend the boundaries of the nation state. Jessop *et al.* (1987), for example, refer to the 'hollowing out' of the nation state to indicate how policy is increasingly being driven by transnational modes of governance, like the World Bank and the International Monetary Fund (see also Dicken 1994).

The issue of national economic policy and its intra-urban effects is complex for other reasons too. Foremost, it is difficult to gauge, with any accuracy, the role of policy in producing specific intra-urban outcomes from other determinate influences. For instance, it is commonly assumed that social polarization has been one of the more visible and evident outcomes of recent rounds of economic restructuring and policy in the cities. Yet, as Pinch (1993) points out, there is a wide range of processes leading to social polarization from changes in family structure to the residualization of state welfare services. From this, he concludes that 'given the multiplicity of such influences it is often difficult to isolate their specific effects, especially as some influences may be mutually reinforcing whereas other influences may work in opposite directions' (Pinch 1993: 780). This point holds in analysing the interrelationships between economic policy and its effects, a complexity that is also compounded by geographical variations and the highly uneven ways in which national economic strategies work themselves out across the UK space-economy.

In the next section, I outline the main lineaments of national economic policy since the early 1980s and develop the contention that the pursuit of neo-liberal strategies, in all of its strands, has purposively cultivated socio-spatial divisions as a cornerstone of economic strategy. The main features of policy to have influenced the fortunes of the cities are, as I shall argue, monetary and fiscal strategy, labour market deregulation, the privatization of the public utilities, the dismantling of systems of income redistribution and the eschewing of a commitment towards the (fiscal) support of domestic industry in favour of inward investment by the leading global corporations. Within the ambit of national economic strategies of this type, spatial policies for the cities have been given a specific role, of recommodifying the seemingly redundant spaces or the propagation of development by seeking to lock the inner cities into wider transnational investment networks. As the third part of the chapter will show, the combination of such strategies has been one of the crucial bases in reproducing and maintaining social divisions within the cities.

THE LINEAMENTS OF NATIONAL ECONOMIC POLICY

As Hutton notes, the period since 1979 has brought to bear 'a uniquely powerful combination of forces . . . to promote the market as the sole organising principle of economy and society' (Hutton 1995: 169). Underpinning this has been neo-liberalism or a particular brand of economic philosophy emphasizing 'individual choice' over 'public good' and 'entrepreneurship' over 'socialised' forms of work organization (see Thornley 1991). The popularization of such conceptions signalled the demise of the post-war commitment to Keynesian economics whereby successive British governments had intervened to manage levels of aggregate demand in the economy. Since 1945, economic policy had revolved around key principles of Keynesian strategy, including the guarantee of a minimum social wage, a commitment to full employment, public support for universal services like health care and social security, and the redistribution of economic surpluses through a whole range of transfer mechanisms. However, the rising costs of seeking to support the welfare state, in combination with the diminution in the competitiveness of British capital, served to undermine political consensus on the efficacy of welfarist, redistributive measures, and by the end of the 1970s neo-liberal ideologies were taking centre stage.

While it would be problematic to talk of a coherent approach to national economic policy since 1979, it is clear that one of the consistent elements has been the pursuit of a broadly monetarist policy, what Margaret Thatcher, on her election victory in 1979, referred to as a 'sound money' policy. In part, this focus reflected the domineering hold of financial values over British industrial structures and the obsession of its main financiers with liquidity and short-term profits. For Hutton, this represented the staple basis of British capitalism where 'the state has developed a rentier culture that complements "gentlemanly capitalism", in which the state stresses the financial virtues of balanced budgets and financial targets over investment and production' (Hutton 1995: 53). The unfolding of macro-economic policies in the early 1980s, then, was purely financial: stringent targets for the money supply; a reduction in public borrowing and spending; and, above all, a low inflation rate. The overall aim of achieving price stability was manipulated by the government's control over both exchange and interest rates and, throughout the 1980s and early 1990s, one of the consistent features of policy was to maintain an overvalued exchange rate, while maintaining high interest rates with which to depress the money supply.

This broader framework has been implicated in supporting both the City of London and select multinational firms in seeking to source profitable spaces and places for investment; and one of the more significant events of the last twenty years or so was the relaxation of exchange controls, an event that was at the root of the subsequent collapse of Britain's manufacturing base and with it the fabric of many of Britain's older industrial cities. Indeed, the abolition of exchange controls in 1979 stimulated international demand for sterling which, in part, was linked to a 17 per cent rise in effective exchange rates of sterling between May 1979 and the autumn of 1980 (Leys 1985). The (overvalued) currency, in turn, precipitated the collapse of exports while domestic markets, due to major import penetration and the collapse of consumer spending in the early 1980s, were more or less non-existent. This, then, set the stage for rampant monetarism, with whole industrial sectors being decimated and the (formerly) staple industrial bases of the older industrial cities being cut away.

Such processes were, in part, connected to the shifting international division of labour and the growing globalization of finance and industry which, in turn, exerted pressure on the British state to facilitate the international competitiveness of British industry. As Cox argues, 'states are becoming the instruments for adjusting national economic activities to the exigencies of the global economy . . . adjustment to global competitiveness is the new categorical imperative' (Cox 1993: 260). For Britain, in the late 1970s, such adjustments were crucial in a context of a profits and labour productivity crisis in its leading industrial sectors, while overproduction and underconsumption were endemic. For some, the response (necessarily) represented the break-up of Fordist forms of industrial organization and the diminution of vertically integrated forms of production towards flexible production systems, while the regulatory nature of the state, in providing social wages and support to many industries through direct subsidy, was seen as inflationary and also as a core element in maintaining the overproductionist and/or underconsumptionist structures of the British economy.

Thus, in responding to the exigencies of the wider global economy, an underlying restructuring of some of the regulatory elements of the state became a key feature of the neo-liberal strategy. In particular, stringent fiscal policy underpinned the demise of investment functions by the state and, as Hutton notes, throughout the last fifteen years the state has abandoned support for industrial investment, R & D and innovation because of the belief 'that growth and investment are determined naturally by technical change and population growth, and that the best governments can do is not to artificially interfere with these processes' (Hutton 1995: 80). Yet, the real effect of this, as Hutton (1995), and others, have documented, was to starve manufacturing industry of much needed investment capital and, paradoxically, to inhibit its chances of regaining international competitiveness. In the early period of the monetarist 'experiment', the effect of such policies was highly uneven and the experiences of the British cities have tended to reflect Jessop *et al.*'s (1987) conception of the 'two nations hegemony' of Thatcherism.

Likewise, Peck and Tickell have argued that Thatcher's economic policies were underpinned by a 'limited hegemonic policy' or one which 'underwrote the growth of incomes in the south at the expense of a withdrawal of public services in the north and by the redistribution of incomes from the poor to the rich' (Peck and Tickell 1995a: 38). Higher rates of income taxation were abandoned, so providing a windfall for the rich, yet forms of indirect taxation, much of it on basic commodities, were increased with the result that tax burdens were switched from richer to middle income and poorer households. As Peck and Tickell (1995a) have outlined, tax cuts in 1988 resulted in 60 per cent of the benefit going to the south-east of England, a region with 30 per cent of the national population and the most affluent households. Yet, even here, there were marked inequalities and, as Thornley (1992) recounts, in London in 1987, the average gross weekly income of the poorest 10 per cent of households was £57, representing an increase of 30 per cent since 1980. In contrast, the richest households had a gross weekly income of £670 or a figure that had increased by 118 per cent since 1980.

Yet, other processes were at work too in undermining the fabric of the socio-economic structures of the cities and one of the crucial episodes since the late 1980s has been the encouragement of a credit boom based upon the deregulation of financial services. For instance, the relaxation of the rules for lending mortgage finance in the mid-1980s led to a flood of mortgage finance being made available for home ownership, yet the collapse in housing markets after 1989 has left a legacy of negative equity and social misery, with many people in the cities unable to 'pay their way'. Moreover, the flood of credit stimulated speculative property developments, from new starter homes through to skyscraper office blocks. The spatial reorganization of neighbourhoods was increasingly underpinned by 'easy money' which, in part, was facilitating new forms of gentrification. For Kleinman, the credit boom was one of the key mechanisms in exacerbating social polarizations between the rich and the poor in the British cities, between those who could access it and those who could not. As he argues, one of the effects was a 'sharp and shocking distinction between a successful urban elite and an impoverished underclass' (Kleinman 1994: 8).

Ironically, while government policy was seeking to facilitate (inflationary) flows of financial capital, it was simultaneously restructuring public expenditure and, as a range of authors have noted, perhaps the most precipitous effect of national economic policy on the cities was the dramatic cuts in their fiscal bases and the diminution in their funds for investment in social infrastructure (see Cochrane 1993, Healey *et al.* 1995). For the central state, cuts in public expenditure *per se* were a prerequisite towards a tax cutting agenda and, in essence, redistributing funds from social welfare programmes to individual consumers was a core element of the neo-liberal philosophy of seeking to 'incentivize' the (seemingly mythical) entrepreneurial basis of society. In effect, this manoeuvre was part of a wider, well-documented, regressive redistributive process from poorer to richer households and, as Hutton notes, the emergent welfare delivery systems were being organized 'so as to mimic markets in the search for economic efficiency, and the welfare system shaped to reinforce a flexible labour market' (Hutton 1995: 307).

The undermining of social distributive mechanisms, while problematical for the poorer populations of the British cities, has been compounded by the central state's avowal to create a 'leaner and fitter' economy by deregulating the labour market and creating the conditions for flexible forms of labour utilization. The period since 1980 has witnessed an increasing exposure of employees to employer power. In part, this has been achieved by a diminution in the power of the trade unions, and the dismantling of a range of legislation governing the conditions of work, including the wage councils.

As Hutton (1995) argues, Thatcher's union reforms, for instance, were crucial to her larger aim of revitalizing British capitalism, while her exhortation to workers to price themselves into a job was a convenient way of blaming them and their (restrictive) practices for the collapsing state of the British economy in the early 1980s. In addition, firms began to utilize more flexible forms of employing people, ranging from part-time, contract labour to a growing casualization of the labour market, all of which were, and still are, being encouraged and, in part, facilitated by strands of national economic policy.

As Hutton notes, such processes are at the heart of society 'dividing before our very eyes, opening up new social fissures in the working population' (Hutton 1995: 106). Thus, 4 million men are out of work, while the new workers are predominantly women working part-time, usually working for minimal wages and with few rights to statutory redundancy payments and/or the means to contest unfair dismissal. Moreover, it is estimated that 30 per cent of the adult population are either unemployed or economically inactive, while even for those in work the realities hinge around fixed term contracts and ever widening job descriptions which result in people having to take on ever increasing work loads. Thus, as Hutton (1995) notes, while in 1975, 55 per cent of the adult population held full-time, tenured jobs, this proportion had fallen away to 35 per cent by 1993, while employment demarcations have more or less disappeared. The segmentation of the labour market in such ways is, according to Hutton, underpinning the 'sculpting of the new and ugly shape of British (urban) society' (Hutton 1995: 108).

THE PARADOXES OF NATIONAL ECONOMIC POLICY FOR THE CITIES

The impacts of economic policy on the cities have been many and profound and range from the increase in absolute and relative levels of poverty, to the emergence of significant mobility inequalities given the reduction of expenditures on intra-urban public transportation systems (Imrie 1994). The effects go well beyond this though and include the emergence of a fiscal crisis within local government, the collapse of urban housing markets and a growth in income inequality and labour market polarization. Fainstein *et al.* argue that the core (economic) policy programmes in the 1980s were also typified by a switch away from compensating the 'vulnerable sections of the population for the diswelfarism of a market-dominated urban system, to those who focused on supporting the further growth of the dynamic sections of the urban economy' (Fainstein *et al.*: 6). In this sense, the underlying structural logic of neo-liberalism was, and still is, predicated on the cultivation of socio-spatial inequalities. In particular, three significant, interrelated effects of neo-liberal economic policy are discernible in the British cities.

The first relates to the shifting modes of local governance in the cities and the central state's attempts to secure both fiscal and political control over potentially rebellious local administrations. The implementation of national economic strategy has required the (political) quiescence of the cities, which, in turn, has been reflected in forms of fiscal control, the de-democratization of the local state and the political marginalization of a whole host of urban communities. In this sense, political quiescence was both a pre-condition of and, in part, an outcome of neo-liberal economic strategy. Such quiescence, in turn, has underpinned a second aspect of change in the cities: the emergence of new (privatized) spaces of production and consumption predicated on a property-led approach to urban regeneration which, in turn, are implicated in the production of places of inclusion and exclusion. A third, related, strand has been the emergence of new patterns of social inequality, or, to use Marcuse's (1995) metaphor, the reproduction of the 'quartered city', characterized by increasing fragmentation of the labour market. Such fragmentation has, in turn, (re)produced particular patterns of inequality, especially by age, gender and race. I will now discuss each of these categories in turn.

National economic imperatives and the restructuring of local modes of governance

One of the more important effects of national economic policy on the cities has been the transformation in local governance structures. In a context of fiscal austerity, the central state has sought to impose market criteria as the basis for the development and delivery of local social and/or welfare services. In effect, the privatization of policy has been utilized as a mechanism to achieve 'value-for-money', based on the assumption that resources are best utilized in a private sector culture. A whole range of policy measures have been co-opted by business leaders who, as Healey *et al.* (1995) note, find themselves bringing their 'vision' and 'leadership' to the Boards of Urban Development Corporations (UDCs), Training and Enterprise Councils (TECs), Enterprise Agencies (EAs) and Regional Development Companies. As Peck and Tickell (1995b) argue, the 'new leaders' seem to constitute an empowered urban elite akin to the philanthropic businessmen of the Victorian era; yet, as the Centre for Local Economic Strategies (1990) and others have noted, their power base is partial and their development objectives narrowly construed around commercial returns on projects.

The redrawing of central–local government relations, which has accompanied the privatization of particular facets of policy, has been multidimensional, yet with the common objective of a diminution in the power of local government. As Cochrane (1993) notes, local government finance has shifted from the introduction of expenditure targets for specific local authorities, with sanctions for overspending, to the utilization of local tax capping with limits placed on the level of local taxes that local government may levy. Local government has lost powers in policy formulation and service delivery to a range of sectional interests, including parents, voluntary groups and, significantly, businesses. The strategy of privatization has involved the withdrawal of key activities from local government, the contracting out of services, and an increase in legal forms of control over local authorities. Moreover, the focus on business elites, quangos and other localized forums for policy delivery has led to a proliferation of non-elected bodies dealing with the socio-economic problems of the British cities. In sum, public policy has gradually come to be dominated by central directions, with implementation heavily influenced by the private sector and market trends.

The centralization of public policy, and a diminution in local government autonomy, is well exemplified with regard, for example, to the financing of urban policy which, as Robson claims, has been 'piecemeal, ad hoc, and subject to the law of one hand taking away what the other was giving' (Robson 1988: 96). For instance, while the overall level of finance in the Urban Programme increased substantially, from £29 million in 1977–8 to £361 million in 1985–6, no additional government spending was actually involved, in that additional spending on the UP was a product of savings made by government cuts in the Rate Support Grant (RSG) to urban councils. Moreover, cuts in the RSG were falling disproportionately on the inner cities and, as the Archbishop of Canterbury's Commission on Urban Priority Areas noted, between 1981–2 and 1984–5, the inner city partnership authorities had their RSG cut by 22 per cent, well above the national average of 9 per cent for the same period. Fiscal controls were also being redrawn and, as Duncan and Goodwin (1988) have noted, the increase in spending on the Urban Programme signalled a shift from local authority controlled Partnership Programmes to the non-elected UDCs and other (privatized) policies. By 1988–9, of the alleged £3 billion that central government had allocated to the cities, only 10 per cent, the Urban Programme, was subject to any form of local government influence (Lawless 1991: 25).

This, however, was an important objective of neo-liberal economic policy, of bringing local state spending into line with monetary targets and reducing overall levels of public expenditure. Yet, as a range of commentators have documented, this precipitated an on-going conflict between the local and central state over fiscal targets; and one of the key effects, as the central state has sought to discipline local states and exercise its authority, has been a process of de-democratization and of social and political closures of wider civil society to the new institutions of governance (Cochrane 1993, Lewis 1993). In particular, it is estimated that 40 per cent of government expenditure is administered by non-elected bodies of one type or another while, for organizations like the UDCs, their constitutional position is 'inadequately defined and regulated' (Lewis 1993: 3). Indeed, UDCs operate in a number of 'closed' ways including the exclusion of the public from their board meetings to the non-disclosure of the recorded minutes. They have also been criticized for their failure to publish corporate plans, while Byrne (1993) likens them to a form of colonial administration, a superimposed body with little or no sensitivity to their localized operating environments.

In part, this reflects the new agenda or one in which neo-liberalism is attempting 'to reduce spending on welfare, helping to create a permanently beleaguered atmosphere of managing cuts within the local state' (Cochrane 1993: 118–19). Part of the price being paid is the recasting of democratic involvement in local governance as a market relation, where the relationship between service providers and users is one in which the latter are conceived of as customers or consumers and where their rights to influence the nature of service provision are being propagated as a function of their 'spending power'. What we have here then is a shareholder mentality or one where those who hold the stakes get to play the game. For the poor, the reality is one of effective exclusion from localized political systems. The objective, of course, is the development of a new welfarism in the cities, a welfarism leading to a neglect of the more basic, yet important, forms of social provision. In turn, the crisis of local government, of rate capping, the truncation of universalistic services and the linking of service provision to consumer sovereignty (some would say solvency) is there for all to see, ranging from the 'sink' council estates, which lack the most basic of amenities, to the crumbling roadways and the collapsing sewer systems.

Property-led regeneration and the commodification of the inner cities

One of the more significant intra-urban implications of government economic policy has been the emphasis placed upon the revalorization of inner city spaces by encouraging the development of new spaces of production and consumption. A range of fiscal measures were introduced throughout the 1980s to encourage developers to take up derelict and semi-derelict sites in the inner cities and to develop them for a range of commercial uses. The underlying idea of a property-led approach to urban regeneration is linked to the notions of 'development', and the necessity for urban government to pursue economic growth goals given the alleged dependence of the local citizenry and state on global capital and labour flows for their (economic) welfare. Such views have been popularized by Peterson (1981) who, in writing about the regeneration of cities in the USA, argues that the pursuit of welfare simultaneously requires cities to maintain people who contribute most to the tax base, while minimizing redistributive policies that act, potentially, as disincentives to capital investment. This perspective suggests that cities are unable to pursue an autonomous pathway and that (urban) political strategy is firmly tied to the wider imperatives of globalized capital accumulation and of seeking an advantageous position within the international division of labour.

This set of neo-liberal views has underpinned urban policy which, in the early 1980s, was restructured in line with national and international economic imperatives to reflect the broader concern with improving the productive

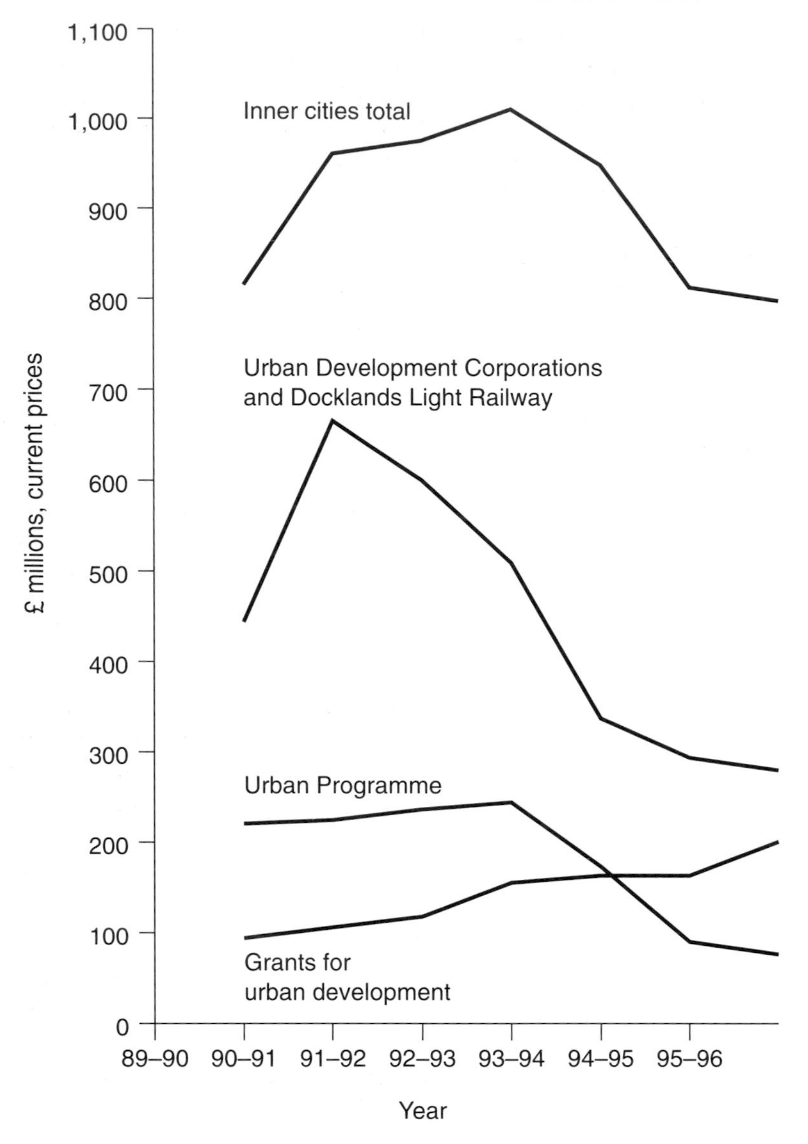

Figure 3.1 The changing nature of urban policy expenditure 1989/90–1995/6
Source: National Audit Office 1989, Department of the Environment 1993

capacity of the cities. As Figure 3.1 conveys, the changing expenditure on urban policy in the UK indicates a shift from broad-based welfare spending programmes towards infrastructure and building projects directly linked to the supply-side ethos of monetarism. Thus, between 1989–90 and 1995–6, welfare-oriented Urban Programme funds were truncated from £223 million to £80 million. In contrast, spending on the supply-side policies of the UDCs and the Docklands Light Railway increased from £477 million in 1989–90 to £607 million in 1990–1 before diminishing to £284 million in 1995–6, still considerably above expenditure on the Urban Programme. Moreover, grants for urban development, which are primarily subsidies to private investment capital, more than doubled over the period 1989–90 to 1995–6 and, as Amin and Tomaney have noted, such policies are being oriented towards a new form of welfarism, one that is 'less concerned with the needs of the unemployed and economically marginalised groups but more with targeting resources towards reproducing the most economically active groups' (Amin and Tomaney 1995: 172).

In turn, such policies have been expressed directly through the transforming nature of the built environment and by the recommodification of particular inner city spaces. In particular, property development has interlocked with wider economic processes in reproducing what Soja refers to as 'a combination of decentralisation and recentralisation, the peripheralisation of the center and the centralisation of the periphery, the city simultaneously being turned inside out and outside in' (Soja 1995: 131). Thus, the plethora of tax breaks, development grants and other fiscal incentives targeted at the cities by government became the staple basis of neo-liberalism in the cities or where particular spaces were being targeted to draw in global capital. As Soja (1995) notes, the new economic imperatives are introducing new and different forms into the built environment, encouraging gentrification in working-class neighbourhoods, and the development of new business centres, while sweeping away older industrial areas with skyscrapers devoted to the emergent service economy. For Fainstein, the new spaces are immoral and richly symbolic of the ethos of neo-liberalism, that inequality should be seen as the price of wealth creation. As she recounts:

> By focusing on the construction of first class office space, luxury housing, and tourist attractions, and short changing the affordable housing, small business and community-based industrial sectors, they prompted developers to engrave the image of two cities – one for the rich and one for the poor – on the landscape. Redevelopment took the form of islands of shiny new structures in the midst of decayed public facilities and deterioration in living conditions for the poor. The symbolic statements made by the new, completed projects are irritating – not because their internal environments are obnoxious in themselves but because of the contrast between them and the rest of the city.
>
> (Fainstein 1995: 133–4)

Such contrasts are everywhere in the British cities, between the opulence of Canary Wharf in London Docklands to the surrounding council estates which house a population with few of the skills to take up the new service sector jobs on offer (Plate 3.1). The emergent landscapes are richly symbolic of the social cleavages engendered by neo-liberalism, seen in the residualization of blue collar workers left high and dry by the decline of the manufacturing base of the cities, and the juxtaposition of rich and poor housing estates (Plate 3.2). New forms of consumption too are reinforcing the emergent socio-spatial juxtapositions between the 'haves' and 'have-nots', and the new environments, like Broadgate in London, Canary Wharf and Glasgow's Merchant City, are infused with a conspicuous consumption that many can never hope to engage in. Keith and Cross provide an apt characterization:

> At one end of Canon Street Road, London E1, you can pay £4 for a two-course meal. At the other end of the street, less than 500 metres away, the same amount of money will buy a single cocktail in Henry's wine bar in a post-modern shopping mall come upmarket residential development . . . the leitmotif of social polarisation is unavoidable. Golf GTIs share the streets uneasily with untaxed Ford Cortinas. Poverty is manifest, affluence is ostentatious. Gentrification sits beside the devalorisation of old property.
>
> (Keith and Cross 1993: 1)

There is also a marked privatization of public spaces emerging in the inner cities, places of enclosure, exclusion, or what Davis (1990) refers to as the emergent panoptican malls where the poor, or those unable to engage in consumption, are moved on. The revalorization of the inner city, then, requires the protection and safe-guarding of the consumer, while those that consume demand that their consumption is not interrupted by the gazes of the poor (so cementing division upon division). For Davis and others, the divide between the rich and the poor, the centred and the peripheral, is also experienced in what he terms 'the security driver's logic of urban enclavization' or the 'insulate of specific home values and lifestyles' (Davis 1990: 244). Indeed, the 1990s has become the period of a collective psychological neurosis, of the emergence of new forms of public surveillance, cameras placed everywhere, while the proliferation of neighbourhood watch schemes only serves to reinforce a siege mentality. Soja, in writing about Los Angeles, refers to the emergence of ungovernable spaces or 'carceral' cities, 'walled-in estates protected by armed guards . . . bunkered-in buildings and fortress-like stealth-houses' (Soja 1995: 134), forms of segregated (social) spaces.

Social fragmentation in the cities

The division of the city into exclusive and inclusive, public and private, realms is a continuation of trends described by many Victorian commentators, and, as the previous section outlined, is inscribed into the emergent physical landscapes of contemporary urbanism. Indeed, from Riis's (1890) conception of 'the other half' to Disraeli's proclamation of 'two nations' (Disraeli 1927),

Plate 3.1 New environments in the city – Canary Wharf, London. A symbol of the realignment of national economic and urban policy which has sought to create the bases for the recapitalization of inner city areas such as Docklands

Plate 3.2 New forms of consumption – Spiller's warehouse, Cardiff. The former site for small manufacturing firms has been converted into prestigious up-market flats in a location close to the impoverished Butetown estate

a recurrent concern with divided and polarized cities has been an undercurrent of urban studies. However, while income differentials began to converge during the Fordist era, leading some to proclaim the end of material inequalities in the UK, the combination of global restructuring and neo-liberal economic policies has underpinned a marked divergence of incomes amongst the citizens of the British cities. The patterns and processes of income polarization are now well documented and clearly they are connected to a period of regressive tax cuts that gave more back to higher income earners than to lower, and to the dismantling of social welfare networks which, since 1945, had been a guarantee of a minimum social wage for the poor.

For Mingione (1994, 1995), and others, such policies are implicated in the emergence of new social cleavages and divisions within the cities, the fragmentation of familial ties, and heightened levels of crime and social disorder. Indeed, the interlocking between global processes of economic restructuring and the truncation of welfare capitalism is, as Mingione (1995) notes, underpinning mass impoverishment and the overall disruption of social life in the cities. The emergence of new forms of work, the increased instability and heterogeneity of work contracts, and the increased numbers of people in precarious and low-paid employment, are all off-shoots of the wider objectives of neo-liberal policy to cheapen labour costs as a mechanism of enabling British capitalism to regain international competitiveness. As Mingione argues, such strategies have contributed to a 'complex map of social inequalities and forms of integration of the social division of labour' (Mingione 1995: 201). Mingione summarizes some of the complexity and variations in the following way:

> the typical result of exclusion from adequate employment, in combination with poor social integration and support, ranges from the work-poor households in England to the impoverished households hard hit by long term youth unemployment in southern Italy and Spain and the growing numbers in every large city of young adult homeless dropouts.
>
> (Mingione 1995: 203)

Hutton (1995) comments on the wider social implications for households, of the possibilities of the fragmentation of particular social and familial ties and of the disintegration of social networks.

Thus, the instability of the labour market has increasingly created dual-earner families, with parents working around the clock while their children are left to their own devices. For Hutton, 'economic liberalism has not provided an effective response to the parenting problem' (Hutton 1995: 225) while poorer households are increasingly stretched to provide for all of their members. Indeed, the break-up of households and familial ties has been linked to the interplay between economic restructuring and policy by a range of researchers, and, as Mingione (1995) notes, the (economic) marginalization of the unskilled and the uneducated is not unconnected to the rise in urban crime child prostitution and the proliferation of informal forms of work. In part, this underpins Lovering's (1995) comment that 'the only substantially resourced element of urban policy which can be said to be targeted on young unemployed males, the marginalised children of the Thatcher years, is policing' (Lovering 1995: 121).

Moreover, one of the more protracted, perhaps less easily measured, symptoms of the crisis of the cities relates to the socio-psychological effects of the wider economic transformations that have been experienced. As the British Medical Association has indicated, life expectancy for the poorest people in the cities has fallen in recent years, while research by Wilkinson (1994) suggests that the self-esteem of the poor falls as incomes become less equal. Indeed, a range of research highlights the increasing psychological problems of the young, especially of young men who have been hit hardest by the collapse of the staple industrial bases of the cities. As Wilkinson (1994) shows, the relative earnings of young men aged between 15 and 24 deteriorated throughout the 1980s, while their unemployment increased and their benefit entitlements were reduced. They, then, were at the sharp edge of the 'new welfarism', a process that according to Wilkinson (1994), has been manifest in the loss of self-esteem, an escalating suicide rate and the disaffection of young men from society more generally.

Such disaffection has also had marked gender and race dimensions, and the racialization of the 'urban problem' has come to the fore (Keith and Cross 1993). Indeed, the Chief Commissioner of the Metropolitan Police force, Paul Condon, provoked widespread outcry amongst ethnic groups by proclaiming in 1995 that they (especially black youths) were primarily responsible for crime in the cities. Thus, a 'blame-the-victim' social pathology has never disappeared and, in part, has underpinned the rise in racial tension in the British cities throughout the last fifteen years. Racially motivated violence has proliferated, while neo-fascist politicians have gained some ascendency in many white working-class neighbourhoods. For Keith and Cross (1993), and others, the economic marginalization engineered by the politics of neo-liberalism has also manifest itself in a 'politics of intolerance, of scape-goating, of seeking to blame those who themselves are the victims of the same or similar processes'. Yet, it is seemingly part of a system whereby 'irrationality, economic efficiency, and mounting unfairness result – all forgiven by a value system based on exclusion and systematic inequality' (Hutton 1995: 225).

CONCLUSION

After sixteen years of neo-liberal economic policy in the UK, the wider objective of drawing the cities into the international division of labour has had profound and, in some instances, precipitous effects. While London has cemented its position as a dominant world city, the older industrial cities, like Glasgow and Liverpool, have been recent recipients of Objective Five European regional funds, or of resources that are given only to the more impoverished and/or economically marginal parts of the European Union. The intra-urban variations in the British cities are, if anything, greater in a context where economic policy is increasingly being driven by global competition or by forces that seek to deregulate financial markets and reduce government support for industrial sectoral policies, while underpinning the injection of new forms of work organization and labour flexibility into the corporate economy (see Dicken 1994, Jessop *et al.* 1987, Pinch 1993). Such policies, combined with the emergence of a 'new welfarism', has, as Keating argues, generated a national economic policy forum in the UK geared to the deconstruction of place, or 'its reduction to a mere commodity in the global market' (Keating 1993: 392).

Indeed, the pursuit of international capital by many cities has often led to a series of 'false dawns' based upon a series of supply-side policies that amount to no more than a testament of faith in the benevolence of corporate capital to take up the bundle of incentives on offer. And, from Baltimore to Glasgow, the efficacy of the market has more than left a legacy, or, as Arlidge comments in relation to Glasgow, 'the can-do spirit and PR glitz, which swept Clydeside in the 1980s have failed to reverse the city's decline . . . and poverty in the city has doubled in the past two decades' (*Independent* 23.8.1995). Moreover, the legacy of redundant unused and underutilized spaces in Britain's cities continues, while a semi-permanent pool of people without a job underpins the reductionism inherent in neo-liberalism, whereby the value of people and places is solely measurable in terms of their (economically) commodifiable status. This highlights the ironic and contradictory nature of the last sixteen years of national economic policies, during which the pursuit of an internationalizing agenda has unleashed powerful disintegrating forces that have fragmented, even dissolved, the socio-economic fabric of the cities.

GUIDE TO FURTHER READING

One of the most readable accounts of the changing state of the British economy and its impact upon people and communities is by Will Hutton (1995). *The State We're In,* London: Jonathan Cape. Healey *et al.* (1995) provide a comprehensive coverage of some of the main theoretical debates about the cities, while focusing on the socio-economic and political changes that affect urban areas. Cochrane (1993) provides a good account of the restructuring of the welfare state in the cities, and provides some understanding of the interrelationships between economic policy, fiscal restructuring in the cities and the consequential socio-economic effects on the poor. Fainstein *et al.* (1993) is also to be recommended for its evaluation of how far, and in what ways, public policy influences the social and economic structures of cities.

REFERENCES

Amin, A. and Tomaney, J. (1995) 'The regional dilemma in a neo-liberal Europe', *European Urban and Regional Studies* 2(2): 171–88.

Byrne, D. (1993) 'Property development and petty markets versus maritime industrialism: past, present and future' in R. Imrie and H. Thomas (eds) *British Urban Policy and the Urban Development Corporations*, London: Paul Chapman: 89–103.

Cochrane, A. (1993) *Whatever Happened to Local Government?*, Milton Keynes: Open University Press.

Cox, R. (1993) 'Structural issues of global governance: implications for Europe' in S. Gill (ed.) *Gramsci, Historical Materialism, and International Relations*, Cambridge: Cambridge University Press 259–89.

Davies, H., Fielder, S. and Watkins, S. (1982) 'Managing urban decline: trends, problems, and policy implications', *Report for the Policy Project Group on Policies to Address Urban Decline*, Geneva: Organisation for Economic Cooperation and Development.

Davis, M. (1990) *City of Quartz*, London: Vintage.

Department of the Environment (1993) *Annual Report*, London: HMSO.

Dicken, P. (1994) 'Global-local tensions: firms and states in the global space-economy', *Economic Geography* 70(2): 101–28.

Disraeli, B. (1927) *The Two Nations*, London: Bodley Head.

Duncan, S. and Goodwin, M. (1988) *The Local State and Uneven Development*, Cambridge: Polity Press.

Fainstein, S. (1995) 'Urban redevelopment and public policy in London and New York' in P. Healey, S. Cameron, S. Davoudi, S. Graham, and A. Madani-Pour (eds) *Managing Cities: The New Urban Context*, London: Wiley: 127–44.

Fainstein, S., Gordon, I. and Harloe, M. (eds) (1993) *The Divided City*, Oxford: Blackwell.

Healey, P., Cameron, S., Davoudi, S., Graham, S. and Madani-Pour, A. (eds) (1995) *Managing Cities: The New Urban Context*, London: Wiley.

Hutton, W. (1995) *The State We're In*, London: Jonathan Cape.

Imrie, R. (1994) 'The policies and paradoxes of "Greening" the motor vehicle in the United Kingdom' in P. Nieuwenhuis and P. Wells (eds) *Motor Vehicles in the Environment*, London: Belhaven: 76–96.

Imrie, R. and Thomas, H. (1992) 'The wrong side of the tracks: a case study of local economic regeneration in Britain', *Policy and Politics* 20(3): 213–26.

Jessop, B., Bonnett, K., Bromley, S. and Ling, T. (1987). 'Popular capitalism, flexible accumulation, and Left strategy', *New Left Review* 165: 104–22.

Keating, M. (1993) 'The politics of economic development', *Urban Affairs Quarterly* 28(3): 373–96.

Keith, M. and Cross, M. (1993) *Race and Racism in the City*, London: Routledge.

Kleinman, M. (1994) 'Inequality, public policy, and cities', paper presented at the conference *Cities, Enterprises, and Society at the Eve of the XXIst Century*, IFRESI, University of Lille, Lille, France, 16–18 March.

Lawless, P. (1991) 'Urban policy in the Thatcher decade: English inner-city policy, 1979–90', *Environment and Planning C: Government and Policy* 9: 15–30.

Lewis, N. (1993) *Inner City Regeneration: The Demise of Regional and Local Government*, Buckingham: Open University Press.

Leys, C. (1985) 'Thatcherism and British manufacturing', *New Left Review* 151: 5–25.

Lovering, J. (1995) 'Creating discourses rather than jobs: the crisis in the cities and the transition fantasies of intellectuals and policy makers' in P. Healey, S. Cameron, S. Davoudi, S. Graham and A. Madani-Pour (eds) *Managing Cities: The New Urban Context*, London: Wiley: 109–20.

Marcuse, P. (1995) 'Not chaos, but walls: postmodernism and the partitioned city' in S. Watson and K. Gibson (eds) *Postmodern Cities and Spaces*, Oxford: Blackwell: 243–53.

Mingione, E. (1994) 'Socio-economic restructuring and social exclusion', paper presented at the conference *Cities, Enterprises, and Society at the Eve of the XXIst Century*, IFRESI, University of Lille, Lille, France, 16–18 March.

——(1995) 'Social and employment change in the urban arena' in P. Healey, S. Cameron, S. Davoudi, S. Graham and A. Madani-Pour (eds) *Managing Cities: The New Urban Context*, London: Wiley: 195–208.

Peck, J. and Tickell, A. (1995a) 'The social regulation of uneven development: regulatory deficit, England's south east, and the collapse of Thatcherism' *Environment and Planning A* 27(1): 15–40.

——(1995b) 'Business goes local: dissecting the business agenda in Manchester', *International Journal of Urban and Regional Research* 19(1): 55–78.

Peterson, P. (1981) *City Limits*, Chicago: Chicago University Press.

Pinch, S. (1993) 'Social polarisation: a comparison of evidence from Britain and the United States', *Environment and Planning A* 25(6): 779–98.

Riis, J. (1890) *How The Other Half Lives*, New York: Hill & Wang.

Robins, K. (1995) 'Collective emotion and urban culture' in P. Healey, S. Cameron, S. Davoudi, S. Graham and A. Madani-Pour (eds) *Managing Cities: The New Urban Context*, London: Wiley: 45–62.

Robson, B. (1988) *Those Inner Cities*, Oxford: Clarendon.

Savage, M. and Warde, A. (1993) *Urban Sociology, Capitalism, and Modernity*, London: Macmillan.

Simmel, G. (1964) 'The metropolis and mental life' in K. Wolff (ed.) *The Sociology of Georg Simmel*, New York: Free Press: 409–24.

Soja, E. (1995) 'Postmodern urbanism: the six restructurings of Los Angeles' in S. Watson and K. Gibson (eds) *Postmodern Cities and Spaces*, Oxford: Blackwell: 125–37.

Thornley, A. (1991) *Urban Planning Under Thatcherism: The Challenge of the Market*, London: Routledge.

——(ed.) (1992) *The Crisis of London*, London: Routledge.

Wilkinson, R. (1994) *Unfair Shares*, London: Barnardos.

Wirth, L. (1938) 'Urbanism as a way of life', *American Journal of Sociology* 44(1): 1–24.

4

NATIONAL SOCIAL POLICY IN THE UNITED KINGDOM

David Byrne

INTRODUCTION

In examining the impact of national social policies on the UK's cities we have to recognize that the form of such impact is a complex resultant reflecting the nature of the individual policies, the way in which formal policies as determined at the national level are actually implemented locally (both by elected councils and by appointed quangos) and the way in which the effects of such policies interact in particular places. Moreover, these complex processes do not occur on a blank landscape, but rather have effects in particular places on local matrices that are the historical products of the interaction in space of economic development and of the politics and policies of previous eras (the core of the idea of 'locality' – see Dickens *et al.* 1985, Bagguley *et al.* 1990, Byrne 1989 for an elaboration). Conventional 'social policy' accounts tend too often to emphasize only the level of the national formation of policy, although the older tradition of 'social administration' never ignored the crucial significance of implementation. However, in a formal sense at least, if not always in particular case studies, both approaches tended, in an interestingly Fordist way, to assume the universality of policy as implemented in a national context. One of the advantages of the interest now taken by geographers in social policy issues is that spatial variation within the nation state has been forced onto the agenda.

This book is concerned with intra-urban, rather than inter-urban, variations – with social divisions in the city – and it is true that there is a general form of variation in our post-industrial cities that is best described by reference to Complexity Theory's concept of the 'butterfly attractor' (Byrne 1996). This notion seems absolutely defensible for all the UK's non-world cities, the large industrial conurbations other than London but including the English shire cities such as Leicester, Hull, etc., which are actually the largest urban form by resident population in the UK. In other words more of us live in places like this than anywhere else. London's 'world city' status (and by London I mean the Greater Metropolitan Area) does complicate its picture somewhat, but even for this massive metropolis the historic significance of its industrial past is such that the notion of a division of the city into two general sorts of places holds largely true. Nonetheless, and this is an important warning, although this chapter will discuss 'National Social Policy' in relation to the general 'affluent' and the 'excluded' wings of the butterfly, the significance of particular local politics, local history, the local in general, in shaping the actual character of any specific place, is immense.

My general argument is this. The effect of the global processes described in the first two chapters of this book has been the transformation of the stepped and gradual inequalities (and, moreover, inequalities expressed over a historically restricted range) of the Fordist/Keynesian/Beveridge – in summary 'Universalist' – city, into one characterized in the sort of society Therborn meant by his expression 'the Brazilianization of Advanced Capitalism':

> At the bottom will be the permanently and marginally unemployed with certain welfare elements which are almost certain to be reduced over time. Some of these people will make a living in the black economy. In the middle will be the stably employed, those with the possibility of re-employment, who will be increasingly divided according to enterprise, sector and geographical position. They will make a fairly decent living, but no more, but will be able to congratulate themselves on the widening distance between themselves and the unemployed. The marginalization of the working class has already gone hand in hand, in the first half of the 1980s, with increasing wealth and incomes of capitalists and top business managers. Politically this ruling class will appeal to the bulk of employees as guarantors of the latter not falling into the abyss of unemployment and they will invite the citizenry at large to a vicarious enjoyment of the success of the wealthy and beautiful.
>
> (Therborn 1986: 32–3)

I would modify Therborn's detailed schemata by arguing that in 'old' capitalist economies the significant difference lies not between the 'unemployed – irregularly employed' and the employed, but rather between the category comprising the unemployed, irregularly employed and 'poorly' employed on the one hand, and those with 'decent' work on the other. Here 'poor' employment is essentially a matter of very low wages. Those with 'decent' work are reasonably paid. They are also somewhat more secure in employment, although the development of casualization as a policy commitment has extended insecurity to 'staff' groups and professionals who before were essentially secure throughout a working life.

The role of national social policy in relation to this is twofold. First, national social policies have played a very important part in shaping the specifically UK form of the divided city – the *creative* mode. Social policies, whilst working in the global context, are never merely responsive to global tendencies, but have a creative effect of their own. The Right wing libertarian ideology of market assertion, coupled with the authoritarian practices of social control of the new paupers brought into being by the combined effects of global tendencies and an ideological programme that worked absolutely to reinforce those tendencies, matters in its own right. So too does the 'new realism' which in effect amounts to an acceptance of the inevitability of globalism, characteristic both of 'New Labour' at a national level and the actual practices of Labour-controlled local authorities over the last decade.

The second role of social policy is *responsive*. Policies are developed that attempt to handle the issues that emerge because urban social divisions exist. The formal rhetoric of such policies is frequently that they are intended somehow to redress the inequalities inherent in such social division. The reality is that often they are policies concerned with the management of the social consequences of division – policies that range from containment to the actual establishment of effectively separate policy regimes in the different social spaces of the divided city.

Let me begin with a summary overview of the policy orientation of UK governments since 1975. I have chosen that date because it was the package of cuts introduced then by Denis Healey as a Labour Chancellor, under pressure from the IMF to which he had turned (as it transpired wholly unnecessarily) to resolve a sterling crisis, that signalled the end of the post-war consensus. This introduced the theme of 'fiscal crisis', or the 'unaffordability' of universal welfare, into the political process. The Conservative government coming into office in 1979 reinforced this by introducing a cash limits control mechanism for the expenditure of central government itself, and a series of increasingly onerous controls on the spending of local authorities (Heald 1983).

The Conservative agenda (or more properly, the agenda of the Thatcherites who were not Conservatives in a traditional sense) went much farther than responding to perceived structural pressures, and emphasized anti-statism, and the use of market mechanisms, both in delivering services and to replace policy planning in the decision-making process (Harris 1989 and Taylor 1992). It is important to realize that the commitment to the market does not simply mean that direct provision of services should be replaced by purchase from a plurality of providers, preferably of course purchase by ultimate individual consumers. Most fundamentally, this approach has profound implications for political forms and the degree of centralization of power in the nation state (or at whatever highest level the proponents of the market have power). As Crouch and Marquand note:

> One of the central assumptions of the new right is that choice is maximized through the market, not through politics: that the frictionless, undisturbed market is a realm of freedom and the polity a realm of domination and manipulation. Grant that and it follows logically that

> the sphere of the political should be curtailed; one obvious way to do this is to limit the scope of subordinate political authorities. On new right assumptions, provided people have the right to vote in national elections and to participate in open political discussion and lobbying, they should find variety, choice and delegation through market activity alone, and not through further political forms.
>
> (Crouch and Marquand 1989: viii–xi)

The Thatcher era was formally hostile to traditional corporatism, seeking to reverse what Middlemas (1979) had described as the UK's general corporatist tendency. In fact the actual institutional forms that have replaced local government in a range of areas, but particularly in urban development and the control and provision of post-16 education and training, have what I see as a 'stripped' corporatist form in that they involve a partnership between government and the representatives of capital, with trade unions and local government now excluded (Cawson 1986 calls this post-corporatism). The significance of the retreat not only from democracy, but even from traditional inclusive corporatism, in the governance of cities and regions, is one of the key features of the Thatcher and post-Thatcher years.

In this chapter I will examine the recent history of the following sets of UK national social policies in relation to their creative and responsive roles in the divided city: *income maintenance and fiscal policies*, (i.e. the combination of what is given and what is not taken away, together); *urban policies*, by which I mean both urban planning in general and the range of specific policy programmes concerned with urban regeneration, most of which date from the 1980 Planning and Land Act; and *educational policy* concerned with schools. I have chosen these three areas because the first is central to the creation of the general forms of income inequality in society as a whole; the second has a crucial role in constituting the form of urban spatial divisions; and the third is both shaped by urban spatial division and is crucial in the intergenerational transmission of inequality. My concern here is with an overview illustrating general principles. Details of the different planes of urban social division are provided in Part II of this book.

Of course these three areas do not constitute the whole of UK social policy over the period under consideration. Even the most traditional and limited definition of social policy would also include the areas of Health, Personal Social Services, and Housing, as well as much else in Education in addition to the issues of secondary schooling discussed here. So far as Health and Personal Social Services are concerned, I have not included any detailed review because these services do not operate in a way that is simply spatially ordered within cities. There is certainly a valid geography of Health and Personal Social Services but it is primarily a geography at the level of locality rather than within localities. Of course there are enormous intra-urban spatial variations in health states, but these are by no means simply the product of policy, and there has been no specific policy set in Health that has had much impact here. The introduction of general management and an internal market in health care, and especially in the secondary health care of the hospital system, is an extremely important development. However, it is not one that is expressed in intra-urban terms.

In Personal Social Services, and in the crucially important area of community care where Health and Personal Social Services intersect, there are intra-urban effects even if policies are not spatially ordered in 'within locality' terms. In effect, the agencies that deliver these services do so on a whole locality basis and changes have happened at that level. Those changes can have neighbourhood level effects. The general abandonment of institutional forms of provision for long-term dependent groups, and especially for the mentally ill (which is the essence of 'community care') has differentially dumped very dependent people with acute needs (and sometimes behavioural problems that have an impact on others) into the poorest areas of our cities. Things are not as extreme as in the US, but homelessness in particular is to a considerable extent a consequence of this major policy shift. However, important as these developments are, they are not policy areas that are generally constitutive of the system of inequality in our society.

Housing is another matter. The absolute shift in support away from social housing and the total commitment to owner-occupation as the tenure for 'normal' people, has had profound effects on our urban systems. Things have become very messy in the aftermath of the recent dramatic reversal in the real prices of houses for owner-occupiers, and the development of significant 'negative equity' among such households, but the general tendency has been towards 'socio-tenurial polarization' (Hamnett 1987). Basically the fiscal support given to owner-occupiers (tax relief on mortgage interest payments, exemption of owner-occupied dwellings from Capital Gains Tax, and most importantly the disregard since the abolition of Schedule A of real housing incomes in kind by the tax system) so massively advantage them that all average and above average income households are much better off as owners rather than tenants. This is a long-term tendency and the effect of it over two generations has been the creation of a social housing system in which there are really only two sorts of tenants, both of whom are relatively poor.

Many social housing tenants, particularly in the best council estates, are pensioners who were exactly the respectable, employed, working-class residents of those estates thirty years ago. Almost all the rest are poor households headed by people of working age who move between total benefit dependency and dependency on a mix of benefits and low wages. In Housing Association stock, central government policies on rent levels have raised these to such a high level that almost all households are benefit trapped, in that they have no access to jobs with wages high enough to mean that working generates a higher income than total benefit dependency. If they had, they would of course become owner-occupiers. In council housing the impact of rent-pooling, which produces a collective benefit for all tenants from the low historic cost of much of the stock, keeps rents to a level where legitimate poor work is an option. It should of course be noted that those social housing tenants whose incomes, whilst still low, are just high enough to put them above benefit levels, receive no assistance with their housing costs whatsoever, in marked contrast to much more prosperous owner-occupiers. General housing subsidies, separated from means testing at a very low income level, no longer have any real function in the UK housing system.

The sale of social housing under right-to-buy legislation has really had very little effect on the internal spatial form of urban areas. It has of course reduced the available good social housing stock and contributed to 'family' homelessness in areas where there is not a crude surplus of housing stock. It has to be remembered that in most of the urban UK outside the south-east and a few other areas, there is a crude surplus of stock if the quality of that stock is disregarded. However, it has not in any way changed the character of urban housing areas.

It is absolutely true that social housing areas are almost without exception now locales in which only relatively poor people live. The converse – that owner-occupied areas are only affluent – is, however, not true. As Forrest *et al.* point out, we need:

> to see home ownership in terms of a continuum that at one end is indistinct from renting. The owner has continuing mortgage commitment and limited choice and autonomy in sectors of the market that do not keep pace with the general rate of accumulation of housing. Low-paid or insecure employment leaves the owner with important periods of dependency on state benefits and external support. At the other end of the continuum, home ownership involves independence, autonomy and high rates of accumulation
>
> (Forrest *et al.* 1990: 198)

The private rented sector is now small in scale, and increasingly specialized in function. Probably its largest component in cities is essentially generationally differentiated in that it involves student renting. There is developing what seems to be a 'very bad' element in the system, some of which consists of housing in multiple occupation, in which landlords obtain very high rental incomes from very poor property let to benefit claimants whose rents are paid for them. This may well be associated with criminality, either through simple benefit fraud or through the recycling of cash from other criminal enterprises (Byrne and Green 1995). Overall, housing tenure is related to spatial differentiation within cities, but I would argue very firmly that recent policies are not constitutive of that differentiation. Rather the socio-spatial patterning of housing tenure existed as a matrix onto which the combination of global economic tendencies and policy have written inequality; so inequality is tenure structured, but housing policy as such has not created that inequality.

In contrast, both the benefit/fiscal policy combination and 'urban policy' have been primarily 'creators' of the divided city, despite claims that aspects of the latter have been about redressing social division. Schools policy has been primarily responsive to division (although there has been a creative element introduced by parental choice), and the medium-term implications of what is becoming effectively a system of social segregation within the state secondary school system are profound. I will take each of these policy systems in order and review their consequences for the divided city.

INCOME MAINTENANCE AND FISCAL POLICIES

The significance of the combination of benefits and fiscal policies is that it is of very great importance in determining the disposable income of the households that are the crucial social units through which people are located in residential space, and the divided city is, to a considerable degree, a matter of segregated residence. Indeed the tax/benefit position of a household in relation to waged income is the key determinant of which sort of residential space a household can locate in. It should be noted that the division is not between households solely dependent on benefits and those not dependent on benefits. There are some 1.8 million households in the UK that receive either or both of Family Credit and Housing Benefit as a subsidy to their low wages. Many of the 500,000 'other' non-pensioner Income Support recipients will be those who are working part-time whilst receiving their main income from the state (Social Security Statistics 1994). In addition to these groups there is the general category of the low paid who either do not claim these benefits (take-up rates for Family Credit are just over 60 per cent) or earn just enough to put themselves outside the eligibility criteria.

If we look at taxation policy over the last sixteen years we can see very clearly that it has cut direct taxation, very much for higher income earners, and increased regressive indirect taxation. According to *Social Trends*:

> Income tax rates have been reduced markedly since 1978–79, particularly for those on higher incomes. The basic rate fell progressively from 33 per cent in April 1978 to 25 per cent in April 1988, and in 1992–3 a new lower rate of 20 per cent was introduced. The higher tax rates, which rose to a maximum of 83 per cent on earned income have been replaced by one 40 per cent rate.
>
> (*Social Trends* 1995: 89)

The regressive effects of these changes are reinforced by the fact that PAYE taxpayers have little choice but to pay their taxes, while many higher earners and business people are able to utilize perfectly legal tax avoidance schemes to pay less as a proportion of their incomes. The end result is that the top group in the UK version of the 'Brazilianization' of advanced capitalism have a great deal more money to spend on themselves than they had in relative terms twenty years ago. The middle group come out about even and the poor are worse off. Indeed if we look at changes in real net income (after direct tax and with benefits added) after housing costs – a necessary qualification because the poorest get their housing costs paid or pay rent for assets that have remained constant in quality – then between 1979 and 1991/2 this fell by nearly 20 per cent for the poorest decile whereas it rose by more than 60 per cent for the most affluent decile (Hills 1995: 31). Only the most affluent three deciles, the best-off 30 per cent of individuals, had a real increase of more than the overall mean increase of 37 per cent. This is in marked contrast to the period 1961–79 when only the poorest decile, whose real incomes increased by 53 per cent, saw a higher than average increase. If we turn to the household picture we find that the poorest 40 per cent of households had 23 per cent of disposable income (net of direct taxes) in 1978 and 18 per cent in 1993. In contrast the best-off 20 per cent of households saw their share increase from 35 per cent in 1978 to 42 per cent in 1993 (Hills 1995: 24). Only this most affluent quintile increased their share of disposable income. The next quintile stayed constant. The other three all saw their share decline.

National policy on benefits has been driven, formally, by a perceived structural necessity to reduce their enormous cost as a proportion both of overall government expenditure and of GDP – i.e. by the conception of a fiscal crisis/welfare burden. In the year 1994–5, from a total government expenditure of approximately £250 billion (of which £176 billion was central government expenditure), the cost of Social Security programmes was £65.8 billion (Cmnd 2821 1995). In fact, through the Tory years the cost of Social Security programmes has risen substantially because of the massive increases in non-employment among people of working age. This includes not only the recorded unemployed but those in receipt of invalidity benefits (now costing, at £6.8 billion, nearly as much as the £8.8 billion paid out in Unemployment Benefits) and the early 'retired' receiving Income Support without a requirement to register for work. The massive increases in costs are a function of rising numbers of recipients, not of more generous payment of benefits. Quite the contrary – the relative level of benefits has fallen, because in 1980 the link between the uprating of insurance benefits in line with changes in wages or prices, whichever was the greater, was replaced by a linking of benefits solely to prices. The abolition of earnings-related supplements to National Insurance benefits, whilst retaining earnings-related contributions, was a particularly harsh cut. However, changes in basic rates are not the whole story. The 'Fowler' review of benefits, which resulted in the 1986 Social Security Act, in effect eliminated the discretionary additional systems of regular allowances and special payments, substituting loans from the Social Fund in place of the latter. The effect of this on the very poor has been dramatic. Other policy changes included the transfer of the administration of housing benefit from central government to local authorities without full funding of the administrative costs. These are now effectively borne by those council tenants who do not receive housing benefit – a group who are themselves generally not far above qualifying income levels. In effect, the moderately poor have to support the very poor.

The real efforts at eliminating the dependency culture through national policy on Social Security have been those involving administrative changes. By far the most important has been the exclusion of almost all those under the age of 18 from any Income Support entitlement. A series of administrative practices, culminating in the proposed replacement of Beveridge's universal, insurance-financed unemployment benefit by a 'job-seekers' allowance, have reintroduced the 'genuinely seeking work' clause of the 1930s to the contemporary treatment of the unemployed. Now the test of availability for work is no longer the refusal of a suitable job, but the demonstration of an active work-seeking strategy by the claimant.

Markets as such are not something that relate to the direct consumers of income maintenance benefits, other than through the government's attack on the State Earnings Related Pensions System as a way of forcing more people into (often wildly inappropriate and grossly exploitative) private pension schemes. Where market systems have come in is through the transfer of almost all the administrative functions of the benefits system to 'Next Steps' Agencies, which are no longer part of the civil service. New recruits, for example to the Contributions Agency Head Office in Newcastle, are no longer civil servants and the culture of security of employment has been done away with. These agencies are currently being subjected to a wave of market testing and are having to bid for their own work load against private sector competitors. This, effectively, has the same social effects as the continuing development of compulsory competitive tendering for a range of local authority services.

What these changes reflect is the importing into official (both political and civil service) culture of a conception of the nature of the post-industrial labour market, and especially of the fragmented and temporary character of employment within that labour market. People have to take insecure and low-paid jobs, and the general line of other policies has been to ensure that there are plenty of such jobs available. In this section I have dealt with the mixed effects of fiscal and benefits policies. However, it is important also to consider the nature of wages in establishing the sort of household resource structure that the system has generated. The combined effect of anti-trade-union legislation and the abolition of wages councils, including the elimination (through compulsory competitive tendering and other means) of the culture of Whitleyism (the corporatist system of managing public sector employment issues at the national, local and regional level which dates from the First World War) in the public sector, has played a large part in disrupting a series of wage relativities that had held pretty well constant from the fifteenth century to the 1970s. The recent Rowntree Inquiry concluded: '. . . the male earnings distribution is now wider than at any time in the century for which we have records' (Hills 1995: 42). Interestingly we now have less gender inequality for earnings than ever before, but more inequalities within genders.

The weakening of the trade union power of the ordinary worker is very important in explaining all this, but transformations in the nature of work are plainly also of great significance. The Rowntree investigation concluded that '. . . between a third and a half of the growth in the earning dispersion over the 1980s can be explained by higher wage premiums commanded for greater levels of experience and/or educational qualification' (Hills 1995: 44). The report also notes that this was of particular importance for young workers. We will return to the very great significance of this in our discussion of education policy.

The divided city, in socio-spatial terms, is about the residential segregation of the haves and have-nots (Byrne 1995b). Increasingly what we have are two nations in one locality but segregated in the spaces of that locality. We now turn to policies specifically directed at 'the urban' itself.

URBAN POLICIES – CATALYSING THE DIVIDED CITY

There is no simple definition of 'urban policy' – here it will be taken to be that set of policies concerned with urban land and urban people that are directed at shaping the future social form of cities. This means that it includes the urban programme as developed and transformed since the translation into legislation of Labour's 1978 White Paper until its culmination in City Challenge, now transformed into a component of 'the single regeneration budget'. Urban policy also includes the range of initiatives including Enterprise Zones, Urban Development Corporations and, most recently, the Urban Regeneration Agency which derive from the Local Government Planning and Land Act of 1980, as well as the planning system as now constituted under the Unitary Development Plan procedure. Although not strictly 'urban policy', the operations in England and Wales of Training and Enterprise Councils, quangos originating as a post-corporatist replacement for the Local Manpower Boards of the corporatist Manpower Services Commission, are also of considerable importance in the urban policy local development and implementation process, as are their Scottish equivalents – the Local Enterprise Councils. These things have an impact on other aspects of policy, notably housing and education, but they also have a prior dynamic of their own.

My intention here is not to provide a detailed chronology of the development of urban policies but rather to emphasize the objectives and effects of this policy set. Urban policy has had two contradictory objectives over the period under consideration. The first is to allow the development of an urban land market to replace a politically located urban planning system. Free market ideologues have always particularly detested the post-war land use planning system which gave an important role in shaping cities to elected

Plate 4.1 The former John Marley Junior High School in Scotswood, Newcastle upon Tyne, an inner city school that has closed and is now used as an Enterprise Centre

local authorities and their officers. They want this process 'depoliticized' by replacing determinative planning with allocation of land according to prices in a market. For this to happen there must be money to be made out of the redevelopment of urban areas. If the conditions for such activity do not exist, then they must be created. The efforts to create such a real market can be summarized by the expression 'catalytic planning'. The second objective has been the management of the populations of inner urban areas, and of that very large part of the inner urban population of all cities other than London that was displaced to public sector 'outer estates' by the slum clearance processes of the 1960s and early 1970s. In the earliest phase this was an effort at reintegrating such people by using the Urban Programme to provide a targeted selective replacement for the universalist provision of core local authority services that had been cut in 1975. Under the Tories efforts at reintegration have been replaced by a combination of efforts at containing disorder. The object here has been incorporation, rather than integration which would have attempted to restore the excluded to a full citizenship status. Incorporation is the process of involving the excluded in the management of their own containment. The other key policy of this period has focused on the continuing training-based renewal of elements of the dispossessed poor as both the actual performers of poor work and as a reserve army for such work. The target of the first set of policies has been urban land, the target of the second has been people.

In reviewing the nature of urban policy, we must pay some attention to the consequences of means as well as ends. UK central government has in effect handed over much of the administration of our cities to non-elected quangos – UDCs, TECs, LECs, etc. The background to this is the free market hostility to political process identified by Crouch and Marquand (1989), but it is worth reflecting on why these elements of the 'New Magistracy' (Stewart 1995) are utilized instead of making all such processes part of a centrally directed administrative machine. Duncan and Goodwin (1988) have pertinently pointed out that a logical implication of the notion of locality is that administrative mechanisms must be both able to respond to specific local conditions and have adequate knowledge about those conditions so that decisions can be made that fit particular places. This predicates a necessary local level in the government process. By restoring former power to local capitalist elites, power that in most UK urban areas they could never now acquire through democratic electoral means, central government can accommodate this without the challenges that democratically elected local government in such areas would mount (and have mounted, such as the GLC) to its general anti-political programme. However, as Lewis points out, this approach is not without its costs.

> one of the most striking phenomena in this area is the sheer complexity engendered by the range of initiatives and actors involved in the field. This produces a number of problems. The first is simply to identify who is doing what at any given moment. An anatomy of the major actors is called for, but is often very difficult to draw. An attendant problem is the likelihood that strategic planning will suffer with a myriad of different schemes in operation. Finally, we are faced with a serious question of constitutional accountability for what is happening in our cities. An increasingly opaque central government machine is exhorting industrial and financial capital, the community and volunteer movement, to a greater or lesser extent local government, and a clutch of quangos to act in the local interest. With local government not necessarily playing the central role it becomes exceedingly difficult to identify responsibility for action taken and not taken.
>
> (Lewis 1992: 1–2)

Let us now turn to actual policy. I first encountered the idea of 'catalytic planning' at the public inquiry concerned with the compulsory purchase of a number of properties on the East Quayside site in Newcastle by the Tyne Wear (Urban) Development Corporation (TWDC 1989). There it was defined by TWDC's expert planning witness, P.W. Jones, a director of Debenham, Tewson and Chinnocks, project advisers:

> There is, in my opinion, a distinction to be drawn between 'regeneration' and 'redevelopment'. Redevelopment of a site will succeed in bringing land and buildings into whatever use the market determines as the most appropriate for that site at that time. Regeneration on the other hand, aims to create new markets by increasing confidence and attracting inward investment. A regeneration project is needed to rekindle economic and cultural vitality of the site itself and also creates similar betterment to its immediate environs. When combined with other such schemes, it will also be a catalyst for sustained improvement and growth in the whole city and indeed the region.
>
> (TWDC 1989: 12)

In other words, the task of the UDCs was to use public resources to get the market going, as TWDC put it, to act as 'A Catalyst for Regeneration' (TWDC 1989: 16) – the catalytic image implying that the potential existed. It was only necessary to inject some energy into the system to initiate a self-sustaining reaction that would proceed without further intervention. The use of the term 'flagship' by UDCs to describe particular developments is also significant. The 'flagships' are the physical representation of the catalytic process – the late twentieth-century equivalent of Gray Street around the development of which the mid-nineteenth-century urban renewal of Newcastle was hinged. The concept is quite well founded in the history of urban development and renewal but has proceeded without any subsidy only in core urban areas (a special exception being Olympia and York's successful development of Welfare Island in the world city of New York).

The policy of Enterprise Zones, as generally implemented (if not always as rhetorically proposed), seems to have had the same sort of objective. Essentially, Enterprise Zones combined the removal of planning restrictions with very favourable taxation regimes as a method of stimulating what was originally presented as a new wave of manufacturing capitalism in urban areas (Anderson *et al.* 1983), but far more often turn out to be peripheral retail and office parks. The UDCs have generally proceeded by creating brown field sites (i.e. former industrial sites brought to the same physical condition as peripheral green field sites through very expensive derelict land reclamation) and selling them on to the private sector at very low or even negative prices. Indeed when transport and other subsidies are taken into account, developers were not so much given a free gift as paid to take these sites away (see Byrne 1994 for an account of this on Teesside).

The present wave of Unitary Development Plans prepared by elected local authorities might be thought to represent a potential challenge to this approach. However, examination of experience suggests that this is not so. The nature of UDPs is succinctly described in the South Tyneside UDP statement:

> This document, the first Unitary Development Plan for South Tyneside, provides guidance on land use and future development in the Borough,

> up to the end of the year 2001. 'Unitary', in this case, means a single Plan prepared for the whole of South Tyneside. The essential role of the Plan is to determine how much development the Borough can accommodate.
>
> The South Tyneside UDP has been prepared under the provisions of the Town and Country Planning Act 1990 as amended by the Planning and Compensation Act 1991. Under these arrangements a metropolitan district council, like South Tyneside, must prepare a plan which combines strategic issues previously covered in the Tyne and Wear County Structure Plan with detailed, site specific proposals previously considered in local plans.
>
> (South Tyneside UDP 1992: paras 1 and 7)

However, in an authority that includes territory under the control of a UDC matters are not so simple. As Lewis has pointed out:

> The DOE has decided that local authorities should not draw up unitary development plans for UDC areas, but rather that unitary development plans should adopt the strategies of the UDCs. Moreover, local authorities are required to take into account in their UDPs the impact of UDCs outside the UDC's area. In effect then UDCs become the plan-making authorities for their areas – and have a right to influence planning matters well outside their areas.
>
> (Lewis 1992: 53)

This point was recognized in the South Tyneside UDP: 'It should be noted that in this Plan, reference to "the Local Planning Authority" should be taken to mean the Borough Council or the UDC as appropriate' (South Tyneside UDP 1992: para 3).

Even in cases where local authorities are planning in circumstances that are not much affected by UDC operations or proposals, they seem to have often adopted the same strategic conception of the purpose of the planning process. This is certainly true for Newcastle City Council's proposals for development in the green belt to the north-west of the city, which is bitterly opposed by the Labour constituency parties in the city acting in opposition to a controlling Labour Council Group. What this seems to reflect is the extent to which central government policy has induced a culture change in local authorities. This is well described in a recent study of developments in Manchester by Randall (1995) in which he provides a detailed account of how the formerly radical New Left Labour leadership of the authority has become implicated in a civic boosterism that is flatly antagonistic to the poor communities that elect them. Randall includes an interesting catalogue of the financial disasters that have generally been associated with this approach.

The City Challenge programme, now incorporated in the Single Regeneration Budget, is the best illustration of 'people' oriented policies. It was introduced by the unusual method of a response to a written question. This is itself an indication of the executive power ministers have acquired through their capacity to develop programmes on the basis of regulations, rather than of legislation which would be subject to parliamentary scrutiny.

> Under city challenge I will invite Local Authorities to draw up programmes of action to tackle their key neighbourhoods. I will expect the authorities, in the preparation of these programmes, to draw upon the wealth of talent and enterprise which exist in their cities: local people and the voluntary sector, the business and academic communities, the training and enterprise councils, and government and other statutory agencies. I will expect them to attract private finance and involve the private sector throughout in managing the programme.
>
> (Heseltine, Sec. of State for the Environment, HC 5th series, vol. 191: 549–50)

City Challenge was an administrative modification of the way in which Urban Programme resources are distributed, and was funded by top-slicing the Urban Programme as a whole. Urban Programme began with the 1977 Labour initiative (Lewis 1992; Lawless 1989) and has been important in terms of specific programme elements, the cleaning up of industrial areas, some housing improvement work and support for many local community and voluntary agency initiatives. However its very bittiness meant that it did not involve a major recasting of planning objectives. City Challenge was different because bids represented co-ordinated approaches to urban problems. The formal objectives and general style of City Challenge initiatives illustrated a shift in the general direction of policy from the regeneration of land to doing something about people. However, even in City Challenge the *development potential* of the Challenge areas was crucial to their designation for the scheme.

City Challenge bids were not formal land use plans, but were plans in a wider sense and they did indicate both the concerns of local government and its 'partners' about social problems associated with de-industrialization, and their general strategic orientation towards those problems. City Challenge bids departed from the UDC's land-led approach. If UDCs are about redundant land, and primarily redundant industrial land, then City Challenge bids were about a mixture of redundant people and the areas in which they live (primarily outer estates), as well as the run-down commercial centres of industrial towns. This is an odd and probably incompatible combination, particularly when, as in the case of Hartlepool, the whole programme is literally the surrounding support of a collapsing UDC flagship, the Hartlepool Renaissance. Randall's (1995) account of the gentrifying implications of such approaches, if they are successful, reminds us also that poor people have traditionally been displaced by developments of this form. However, City Challenge did have people in view, even if the TEC-led approach to them is highly questionable. In other words, it is probably right to think of City

Challenge as some sort of recognition of the social problems that stem from redundant and disorganized populations. It is probably too early to evaluate City Challenge as a whole. The scheme is by no means complete, and despite the rhetorical similarity of bids, different local authorities have done very different things, ranging from the straightforward palliative measures emphasized by Middlesbrough District Council to the grandiose development suggestions that constitute the core of Sunderland's approach.

EDUCATIONAL POLICY – SECONDARY EDUCATION IN DIVIDED CITIES

Despite the removal of tertiary and post-16 further education colleges from local authority control, education remains the most important locally provided service in the UK. However, the nature of the governance of schools has been changed in a way that fundamentally transforms the social relations of schooling. Essentially, local education authorities now no longer control the schooling system in their areas; instead, the introduction of devolved school budgets (Local Management of Schools – LMS) coupled with formula funding based, essentially, on pupil numbers has created a quasi market in which parents use their 'choice' to seek to send their children to 'good' schools (good being measured by success in public examinations) in a way that reinforces the existing social advantages of schools located in 'better' residential areas. The legislative bases of these changes include the Education Acts of 1980, 1986 and 1988. It is important to note that the much discussed 'opting out' procedures are not all that important here, although extensive 'opting out' – the process whereby schools quit local authority control in favour of a direct relationship with the responsible central department – would exacerbate the situation. The significant divide is the residentially based one within the main state system.

Before these changes, local education authorities controlled the form of state schooling in their areas. In particular they were able to determine the character of catchment areas and of admission policies, and to plan the organization of secondary schooling in a way that could attempt to achieve criteria of universalist social justice. The changes have not occurred in a vacuum. Labour market transformations have produced a significant increase in 'staying on' past the compulsory schooling years with an increase in the proportion of 17 year olds in full-time education. Nonetheless the policy changes, by particularizing schools, have contributed both to the form of the unequal city and to its perpetuation through the lifetime consequences of grossly dissimilar educational experiences. This issue has been much discussed in racial terms in the US (Massey and Denton 1993), but it has equal salience here where the basis of social division is essentially class.

If we consider the situation of English secondary comprehensive schools as they now are, we find a system of neighbourhood schools drawing their pupils from a specific set of feeder primary schools and topping up from

Plate 4.2 Emmanuel College, Gateshead, a City Technology College drawing on inner city and suburban areas

parental choices made by 'out of catchment area' parents. The schools are governed by the governors. This seems like a truism, but it is in marked contrast to the pre-1980 situation when governors had very little power *vis-à-vis* the LEA. Now they run the school, and in practice this means that any coalition of parent and teacher governors can dominate a school by controlling co-option of further governors in a way that reinforces their own intentions as to the school's trajectory. Middle-class parents have been substantially empowered in middle-class areas. Poor working-class parents have in principle the same access to the governance of their children's schools, but are often in awe of the system and lack the experience of control that so many middle-class people possess.

Devolved budgets and formula funding in a market context mean that schools seek to attract pupils. In general the key performance indicators by which schools are judged are their crude (i.e. not adjusted for social background) success rates at GCSE and A level examinations. Middle-class schools do well and attract incoming parental choice. Schools in poor working-class areas do not do well and lose pupils (precisely those pupils whose parents have some knowledge of the educational process) by the same process. The effects are marked, as is demonstrated in Middlesbrough (Byrne 1995a). In the secondary schools in poor, working-class East Middlesbrough only 6 per cent of pupils achieved five or more A–C grades at GCSE in 1994, while in suburban Middlesbrough schools the corresponding figure was 38 per cent.

This simple indicator does describe a system in which internal school regimes become quite different. In the one sort of place the orientation is towards examination success and mobility; in the other it is towards control and management. Given the implications for life-time incomes of early educational qualifications, this means that a spatially ordered system of distinctive, internal school policy regimes has profound continuing implications for social division over the life experiences of entire generations. This situation is not the fault of teachers. As Massey and Denton say:

> The organization of public schools around geographical catchment areas . . . reinforces and exacerbates the social isolation that segregation creates in neighbourhoods. By concentrating low-achieving students in certain schools, segregation creates a social context within which poor performance is standard and low expectations predominate.
>
> (Massey and Denton 1993: 141)

CONCLUSION

This chapter has described a kind of vicious circle of central government policy with regard to the creation and perpetuation of 'divided cities'. It has shown how national policies with regard to the labour market and the tax/benefit systems have exacerbated income differentials and played an important part in the creation of a class of dispossessed poor containing both workers and non-workers. These policies have not been simple, desperate responses to inevitable global tendencies. Rather they have been far more the product of the ideologically driven triumphalist assertion of the logic of those tendencies – not so much a matter of going with the prevailing wind as of hoisting every possible sail and switching on the motor as well. The 'urban policy' set has had much the same consequence. It has involved the triumphalizing of consumption based on inequality in resource distribution. Education shows both the consequences of such developments, and the way in which the positive feedback to them perpetuates and exacerbates such inequalities across the generations.

Perhaps the most important thing to realize is that these developments, although in accord with the implications of the development of a globalized post-Fordist world system, are by no means inevitable. Contrary to even the most informed of 'regulation theorists' (e.g. Lipietz 1992) the changes in the world system were not the product of inevitable systemic tendencies – those things were made to happen. Even more, the policy regime described above is a thing made by people with faces to which names can and should be put. The crucial fact is that policy is not, and never will be, only responsive. In the deadly jargon of these dreadful times it is proactive. The results include the divided cities in which we live – some of us well, and some of us miserably.

GUIDE TO FURTHER READING

Blackman, T. (1995) *Urban Policy in Practice*, London: Routledge, especially Part I. This provides an excellent general account of the recent course of urban policy.

Imrie, R. and Thomas, H. (1993) *British Urban Policy and the Urban Development Corporations,* London: Paul Chapman.
This contains a number of very useful studies of the activities of a key set of agencies operating in the 'post-industrial' city.

Hills, J. (1995) *Inquiry into Income and Wealth* vol. 2, York: Joseph Rowntree Foundation.
A very comprehensive survey of the nature of resource inequality in this country today.

Byrne, D.S. (1995) 'Deindustrialization and dispossession: an examination of social division in the industrial city', *Sociology* 29: 95–115.
This combines an account of an actual place – the Teesside conurbation – with a summary of contemporary debates about the dispossessed poor.

REFERENCES

Anderson, J., Duncan, S. and Hudson, R. (eds) (1983) *Redundant Spaces in Cities and Regions*, London: Academic Press.
Bagguley, P., Mark-Lawson, J., Shapiro, D., Urry, J., Walby, S. and Warde, A. (1990) *Restructuring: Place, Class and Gender*, London: Sage.
Blackman, T. (1995) *Urban Policy in Practice*, London: Routledge.
Byrne, D.S. (1989) *Beyond the Inner City*, Milton Keynes: Open University Press.
——(1992) 'The city' in P. Cloke (ed.) *Policy and Change in Thatcher's Britain*, Oxford: Pergamon: 247–68.
——(1994) 'Planning for and against the divided city' in R. Burrows and B. Loader (eds) *Towards a Post-Fordist Welfare State?*, London: Routledge: 136–53.
——(1995a) 'Deindustrialization and dispossession: an examination of social division in the industrial city', *Sociology* 29: 95–115
——(1995b) 'Radical geography as "mere political economy" ', *Capital and Class* 56: 117–38.
——(1996) 'Chaotic places or complex places? Cities in a postindustrial era' in S. Westwood and J. Williams (eds) *Imagining Cities*, London: Routledge.
Byrne, D.S. and Green, J. (1995) 'Les proprietaires-occupants marginalises dans les villes anciennes', *Les Annales de la Recherche Urbain*, 65: 91–8.
Cawson, A. (1986) *Corporatism and Political Theory*, Oxford: Blackwell.
Cmnd 2821 (1995) *Public Expenditure: Statistical Supplement to Financial Statement and Budget Report 1995–6*, London: HMSO.
Crouch, C. and Marquand, D. (1989) *The New Centralism: Britain out of step with Europe*, London: Political Quarterly Monograph.
Dickens, P., Duncan, S., Goodwin, M. and Gray, F. (1985) *Housing, States and Localities*, London: Methuen.
Duncan, S. and Goodwin, M. (1988) *The Local State and Uneven Development*, Cambridge: Polity Press.
Forrest, R., Murie, A. and Williams, P. (1990) *Home Ownership*, London: Unwin Hyman.
Hamnett, C. (1987) 'A tale of two cities', *Environment and Planning A*, 19: 537–56.
Harris, C. (1989) 'The state and the market' in P. Brown and R. Sparks (eds) *Beyond Thatcherism*. Milton Keynes: Open University Press: 1–16.

Harvey, D. (1987) 'Flexible accumulation thru urbanization', *Antipode*, 19: 260–86.
Heald, D. (1983) *Public Expenditure*, Oxford: Martin Robertson.
Hills, J. (1995) *Inquiry into Income and Wealth*, vol. 2, York: Joseph Rowntree Foundation.
Imrie, R. and Thomas, H. (1993) *British Urban Policy and the Urban Development Corporations*, London: Paul Chapman.
Lawless, P. (1989) *Britain's Inner Cities*, London: Paul Chapman.
Lewis, N. (1992) *Inner City Regeneration*, Milton Keynes: Open University Press.
Lipietz, A. (1992) 'The regulation approach to capitalist crisis' in M. Dunford and G. Kafkalas *Cities and Regions in the New Europe*, London: Belford: 309–34.
Massey, D.S. and Denton, N.A. (1993) *American Apartheid*, London: Harvard University Press.
Middlemas, K. (1979) *Politics in Industrial Society*, London: Andre Deutsch.
PA Cambridge Economic Consultants (1989) *An Evaluation of the Enterprise Zones Experiment* (for the DOE Inner Cities Research Programme), London: HMSO.
Randall, S. (1995) 'City pride: from municipal socialism to municipal capitalism', *Critical Social Policy*, 43: 40–59.
South Tyneside UDP (1992) *Draft for Public Consultation*, South Shields: UDP.
Stewart, J. (1995) 'Appointed boards and local government', *Parliamentary Affairs*, 48: 226–41.
Taylor, P. (1992) 'Changing political relations' in P. Cloke (ed.) *Policy and Change in Thatcher's Britain*, Oxford: Pergamon: 33–54.
Therborn, G. (1986) *Why Some Peoples are more Unemployed than Others*, London: Verso.
TWDC (1989) *Proof of Evidence of P.W. Jones – Debenham, Tewson and Chinnock Ltd – East Quayside Compulsory Purchase Order Inquiry*, DOE Reference N/5038/12P/9.

5

URBAN PLANNING IN THE UNITED KINGDOM

Tim Blackman

INTRODUCTION

This chapter considers the role of land use planning in Britain's cities, locating it in terms of one of a number of planning activities undertaken by local government. The chapter discusses the value of planning and its relationship to social and economic changes, as well as issues specific to planning and regulating urban development. There are two key themes: the question of the strategic significance of land use planning and, more generally, the strategic role of local government itself.

THE ROLE OF PLANNING

All market or mixed economies rely to some extent on state planning to provide public services that meet social needs and that ensure that the externalities of economic and social activities, such as pollution and traffic, are controlled 'in the public interest'. In Britain, land use planning has a very limited role in directly meeting social needs, although this was more prominent in the first three post-war decades, notably in orchestrating the development of New Towns. It has an extensive role, however, in regulating the externalities of development and changes in land use.

Property owners possess the existing use rights of their land, but since the 1947 Town and Country Planning Act the state has owned most of the development rights. The state's share of property rights could extend much further to the ownership of the land itself, or to the right to a share in any profits that arise from developing it (the latter was the case for three periods under post-war Labour governments). Opinions about how large this share for the state should be range from the Right-wing view that governments should have only a minor role in determining what is done with private property, limited to arbitrating in disputes between private owners, to a Left-wing position that the state should have a major role in determining the detail of how property is used. The former is justified in terms of arguments about liberty and the benefits of free enterprise; the latter in

terms of meeting social needs and social planning. Local planning authorities have generally favoured a more interventionist use of planning powers based on the needs of their area as they see them, whilst recent Conservative Secretaries of State for the Environment have imposed a more *laissez-faire* regime with the aim of lessening how far planning interferes with the free operation of the market.

Although the political nature of planning may seem obvious, Goodchild (1990) points out that for much of the history of British planning it was presented by practitioners and academics as a neutral process of rational decision-making. During the 1970s a series of Marxist and neo-Marxist studies challenged this and sought to demonstrate that, far from being neutral, urban and regional planning served to reproduce conditions for capital accumulation by the private sector and often inflicted costs upon the working class (for a review, see Lebas 1982). Thus, for example, working-class communities in inner cities were broken up by redevelopment, releasing the land for more profitable uses, and people displaced to new peripheral housing estates planned as labour pools for the factories of multinational companies. There was also a substantial amount of non-Marxist work which was critical of planning from a liberal perspective. Dennis (1970) and Davies (1972) made early attacks on the planning system and its ideology of 'futurism' which advocated sweeping away the old neighbourhoods of British cities to create modern housing estates and road networks, whatever the cost for the people disrupted in the process.

Hannigan describes this era of modernization as one dominated by the 'interventionist and frequently autocratic planning notions of Le Corbusier and of other European and Anglo-American pioneers of urban planning' (Hannigan 1995: 166). The large-scale planning strategies of the time were, however, relatively successful in improving housing standards and attracting new employment in peripheral regions such as the north of England and Northern Ireland (Townsend 1983). But many of the new jobs disappeared during the recessionary years of the 1980s, and much of the new housing built for the public sector in 'package deals' by large construction companies proved unpopular and often defective due to structural problems and difficulties such as the cost of heating.

This was also an era of community action as local groups formed to defend existing older neighbourhoods from redevelopment or to campaign for improvements in the new mass housing estates (Blackman 1987). In some areas 'planning aid' and 'community architecture' initiatives developed, representing approaches to redevelopment and refurbishment that involved local residents in decision-making about their local environments. This action also had a wider significance as a 'rejection of totality and uniformity' and an assertion of local needs and views (Goodchild 1990).

By the beginning of the 1980s the nature of urban development had changed, reflecting the emergence of post-industrial cities and the New Right

Plate 5.1 Ironic graffiti on this house in Northern Ireland's new town of Craigavon indicates local feeling about the imaginative design solutions used on this and other estates in the early 1970s

policies of Thatcherism. There has been a shift to a *post-modern* phase of development and planning in which the market is a much more significant player than the state. Architectural styles are also different, with a preponderance of neo-vernacular and pastiche designs. In depressed areas designated for regeneration in inner cities, the scale of new development is small compared with the modernization era, the impact on living conditions and employment has been much smaller, and the finance has largely come from

the private sector, 'levered in' by initial public subsidy. River-front sites have been particularly attractive to developers wanting to reuse old industrial land for new housing, shopping, office and leisure projects, subsidized by publicly funded land reclamation schemes, grants and tax-breaks.

The post-modern phase has been accompanied by a visionary language similar in tone, although not in content, to that of the earlier modernizers. Today, the purveyors of these visions are businessmen rather than public sector professionals, exemplified by the north-east's John Hall, who was behind the development during the 1980s of Europe's largest shopping complex and symbol of retail culture, the MetroCentre on Tyneside:

> We built the biggest ships, the biggest engineering products, the biggest armaments in our industrial time. Now in a sense we have to bring ourselves forward and do the same things in the new way. . . . We will create in a sense the new North East out of the ashes of the old – and that's linked in with shopping, with leisure.
>
> (BBC Television, *40 Minutes*, 1989)

A particular feature of post-modern development has been its targeting of the affluent segment of an increasingly divided society. Typical projects have been new working environments for the financial and business services sectors, large shopping developments, bright new housing often protected by intensive security, and restaurants, hotels and private leisure complexes. Whilst the modernization era was a reflection of one-nation politics which included narrowing social and geographical inequalities, present times are marked by an absence of any strong policies in these areas and deep social and economic divisions between households and neighbourhoods.

Cooke argues that post-modernism is essentially a critique of, rather than a coherent alternative to, modernism and its 'centralized solutions . . . imposed upon diverse localities' (Cooke 1990: 341–2). Others have argued that it is in fact a new social condition and detect a shift in people's derivation of social meaning and identity from *production* to *consumption* (Hannigan 1995). The industrial city has been replaced by 'consumption cities' and 'tourist cities' (Zukin 1992). This reflects of course both the de-industrialization of many Western cities and the rise of a 'retail culture' for those who have been able to stay in employment. There is nothing particularly new about these relationships between economic and social changes; what is distinctive about the 'post-modern' era is the politics of deregulation and the widespread denial of the possibility of planning for a better future.

There also appear to be distinctly social changes at work. Bauman (1988) has described the decline of professional and intellectual authority in these terms. Increasingly, the notion that the doctor, teacher or planner knows best is questioned. On the political Right, it is argued that market forces rather than 'experts' should determine how resources are used, whilst on the Left, arguments are made for greater democratization and grassroots participation

in decisions. In public policy, these developments are reflected in the demise of state regulation based on planning blueprints and the rise of flexible decision-making often based on partnerships between a variety of 'stakeholders'. The next section considers how the objectives of planning have been influenced by these changes.

THE OBJECTIVES OF PLANNING: FROM REGULATION TO PARTNERSHIP

Local authority planning departments produce advisory land use plans for their localities and regulate development within these planning frameworks. They have increasingly been expected to work with other organizations in undertaking this work, and the nature of these partnerships has become an important influence upon the role and objectives of land use planning.

In areas of England with two-tier local government, planning functions are split between county councils and smaller district councils. County councils prepare structure plans that contain broad planning policies, whilst district councils prepare more specific local land use plans. In areas of England with single-tier unitary authorities, and in Scotland and Wales where the unitary structure is now universal following the recent reorganization of local government, there is only one local planning authority, and the features of both structure and local plans are incorporated in unitary development plans.

Planning practice at local level is strongly guided by central government policy instruments (Planning Policy Guidance and Regional Planning Guidance in England and Wales, National Planning Policy Guidelines in Scotland, and various circulars). With the election of the first of a succession of Conservative governments in 1979, the planning powers of local authorities began to be reduced, and in some areas significant responsibilities were transferred to new *appointed* bodies such as Urban Development Corporations (UDCs).

Most of the day-to-day work of land use planning is development control. Applications to develop or change the use of land are regulated with regard to advice set out in land use plans. This consists of processing planning applications and making decisions about whether to permit development or a change of use at a particular site. These planning controls are exercised by the state because of the need to have some degree of coordination and regulation of development to protect environmental quality. Practical planning considerations relate mainly to size, siting, visual appropriateness and traffic impact.

Land use planning works largely through the exercise of *negative* powers. Land may be zoned for particular uses but the local authority has few resources to bring forward development itself; it relies largely on the private sector and thus has to be realistic about what types of development to zone for particular areas. In many inner cities, industrial land has been rezoned and developed for housing or retailing in areas with many redundant indus-

trial workers, apparently denying any prospect of a recovery of industry in these areas (Byrne 1989).

Without the resources to plan positively, local authorities have turned to their powers to regulate in order to achieve positive benefits from private development beyond those that might otherwise occur. Under section 106 of the Town and Country Planning Act 1990, they can seek a planning gain by obliging 'within reason' a developer to provide an additional benefit as a condition for receiving planning permission. In most cases this takes the form of contributing towards off-site infrastructure costs associated with the development, but in some cases benefits such as social housing provision are obtained. Land and housing may be passed by the developer, often at nil cost, to a housing association in return for an agreement by the local authority to modify land use, density or other planning constraints affecting the site of a proposed private development.

Controversy surrounds planning gain because, despite its basis in local authorities' regulatory powers, it is achieved through negotiation and flexibility. The bargaining power of a local authority will depend upon its expertise and on property market conditions. It is an activity that takes place behind closed doors, leading some objectors to challenge decisions where they feel a local authority has been too willing to ignore negative aspects of a development because of a promised planning gain. However, recent governments have encouraged the practice as a way of relieving the public sector of infrastructure costs such as road building. Many local authorities have embraced the principle by amending their land use plans to state that the willingness of developers to provide some social housing will be treated as a material consideration when applying for planning permission.

The theory of development control is that decisions are made on the basis of development plan policies, past precedents and government guidance, but in practice these decisions are also made within the political context of a given local authority which can mean that the emphasis on different factors can vary. Although cases of malpractice and even corruption in development control decisions are far from unknown, the question of political context is essentially one of different 'framings' of planning issues that may or may not coincide with the 'official' scope of land use planning (Tewdwr-Jones 1995). Cases have come to light of small rural authorities favouring long-established local residents in granting planning permissions. In the larger urban authorities, the conflicting 'framing' of issues tends to be expressed in more public and political terms. For example, in the 1980s a number of UDCs were set up by central government in designated areas of inner cities, displacing the local authority's planning responsibilities on the grounds that what was needed were new, entrepreneurial and single-purpose agencies to revitalize these areas. Such a framing of the problem unsurprisingly received a hostile reception from local authorities that nevertheless have had to work with the UDCs. The local authority will often frame planning issues within the wider sphere of the city as a whole, whilst UDCs are focused on bringing development into their designated zones. This can cause conflict when, for instance, a local authority is pursuing a strategy of regenerating its city centre whilst a UDC is actively attracting development to another area in the same city (Newcastle City Council 1995).

These issues extend beyond the *how* of planning to wider questions about its purpose. In particular, how far is it the role of planning to co-ordinate development to achieve social or environmental goals beyond narrowly defined land use planning matters? As already noted, there have been instances in the past when a social purpose was very prominent, such as in the planned communities of post-war New Towns and in short-lived legislation in 1947, 1967 and 1976 to tax land development profits that arose purely from the granting of planning permission. Today, however, social goals have a minor role in planning. This can lead to considerable frustration when local people participate in consultation and public inquiry exercises only to find that most of their concerns are excluded because they are non-planning matters. For example, during consultation about Belfast's Urban Area Plan during the late 1980s, community groups raised many concerns about the future of their neighbourhoods. This brought the following clarification of what constituted a planning issue from the Department of the Environment:

> The observations of the Department can only properly address the land use planning issues raised. It does not attempt to deal in detail with the non-planning issues which should be raised directly with the appropriate agencies with legislative responsibility for these matters. These non-planning issues include, for example, the actual provision and creation of jobs, the provision of shops, housing, finance, health and education.
>
> (Quoted in Blackman 1991: 190)

Table 5.1 shows how the arguments between community groups and Department of the Environment planners at the ensuing public inquiry into the Belfast Urban Area Plan crystallized into two conflicting views of how the plan should serve the city.

Local voices such as those of Belfast's community groups are often marginal to major planning decisions. More progressive local authorities have sought to work in partnership with community groups, providing support for them to assess their communities' needs and involving them in making community plans that address these needs comprehensively (Association of Metropolitan Authorities 1993).

Although partnerships are increasingly common in planning, these are generally between organizations with the powers and resources to deliver development projects rather than community groups. Because local authorities have neither the powers nor the resources to deliver large projects themselves, they enter into partnerships with private firms and other public

Table 5.1 'Officialdom' versus 'community': opposing paradigms at the Belfast Urban Area Plan public inquiry

	Official paradigm	*Community paradigm*
Key concepts	Growth of urban system and city image Relating land use to market forces	Quality of existing localities Intervention in land use: equity and stability
Key categories	Order and efficiency City centre boosterism Private transport growth	Community concepts; local facilities; accessibility and public transport
Key relationships	Property market; growth; commuting	Land use; social policy; community development
Key methods	Plan emerges from officialdom within government policy parameters	Plan emerges from public consultation and systematizes local needs

Source: adapted from Blackman 1991

sector agencies. These are now well-established in the housing field, where since the mid-1980s the Estate Action scheme has required private sector involvement in renewing and diversifying local authority housing estates. New social housing development is now through housing associations, making the local authority's role as a major housing developer a thing of the past. Local housing plans, however, remain a local government responsibility.

The City Challenge initiative, launched in England in 1992, required local authorities to have urban regeneration partnerships of the statutory, private and voluntary sectors in place as a condition for receiving government funding for regeneration projects. The initiative marked a shift away from the property-led regeneration that typified government urban policy during the 1980s, and has attempted to target training, job creation, housing and social projects so that disadvantaged local residents benefit directly. There were no new City Challenge projects after the second round in 1993, but the principle has continued of linking urban regeneration funding to the existence of local partnerships working with specific targets for which they are accountable to central government.

This has also been the case in respect of City Challenge's other important feature, that of competitive bidding for resources. Local authorities competed on the basis of their regeneration proposals for City Challenge status and the funding that went with it. This was a process in which there were winners and losers, the outcomes having no direct relationship to the level of local social need. The principle carried over into the Single Regeneration Budget (SRB), which was an attempt to improve the co-ordination of urban policy by absorbing the many separate schemes into a national £1.4 billion pot. Local authorities now bid annually for allocations from this budget to fund partnership projects aimed at regenerating depressed areas – an activity in which local authority planning departments are usually centrally involved.

Partnerships are likely to become an increasingly important method for funding major capital projects via the Public Finance Initiative (PFI). This involves private companies in financing and operating public projects, such as transport infrastructure and hospitals, or in leading joint venture companies to develop urban regeneration projects such as shopping centre improvements, with local authorities as partners in the company. The PFI is controversial because it is being used to substitute private investment in place of public capital expenditure. The approach was pioneered with housing associations during the 1980s. In 1991/2, government grant to housing associations totalled £2.3 billion, or 77 per cent of their budget. By 1995/6 it had fallen to £1.2 billion or 58 per cent of their budget. The difference is accounted for by housing associations raising funds on the financial markets. The scheme has been widely criticized for forcing up rents and reducing new housing starts.

Labour Party policy is to continue the PFI approach should it win the 1997 general election, although Labour has opposed its use in the National Health Service as a means of cutting public capital expenditure. Thus for example, a Labour government would not increase the SRB, but 'expect greater private sector involvement and more private/public partnerships' (Burton 1995: 8).

In the economic field, local authorities have long been keen to work in partnership with incoming investors. A recent example is the decision by Siemens to locate a new semi-conductor plant in North Tyneside, bringing 1,800 jobs to a high unemployment area. This was supported by the local council providing technical data on the proposed factory site, information on the local economy, rates relief, fast-track approval of the planning application and assistance with a project office to find suppliers and contractors for the new factory (*Municipal Journal*, 11–17 August 1995). In some areas local authorities work together in consortia to attract investment from overseas.

Many British cities now have *strategic* partnerships in place (see Table 5.2). These are similar to the growth coalitions seen in North America. Whilst important and pragmatic responses to the problems faced by cities, they have the same issues associated with them of potentially wasteful competition between cities and a side-stepping of fundamental political issues about the distribution of power and resources in an urban society of widening income and wealth inequalities.

Many regeneration partnerships are led by social housing organizations rather than planning departments because of the strong link between social rented housing and high unemployment. The concept of 'housing plus'

Table 5.2 The Leeds Initiative

Leeds Initiative objectives

- Promote the city as a major European centre
- Ensure the economic vitality of the city
- Create an integrated transport system for the city
- Enhance the environment of the entire city
- Improve the quality and visual appeal of the city
- Develop the city as an attractive centre for visitors

Partner organizations:

- Leeds City Council
- Leeds Chamber of Commerce and Industry
- Leeds Training and Enterprise Council
- Leeds Chamber of Trade
- University of Leeds
- Leeds Metropolitan University
- West Yorkshire Playhouse
- Yorkshire Post Newspapers
- Regional Trade Union Congress
- Government Office for Yorkshire and Humberside
- Leeds Civic Trust
- West Yorkshire Police
- Railtrack
- West Yorkshire Passenger Transport Executive

Initiative projects:

- The Leeds Flower Initiative
  – the North's 'City of Flowers'
- Education Business Partnership
  – targets for local education and training
- The Leeds European Initiative
  – promotion of Leeds as 'a major European city'
- The Leeds Retail Initiative
  – marketing and promotion of shopping
- The Leeds Financial Services Initiative
  – 'the UK's leading financial centre outside London and Edinburgh by the year 2000'
- The Leeds Engineering Initiative
  – awareness raising, training and marketing
- Gateways and Corridors
  – transport infrastructure rebuilding, lighting and landscaping
- The Leeds Lighting Initiative
  – floodlighting schemes and lighting improvements
- Leeds City Station
  – redevelopment
- The Leeds Printing Initiative
  – training and promotion
- Leeds Environment City Initiative
  – a range of environmental action schemes and a Business Forum
- Opp2k – Working for Women Working in Leeds
  – initiatives to advance the position of women in the labour market
- Leeds Media Initiative
  – development of the city as a media centre
- Leeds Architecture and Design Initiative
  – promotion of the city as a 'centre of design excellence'
- Leeds Initiative Regeneration Board
  – £17m 5-year programme to increase employment in inner city areas

Source: The Leeds Initiative *Annual Report* 1994–5, Leeds City Council.

goes beyond simple housing development and management to include the local provision of social facilities, services and jobs that promote community development and the overall stability of social housing areas (Page 1994). This approach is being promoted particularly for new social housing developments.

One example is the Blackbird Leys development in Oxford. At the time of writing in 1996, this is the largest new social housing scheme under development in Britain (see Plate 5.2). The development partners consist of the local authority owner of the land, a large regional housing association and its associated company building for shared ownership, a medium-sized community-based housing association, another local housing association specializing in special needs housing, a black housing association, and speculative builders of starter homes for sale.[1] These partners, and especially the largest housing association that is taking the lead during development, are attempting to implement the housing-plus approach by creating and supporting a social infrastructure as well as building houses. This is reminiscent of social development policies in the post-war New Towns, but it remains to be seen how successful this diverse partnership will be in creating a new community.

Although local government's range of responsibilities gives it substantial influence at local level, the positive powers of local authorities have been severely eroded since the end of the 1970s. In the 1960s, a development like Blackbird Leys would have been entirely planned and delivered by the local authority. Legislation has removed or limited some functions, but the main check has been tightening controls on local government expenditure. For local authorities to achieve positive change in their localities, they have increasingly had to work with other agencies and the private sector.

Partnerships can have the benefit of achieving very focused, jointly funded work, but can blur the process of democratic accountability because no single player has overall responsibility and often most of the resources are from the private sector, giving primacy to commercial considerations such as the rate of return from projects. This orientation towards market criteria is not a feature only of regeneration projects but also of day-to-day planning practice, the subject of the next section.

PLANNING IN PRACTICE

The series of Conservative governments since 1979 has introduced legislation and regulations that have changed the nature of British planning and its associated practices. The planning system has been simplified, its processes speeded up and the powers of local planning authorities reduced in order to 'lighten the burden' on entrepreneurial activity and wealth creation. A 'customer care' ethos has been promoted throughout the public services, ostensibly to improve their quality and cost-effectiveness, such as publishing

Plate 5.2 The edge of a major new social housing development at Blackbird Leys in Oxford, with the much 'downsized' Rover car plant in the distance

the time taken to arrive at decisions about planning applications and the cost of the planning service per head of population.

Moves to make planning more 'business friendly' began with the Department of the Environment's Circular 22/80. This asked local authorities to prioritize planning applications that contributed to national and local economic activity. There was a presumption in favour of granting permission for development: in other words, exercise of the local planning authority's power to refuse permission would be expected only if demonstrably greater harm would be done by granting it, and authorities were reminded that the development plan should be only one of many possible considerations that could be relevant.

An effect of the pro-market stance of Secretaries of State for the Environment during this period was that developers increasingly challenged decisions by local planning authorities that went against them, decisions usually based upon the development plan. By making formal appeals, or formal objections to planning policies at public inquiries, developers often persuaded the Secretary of State to use his powers to reverse or amend local authority decisions or plans. The development of new out-of-town shopping centres is a case in point. Local authorities often sought to prevent such developments in order to protect the commercial viability of existing town centres. Many appeals against such decisions have been won, whilst other developments have gone ahead in government-designated Enterprise Zones. The situation led Guy (1994) to observe that the geographical restructuring of retailing in Britain occurred almost entirely in defiance of established local planning policies.

Further circulars during the 1980s developed the pro-market emphasis, but in the early 1990s this was modified by advice that local authorities should pursue the joint objectives of a high-quality environment and economic growth. The basis for this was the Planning and Compensation Act of 1991. Conflicts in Tory parliamentary constituencies about the continuing spread of development in green field areas, growing concern about the impact on town centres of out-of-town shopping developments, and administrative bottlenecks caused by the increase in appeals, led to greater central government support for local authority land use plans. The 1991 Act gave moderately stronger legislative force to development plans and put more emphasis on conservation and environmental issues.

A plan-led approach to development control is now favoured, whereby suitable sites for housing, shopping and other developments are identified in structure and local plans (Packer 1995). However, for decisions to be based on them, development plans must be up-to-date and relevant to the particular planning application, and policies in a local authority's development plan can be overridden by contrary government policies stated in circulars (MacGregor and Ross 1995). These factors introduce the flexibility that is regarded as necessary in today's more uncertain economic conditions so that landowners and developers do not find themselves constrained by fixed plans that are reviewed only periodically.

There also appears to have been some relaxation of the long-standing position that land use planning cannot be used as an instrument of social policy. Guidance issued in 1994 combines the emphasis on market-led development with the need to regulate it in the public interest, and states that the *land use aspects* of social needs and problems may be addressed by local planning policies, such as the impact of development on minority ethnic groups and people with disabilities (Department of the Environment 1994a).

This acknowledgement of the social aspects of planning is not reflected in any greater say for the public; just the opposite in fact. Local authorities have been encouraged to consult the public about outline or strategic proposals, rather than the full draft plan, and urged to make greater use of written representations rather than use costly public inquiry time to consider objections (Department of the Environment 1994b).

The individual citizen, seeking a service rather than an influence over the development of their neighbourhood or city, is the model reflected in the Citizen's Charter, launched in 1991 to improve the performance of the public sector, including local authority planning services. The key performance indicator for planning relates to the number of planning applications decided within eight weeks. This has meant that planners are under pressure to keep any negotiating to the minimum, an imperative that is compounded by staff shortages in planning departments.

Thomas (1994) argues that the emphasis on speeding up decision-making and containing costs disadvantages people who are not informed or resourced sufficiently to make their voices heard in the planning system. Minority ethnic groups are particularly likely to be excluded in this way. It also appears that local authorities have been quick to adopt devices to speed up the plan-making process. One of the costs of this may be the presumably unintended consequence of frustrating the participation of developers as well as the public in planning debates, thereby risking a return to the anti-planning ethos of the 1980s that led to proposals for banning local authority planning committees from dealing with anything other than outline applications (*Planning* 1995a; Vickery 1995).

PLANNING AND ECONOMIC DEVELOPMENT

A less obviously politically driven change in planning practice has been the orientation of planning activity in many local authorities towards economic regeneration in response to the spatial concentration of Britain's economic problems. The supply of small industrial sites and factory premises, infrastructure provision, small grants, the promotion of the area to potential investors and public relations are long-established activities, especially for larger authorities. However, it is not necessary to look far to find political disagreement about how authorities should be able to respond to economic problems in their areas.

For a period during the 1980s, some authorities adopted policies of intervention in key sectors of the local economy. Enterprise boards were established as vehicles to achieve this, carrying out research and developing local economic strategies and initiatives. Selective investment from council funds was aimed at supporting economic activities that had prospects for growth and diversifying away from activities that were clearly in decline. New approaches were encouraged for traditional industries such as footwear and clothing, particularly flexible specialization by small and medium-sized enterprises, which local authorities had the resources to assist.

The abolition of the Greater London Council (GLC) and the English metropolitan county councils in 1986 largely brought an end to this phase of local economic development. The politicization of economic policy pursued notably by the Labour GLC administration contributed towards Conservative government hostility to this tier of local government. Labour's 1981 GLC election manifesto, for example, stated that, 'Our vision of the future is a city in which elected representatives take the lead in economic planning . . . . We shall set out to increase the element of democratic control over industrial decisions . . . ' (quoted in Mackintosh and Wainwright 1987: 2). Regardless of the politics, the efficacy of the approach was also questioned:

one critique comparing it with 'trying to drain an ocean with a teaspoon' (Cochrane 1983: 285).

If the Labour GLC era was a period of attempting to restructure in favour of labour rather than capital, then the phase since the early 1980s of public/private partnerships between local authorities, or agencies sponsored by them, and private firms is much closer to the latter. These partnerships have brought local authority functions in the areas of land, infrastructure and training into the service of inward investors, especially Japanese and American firms seeking to position themselves within the Single European Market.

Some local authorities have attempted to use partnerships to link local unemployed people with new job opportunities by, for example, running customized training in close collaboration with an expanding or incoming firm. This has been a feature of many City Challenge projects. Such initiatives are beset by problems of displacement and low pay, with some critics seeing them as little more than low-level and 'work habit forming' training that keeps a reserve army of labour in British cities constantly in being and 'up to the mark' (Byrne 1996).

Local government's role in local employment training is in fact very limited, this function being the prime responsibility of Training and Enterprise Councils (TECs) in England and Wales and Local Enterprise Companies (LECs) in Scotland. The bulk of TEC budgets is devoted to funding training programmes for young people and adults, but they seem set to adopt a wider role of planning training programmes within a local economic development framework (Blackman 1995a). TECs are quite unlike local authorities. They are publicly funded private companies with boards appointed by the Department of Employment. Whilst they are not strictly speaking quangos, they are an aspect of the fragmented and locally unaccountable reform of the local public sector that has taken place in recent years.

It is, however, in the field of economic development that planning departments have increasingly justified their role beyond routine planning functions. Many chief planning officers have used this justification to defend their departments against cuts in central government grant to local authorities in the face of greater political support among elected members for protecting functions such as education and social services. Chief planning officers have found themselves defending their departments against proposals for a disproportionate share of cuts in their authority's total spending. Indeed, planning has not been immune from the continuing downward pressure on costs in all public services. Efficiency improvements have been sought, particularly in information processing. Planning departments' information systems have traditionally been based upon a mixture of property files, card indexes and hand-annotated maps. Geographical information systems (GIS) based upon the Ordnance Survey's computerized maps have begun to be introduced as comprehensive repositories of planning information, supporting in particular the development control process and land charges administration.

Investment in information technology has not just been stimulated by the promise of cost reductions. There has been growing interest among some authorities in the concept of the 'information city'. At a European level, the Telecities organization has been formed as 'an open network for a concerted development of urban areas through Telematics' (Telecities 1995). Initiatives range from teleworking to distributed information services using new information and communication technologies (telematics). In the UK, Manchester is a leading telecity. Among many of its projects is the Electronic Village Halls (EVH) initiative through which a network of neighbourhood-based telematics centres is being developed (Manchester Telematics Partnership 1995). These provide facilities for training, teleworking and the dissemination of information and advice. They are located in community-based centres offering social facilities such as child care, personal advice, meeting space and educational programmes. Three of the EVHs specialize in supporting groups facing social exclusion in the labour market – the Women's EVH, the Manchester Bangladeshi EVH and the College of the Third Age (which works mainly with people over 55). There are another six EVHs based in local areas within Manchester.

Some authorities see cable television as a way of developing community-based media, distance learning and local information services. Yeomans (1993) portrays a picture of the 'intelligent housing estate', with telecom links bringing residents distance learning opportunities, employment information and improved access to services. Investment in advanced telecommunications infrastructure has the potential to overcome geographical marginalization in peripheral regions and promote them as efficient and attractive locations for inward investors. In 'core regions' such as south-east England there is also an interest in using telecommunications to reduce commuting. Whilst these technologies are dominated by powerful commercial interests, their application for public purposes is being strongly promoted by the European Commission, with major investments in research and development.

CURRENT CHALLENGES FOR URBAN PLANNING: RESPONSIBILITY WITHOUT POWER

Urban planning in Britain's cities faces many challenges including environmental damage arising from traffic growth; maintaining employment and generating or attracting new jobs; the quality of life in neighbourhoods, including problems of environmental quality, social polarization, crime and access to services and facilities; and constant pressure from central government to reduce its costs. If planning is to respond successfully, it not only needs appropriate policies but also the powers to put them into practice. After years of deregulation, this is now a fundamental issue for local government and the efficacy of its role in cities generally. Traffic planning is a prime example.

A 1994 comparative study of urban traffic planning in Heidelberg, Bautzen, Montpellier and Cambridge exemplifies the difficulties that Britain's cities face in tackling the problems of traffic growth (Plate 1994). The traffic problems of these four cities are basically similar, and all the cities are following broadly similar objectives. Cambridge, however, is at a major disadvantage compared with the other cities because the local authority has only a weak influence over the planning and operation of public transport. In Britain, these services are privately owned within a deregulated environment. Thus, a key instrument of policy is not available to planners in Cambridge. This problem is repeated in many areas of British public policy. The ideology of a minimalist role for the state embraced by recent governments has seen a dismantling of much of the capacity of a local authority to intervene in the problems of its area. The headlong rush towards gridlock on many of Britain's roads is just one of the more visible manifestations of this.

It must also be recognized, however, that although fairly strong policy instruments were available to local authorities for much of the post-war period, this did not necessarily mean that problems were tackled successfully. The defects of the systems-built mass housing schemes of the 1950s and 1960s have been well documented; some have interpreted the disaster as resulting from the way planning interfered with the market's ability to signal to builders what consumers wanted (Coleman 1985). During the same period massive investments were made in road building in the UK's peripheral regions, based on the belief that new roads stimulate economic development – a premise for which there is still no conclusive evidence (Whitelegg 1994).

Thus, having the powers of intervention does not by any means guarantee effective results: cities are too complex for planners to have more than probable knowledge about the likelihood of either intended or unintended consequences of intervention. Research has an important role, nevertheless, in identifying conditions for success. Plate's (1994) comparative study, for example, identifies these conditions in the case of traffic planning to be firm political direction at local level, public support, and a sense of common purpose between all levels of government.

Getting these local factors right is still no guarantee that a city can in any significant way control its future because this depends so much upon its position within a globalized economy and global competition for investment and markets. The uneven development of a world economy in which nation states have very limited influence not only creates successful cities at the cost of unsuccessful ones, but also causes sharpening social divisions *within* cities (Byrne 1996). Whilst the European Union offers the kind of scale at which political intervention such as regional policy and employment rights is less likely to be undermined by market forces, initial indications are that completion of Europe's single market is leading to fiercer competition between cities and regions (Wiehler and Stumm 1995).

Competitive pressures are likely for the foreseeable future to be one of the prime shapers of policy for Britain's cities. Cities with developed infrastructures, dynamic industrial and service sectors and well-trained labour forces are more likely to have a long-term future, whilst other cities have much more uncertain prospects. In most European countries, it is regional and local authorities that are expected to respond to this challenge. The EU's principle of subsidiarity, whereby decisions are to be taken as closely as possible to the citizen, reinforces this. However, European regional and local authorities vary greatly in their powers and budgetary resources, with the UK's very centralist state entailing very little policy of budgetary autonomy for local authorities and having no *elected* regional tier of government.

The UK had relatively strong regional policies during the 1960s and 1970s, although only Northern Ireland had regional government. Regional policy was dismantled during the 1980s by the decade's non-interventionist governments but, towards the end of the decade, the lack of a regional planning framework began to create political problems for the ruling Conservatives. Abolition of the GLC and metropolitan authorities had created a strategic planning vacuum, and this was compounded by political pressure to contain market-led development in green field areas of south-east England, business pressure for a more certain planning framework for major investments, and the need for a regional framework to relate to European Union policy and structural funding for the regions (Thomas and Kimberley 1995). The result has been the development of a limited regional policy in the form of regional planning guidance that is issued for the English regions by the Department of the Environment, with input by regional groupings of local authorities.

The weaknesses of local government and narrow scope of regional policies in Britain are compounded by fragmentation and privatization in the local public sector, caused by the growth of semi-autonomous public bodies and the privatization of services brought about by the introduction of compulsory competitive tendering for contracts to run public services. Quangos have acquired responsibilities previously held by local authorities and are contracted to or directly instructed by central government departments. There is growing evidence that fragmentation between TECs, housing associations, NHS Trusts, government agencies and local authorities is undermining the ability to plan comprehensively to meet the social and economic development needs of localities because these organizations are poorly networked together and lack common strategies (*Findings* 1994).

CONCLUSION: TOWARDS THE 'INTELLIGENT' LOCAL AUTHORITY

Urban planning is part of the local co-ordinating and regulating machinery of government. Its traditional land use role remains significant and indications are that it will be so increasingly. Recent estimates that England will see a growth of 4.4 million households by 2016 are stimulating discussion about how to plan for development on this scale. It is recognized increasingly that such planning needs to have greater cognizance of social trends and sustainability, considering the type and tenure of the new dwellings that will be needed as well as the numbers required. Planning ideas include new housing for students, young professionals and older people on brown field sites in cities, and new energy-efficient settlements on green field sites (*Planning* 1995b).

Issues concerning new development, however, are relatively marginal compared to the scale of issues in Britain's existing urban areas. There is a huge backlog of rundown housing, school and hospital buildings, and inadequate or deteriorating infrastructure. Many commentaries on the contemporary state of the UK argue that this is a facet of a wider failure to invest in infrastructure and public services generally, perpetuated by the short-term horizons of British financial institutions and governments (Borrie 1994; Hutton 1995). These arguments for public spending as investment rather than just consumption have been influential. Education, for instance, is increasingly promoted as vital to Britain's economic prospects. Its quality may be a more important focus for urban policy, in terms of achieving an economic return for local populations, than conventional local economic development activities such as providing premises and promotion of the area to potential investors.

In addition to the direct economic returns of public spending, it can also have indirect returns by, for example, reducing the costs of ill-health or energy consumption. Land use planning is potentially in a position to use the state's share in property rights to realize a return for the community, whether preventive, such as requiring designs that reduce the risk of accidents, or positive, such as various types of planning gain. In the case of direct public development, returns can be increased by planning more holistically. For example, housing improvement projects can have designed into them measures to reduce energy use and fuel poverty, generate jobs for local unemployed people, improve access for people with disabilities, provide child care facilities and improve health.

This type of 'intelligent' planning demands the co-ordination of various specialist functions such as valuers, surveyors, engineers and social scientists, as well as a 'people orientation' which means opportunities and support for consultation and political accountability. These elements already exist in local authorities. It is possible for their various services to work with common objectives so that they are more than just routine administrative functions but have *strategic intent* such as maximizing local employment, reducing health inequalities or promoting sustainable development.

Although the powers and resources of local authorities greatly constrain their ability to intervene in urban problems, they nevertheless have major mainstream services that can lack strategic direction because of a preoccupation with day-to-day management. Land use planning practice is an example of where this is less likely to happen because of the nature of its processes, and it is an important model for strategic planning generally. Routine functions are guided by strategic plans and particular issues are approached in terms of interrelationships with the wider environment.

Just as the current constraints on local government need not imply a retreat from strategic planning, there is little reason why local authorities should not lead partnerships with strong strategic policies. A 'healthy city' is an example of an outcome emerging from strategic partnerships based on organizations deploying, sharing and redeploying their resources in ways that achieve health returns (Blackman 1995a). If research indicates that the health problems of a neighbourhood are significantly connected with poor housing conditions, there could be an increase in spending on housing or, if no growth in resources is possible, virement from health care spending if a greater health gain is likely from the investment in improved housing conditions (Blackman 1995b).

For partnerships to work on this basis, they need to be informed by a common analysis of urban problems. Whilst the lead role for a local authority in local partnerships might be based on its democratic legitimacy and the many services for which it is already responsible, this role is considerably strengthened if it is also based on the quality of its 'intelligence': information about needs, capacities and trends derived from local research and consultation. This has also been the land use planning approach, and it is significant that in many local authorities the only research capacity is located in the planning department. There is a case for this capacity to be more extensive and at the heart of corporate management and partnership building in local government.

In conclusion, this chapter has sought to show how the role of urban planning has been influenced by economic and political changes, but has continued to be an important aspect of the governance of cities. It is difficult to envisage its role in regulating development ever disappearing, although the extent to which this is based on detailed plans and guidelines will remain contested because of issues of power and control. Strategies and partnerships rather than plans and regulations currently constitute the dominant paradigm for planning and local government generally. The question is whether local authorities can organize their resources in ways that can make this paradigm work and have an impact on the serious social and economic problems of Britain's cities.

NOTES

1 I am grateful to Jane Darke for drawing my attention to this development.

GUIDE TO FURTHER READING

Blackman, T. (1995) *Urban Policy in Practice*, London: Routledge.
This text reviews and analyses the range of British urban policies from a local government perspective.

Cullingworth, J.B. and Nadin, V. (1994) *Town & Country Planning in Britain*, London: Routledge.
The eleventh edition of this well-known textbook remains one of the best guides to the British planning system and planning policy.

Hannigan, J.A. (1995) 'The postmodern city: a new urbanization?', *Current Sociology*, 43(1): 151–214.
This extended article reviews the many dimensions of the debate about postmodernism and urbanism.

Hutton, W. (1995) *The State We're In*, London: Jonathan Cape.
This book is essential reading for understanding the economic and constitutional causes of Britain's increasingly divided society.

Taylor, A. (ed.) (1990) *The Renaissance of Local Government*, SOLACE.
This series of papers, published by the Society of Local Authority Chief Executives, brings together a range of contributions covering local democracy, local taxation, governance, regionalism and the European Union.

REFERENCES

Association of Metropolitan Authorities (1993) *Local Authorities and Community Development*, London: AMA.
Bauman, Z. (1988) *Legislators and Interpreters*, Cambridge: Polity Press.
Blackman, T. (1987) *Housing Policy and Community Action in County Durham and County Armagh: A Comparative Study*. Unpublished PhD Thesis, Faculty of Social Sciences, University of Durham.
——(1991) *Planning Belfast: A Case Study of Public Policy and Community Action*. Aldershot: Avebury.
——(1995a) *Urban Policy in Practice*, London: Routledge.
——(1995b) 'Recent developments in British national health policy: an emerging role for local government?', *Policy and Politics*, 23: 1–18.
Borrie, G. (chairman) (1994) *Social Justice: Strategies for National Renewal – The Report of the Commission on Social Justice*, London: Vintage.
Burton, M. (1995) 'Labour will stick to current £1.4 billion SRB budget says Vaz', *Municipal Journal*, 42: 8.
Byrne, D.S. (1989) *Beyond the Inner City*, Milton Keynes: Open University Press.
——(1996) 'Chaotic places or complex places?: Cities in the postindustrial era' in F. Westwood and J. Williams (eds) *Imagining Cities*, London: Routledge.
Cochrane, A. (1983) 'Local economic policies: trying to drain an ocean with a teaspoon' in J. Anderson, S. Duncan and R. Hudson (eds) *Redundant Spaces in Cities and Regions?*, London: Academic Press: 285–311.
Coleman, A. (1985) *Utopia on Trial: Vision and Reality in Planned Housing*, London: Hilary Shipman.
Cooke, P. (1990) 'Modern urban theory in question', *Transactions of the Institute of British Geographers*, 15(3): 331–43.
Davies, J.G. (1972) *The Evangelistic Bureaucrat*, London: Tavistock.
Dennis, N. (1970) *People and Planning*, London: Faber & Faber.

Department of the Environment (1980) *Development Control – Policy and Practice*, 22/80, London: HMSO.
——(1994a) *Development Plans and Regional Planning Guidance*, Planning Policy Guidance No. 12, London: HMSO.
——(1994b) *Consultation Paper: The Implementation of Compulsory Competitive Tendering for Professional, Construction and Property Services*, London: DoE.
Findings (1994) 'The governance gap: quangos and accountability', Local and Central Government Relations Research 30, September, York: Joseph Rowntree Foundation.
Goodchild, B. (1990) 'Planning and the modern/postmodern debate', *Town Planning Review*, 61(2): 119–36.
Guy, C. (1994) *The Retail Development Process*, London: Routledge.
Hannigan, J.A. (1995) 'The postmodern city: a new urbanization?', *Current Sociology*, 43(1): 151–214.
Hutton, W. (1995) *The State We're In*, London: Jonathan Cape.
Lebas, E. (1982) 'Urban and regional sociology in advanced industrial societies: a decade of Marxist and critical perspectives', *Current Sociology*, 30: 1–27.
MacGregor, B. and Ross, A. (1995) 'Master or servant? The changing role of the development plan in the British planning system', *Town Planning Review*, 66(1): 41–59.
Mackintosh, M. and Wainwright, H. (1987) *A Taste of Power: The Politics of Local Economics*, London: Verso.
Manchester Telematics Partnership (1995) *Manchester: The Information City*, Manchester: Manchester City Council.
Newcastle City Council (1995) *Regeneration Sub-Committee Wednesday 29th November* (minutes).
Packer, N. (1995) 'Surveys predict further gloom for town centres', *Planning Week*, 3(44): 1.
Page, D. (1994) *Developing Communities*, Teddington: Sutton Hastoe Housing Association.
Planning (1995a) 'Deregulation back on the agenda', *Planning*, 1144: 3.
——(1995b) 'Hall drawn back to the three magnets', *Planning*, 1125: 8–9.
Plate, K. (1994) *From EGO-Mobility to ECO-Mobility*, Dusseldorf: GEMINI.
Telecities (1995) *Presentation of the TELECITIES network*, Brussels: Telecities Coordination Office.
Tewdwr-Jones, M. (1995) 'Development control and the legitimacy of planning decisions', *Town Planning Review*, 66(2): 163–81.
Thomas, H. (1994) 'The New Right: "race" and planning in Britain in the 1980s and 1990s', *Planning Practice and Research*, 9(4): 353–66.
Thomas, K. and Kimberley, S. (1995) 'Rediscovering regional planning? Progress on regional planning guidance in England', *Regional Studies*, 29(4): 414–22.
Townsend, A. (1983) *The Impact of Recession*, London: Croom Helm.
Vickery, D. (1995) 'Selective short cuts', *Planning Week*, 3(48): 15.
Whitelegg, J. (1994) *Roads, Jobs and the Economy*, London: Greenpeace.
Wiehler, F. and Stumm, T. (1995) 'European briefing: The powers of regional and local authorities and their role in the European Union', *European Planning Studies*, 3(2): 227–50.
Yeomans, K. (1993) 'Estate of the art', *Housing*, October: 40.
Zukin, S. (1992) 'Postmodern urban landscapes: Mapping culture and power' in S. Lash and J. Friedman (eds), *Modernity and Identity*, Oxford: Blackwell: 221–47.

6

THE VIEW FROM THE GRASSROOTS

Aram Eisenschitz

INTRODUCTION

The years since the mid-1970s have seen catastrophic changes for much of the British population, changes that data on income distribution and economic growth cannot do justice to. Just as the terror that was inspired by the French Revolution gripped the property-owning classes in Britain and transformed the Napoleonic wars into a civil war against the radical politics of Jacobinism (Thompson 1980: 215), so these twenty years have witnessed a comparable degree of economic exploitation and political oppression that has affected a large section of the population. We are, however, faced by the same question that has bedevilled historians in assessing the impact of the Industrial Revolution. By reducing that question to the consumption of an arbitrary quantity of goods, many of them could argue that industrialization had been, for most, a positive experience, despite qualitative evidence to the contrary. Although figures for the numbers in poverty or the degree of income inequality cannot arrive at such a sanguine conclusion, the feeling of catastrophic dislocation in the way of life for much of the population – what Bodington *et al.* (1986: 39ff) call a political crisis of human need – cannot be conveyed. Accounts of changes in particular localities, despite their specificity, may provide a better feel for these changes (Beynon *et al.* 1989, Widgery 1993), although even at that level there are difficulties in interpretation. The debate over whether London is undergoing professionalization (Hamnett 1994) or polarization along the lines of New York (Harloe and Fainstein 1992) epitomizes the problem.

The complete picture is also obscured because there are so many dimensions of deprivation. As Donnison (1975: 31) argues, the poor cannot be defined by income alone: equally important is their command over resources in kind such as education, the way they are treated by those on whom they depend, such as landlords or bureaucrats, their power over their environment and their life, and, of course, the security of these conditions. Although a wider view of deprivation may deprive us of easy indicators, it does illustrate the complexity of the processes that produce it.

Disinvestment, exclusion and discrimination have created areas of concentrated poverty, the loss of traditional ways of life, and the replacement of skilled, male, high-wage jobs by less desirable service jobs. The network of collective institutions that provided some, if patchy, support and protection from the market – trade unions, local authorities, welfare state, community and extended family – have been weakened. Insecurity at work has been experienced by a large part of the labour force in the last two recessions, aggravating the general precariousness of life in Britain. Deprived communities, as van Rees states (1993: 99), are increasingly on a par with asylums and prisons, while the weakest groups – among them the unemployed, low-income earners, working-class school leavers, many pensioners, certain non-white groups, lone parents and the disabled – are increasingly marginalized.

While there are many more people in poverty it is also more difficult for them to cope with it. The poor cannot escape into consumption nor is there an alternative culture for them at work, in politics, in community or in leisure. Their sense of alienation is heightened by the absence of a meaningful democracy in the economic and political spheres that could provide them with a feeling of belonging either to locality or society. Unemployment and unskilled low-waged work deprives individuals of their self-definition which is given by work and, increasingly, by consumption. Indeed consumption may be seen as individuals trying to buy back the parts of themselves that have been lost in a dehumanized society (Marcuse 1992).

In this chapter it is argued that the wider view of deprivation that has emerged in recent years has underpinned imaginative grassroots policies that tackle poverty by trying to introduce democratic forms of organization to different sectors of society. Such an approach marks a significant break with traditional anti-poverty thinking, particularly if it generates pressure to introduce democracy to labour as a whole. However, it can be a double-edged approach that reproduces dependence and deprivation according to the conditions under which it is implemented. In the first part of this chapter some of the dimensions and processes that underlie deprivation are examined. This is followed by a look at Conservative urban policy in order to situate the bottom-up reactions that it generated, reactions that include attempts at community regeneration, control over the local economy, socially useful production and initiatives for cultural democracy.

POVERTY AND NEO-LIBERALISM

Urban problems and poverty are part of a wider process of exclusion and containment (Gaffikin and Morrissey 1994: 105) that requires an understanding of the neo-liberal project. Neo-liberalism is a response to the economic crisis that emerged in the late 1960s and which is represented by overproduction, the fall in the rate of profit and the growth in the political strength of labour (Mandel 1980, Glyn and Sutcliffe 1972). Neo-liberalism has successfully dismantled the post-war settlement between labour and capital, namely the welfare state and the high-wage economy, with a mixture of coercion and persuasion. The deliberate use of poverty as a means of subordinating labour as part of their programme of savage deflation, mass unemployment, wage reductions, the reinforcement of disciplinary labour relations and the destruction of inefficient capital has been matched by the skill with which poverty has been legitimated. Conservative urban policy has constructed new ideas about welfare and poverty that make this process acceptable; so acceptable indeed, that the political centre's ideas and responses to poverty have been coloured by them.

Anti-poverty policies are a response to the fear of the poor as well as diffusing a fear of being poor. The re-establishment of capital's right to manage has been closely associated with the threat of a descent into destitution, particularly as that destitution is increasingly visible. Weakening the safety nets with which the state has protected people from unemployment, low wages and casual employment means that loss of income translates more readily to poverty. Many more people are at risk of drifting into it at some stage of their lives (Beresford and Croft 1995: 14). Conservative policies expose individuals directly to the economy by introducing market principles in the supply of subsistence goods such as water or housing and by individualizing welfare (van Rees 1993). Their strategy of increasing capital's access to the physical and social infrastructure, to health, education, housing, leisure, culture and child care, reduces individual choice and autonomy for those who are unable to consume these goods on the market. By undermining the collective provision of welfare goods, the Conservatives abandon individuals to the market.

The welfare state had virtually no impact on equality of opportunity; even in its golden years the relative chances of reaching the top for those born into different social classes was unchanged (Goldthorpe 1987). Neo-liberalism has increased those inequalities throughout all aspects of society, imposing even greater barriers between social classes. Levels of poverty and inequality of income have grown further and faster in Britain than in any advanced industrial country. The combination of wage increases and tax cuts at the top of the social scale and unemployment, downward pressure on wages and reductions in the real value of welfare benefits at the other end has increased inequalities in the distribution of income. When poverty is defined as below half the average income (after housing costs) there has been spectacular growth in the numbers of the poor, rising from 7 to 24 per cent of the population between 1977 and 1990 (Barclay 1995: 15). The gap between the top and the bottom decile has increased to levels not seen this century; between 1979 and 1991 the real income of those in the top decile has risen by nearly two-thirds, while those in the bottom decile have seen a 17 per cent fall in their incomes, after housing costs (*Poverty* 1994). Over that period the poorest fifth of the population have not gained in real terms; since 1979 the shares of all those in the bottom six deciles of the distribution of income have fallen. Real income data, however, is problematic because it does not take into account the fact that inflation rates differ by income group. Nonetheless, the popular notion of the one-thirds–two-thirds society remains a rough approximation. That this bottom third of the population has not benefited from economic growth is apparent from such figures as the 17 per cent of all children in households with no earners, the 1.5 million families that are unable to feed their children on an 1876 workhouse diet (Campling 1994: 406), or the 2.7 million children living on benefits that provide an income 25 per cent less than is needed for subsistence.

There is, however, no linear relationship between income and quality of life. Those on incomes above the poverty line may experience as much disadvantage as those below it, particularly for those becoming poor through redundancy, divorce, illness or repossession, while figures of those in poverty often understate the discomfort associated with low incomes. Focusing on the numbers in poverty gives no indication of the sum of the *efforts* of keeping afloat in rapidly changing labour markets – such as multiple job holding or long hours of work – and the costs of that effort as marked by illness, neurosis, stress or premature death.

The history of capitalism demonstrates its scant regard for human life. The truth of this statement may be seen in contemporary Britain where being poor means increasing constraints in access to the basics of life and in the ability to participate in ordinary society. One can illustrate this virtually at random, with such figures as the 20,000 annual water disconnections, or the 4 per cent of boys in Glasgow's Easterhouse estate so seriously injured as a result of poor living conditions that they had to be detained in hospital in 1988 (Holman 1991: 6). Being poor means a greater likelihood of death by hypothermia: a third of households cannot afford adequate warmth and 13,000 people will die each year who would have survived had they lived in France or Sweden where housing is better (Hanlon 1995: 10). Or being poor means a greater chance of becoming homeless, with a fivefold increase in household-days spent in bed and breakfast between 1979 and 1991 (Ambrose 1994: 174).

What these figures show is that privilege is reproduced on many fronts. Policy aimed at one form of deprivation may well be met by greater differentiation elsewhere; greater equality in income may be met by increased inequality in housing markets or in the distribution of environmental costs. Jobs for which wages improve may become more difficult for the poor to enter, while welfare gains for particular groups in Britain may lead to an increase in child labour in, say, Pakistan, as labour-intensive, low-skilled products are increasingly sourced from such countries. That is why stratification in cities and society needs to be treated as class divisions in order that connections be made between them and so that policy is not focused on particular inequalities rather than the concept of inequality itself.

SPATIAL ASPECTS OF DEPRIVATION

How do spatial processes and structures contribute to the formation and alleviation of poverty? Cities do not just reflect aspatial economic and political features, but actively recreate social stratification, constructing deprivation and vulnerability in a diverse number of ways. As the chapters in Part II of this book demonstrate, exclusion and marginalization through class, race and gender relations will vary by locality which reflects their different political and economic histories (Massey 1984: 194–226; Gordon 1981).

The housing market is a central element in the reproduction of class relations. It is, for instance, the principal mechanism for distributing environmental costs and benefits, which are an important component of poverty. Since the environment is a conditional good, the actions of those with power will often block others from gaining it. This process is able to continue because the components of a good environment are so subjective, because the mechanisms with which it is distributed are opaque, because of the hereditary nature of housing classes and because positive externalities are underpriced. Variations in house prices do not accurately reflect the impact that environmental conditions can make on people's lives; the benefits that a high-quality environment can bring are disproportionately large in comparison with those who are excluded from it. Owner-occupiers above a price range determined by competition, for example, have always been able to buy public services, in particular amenity and education. Giving the poor more money would therefore not allow them access to these environments but would simply raise the entry threshold price.

Virtuous and vicious circles develop in good and bad environments, intensifying their differences and making physical exclusion an important element in the reproduction of class relations. Indeed as Cox (1976: 88) argues, most people live in 'purified communities', areas that are 'purged of the disturbing presence of outsiders' – a form of class and ethnic cleansing. Those who are not accepted there and who have little political muscle are dumped in 'soft' areas, areas that are set apart by institutional segregation such as the redlining by financial institutions. The state has always separated the poor from the rest of the working class (Stedman Jones 1971), through its policies for housing allocation, New Towns or suburban expansion, to take a few examples. Their visible concentration produces feelings of insecurity in the rest of us, 'there by the grace of god . . .' which encourage labour's compliance.

Concentrations of the poor are also politically important because they give rise to the linked ideas of social polarization and the underclass. As the middle-income groups leave for the suburbs, the cities are thought to become polarized between the high-income earners and the underclass. Notions of an underclass concentrated in the inner city marks a return to the idea of the city as the source of corruption. This is, however, a political response to structuralist explanations of poverty, which denies the significance of capitalist social relations in creating urban and social problems. The idea of the underclass, however, manages the poor by marginalizing and segregating them so as to cope with the fears of the social and political impact of mass impoverishment that spring from youth unemployment of 30 per cent, or more than double that among young black males in some places. By 'blaming the victim' the behaviour patterns of the poor, which are often symptoms of poverty, are regarded as causes (Loney 1983). The extended family and the female-headed household, for instance, are successful adaptations to hostile conditions (Miller 1993), yet the social pathology model of poverty identifies such successes as problems, using the concept of the underclass to scapegoat women as breeders of poverty and criminality (Medoff and Sklar 1994: 206).

The state has tended to reinforce this politics. Social housing, for instance, has always been a way of separating the 'respectable' from the 'undeserving' poor, while John Major's pronouncements on 'yob culture' renew another stereotype to join the demonology of the Right. The underclass is often code for the group politicians are most frightened of, an example of cultural engineering that creates an image of deviancy, 'otherness' and dependency to justify the neglect, segregation and repression of the poor. By blaming them for their predicament, the state is justified in leaving them as a warning to others (Boyer 1993). The treatment of the inner city in the film *Fatal Attraction* – associating it with madness, the elemental forces of fire and ice and the slaughter of the abattoir – illustrates the social significance of physical and social segregation in constructing images of the poor.

A further dimension of exclusion is to be found in the way that the city itself is becoming a focus of consumption and a means to establish the social status of the new service-based managerial class (Knox 1993) which is assuming the prestige and power once associated with the professions. The consumption patterns associated with the new inner city waterfronts, for instance, demonstrate this. The city is increasingly designed to exclude the poor, to make them invisible, and to provide security for consumers and residents (Boyer 1993). The enclosure of public open space and the replacement of the street by the institutionally owned shopping centre for the affluent (Worpole 1991) marks the privatization of the city and another stage in the management of the poor (Davis 1992).

Segregation also extends to the experience of the past. The sense of a shared continuity and common experience is part of psychic well-being, the glue that binds many communities. The notion of community allows individuals to cope with change, as ties of locality, familiarity and collectivity compensate for class inequality, particularly among those with the bleakest prospects (Callaghan 1992: 31). De-industrialization, physical renewal and the devaluation of working-class history affects those who would benefit most from those ties. In a world in which history ceases to exist, individuals are disconnected from their roots and cast adrift. In areas that are strongly associated with militancy, 'employers by-pass the last generation with a collective memory, a history and a notion of them-and-us' (Campbell 1989: 282). In one of the cradles of the labour movement, East London, the past has been obliterated, not least with the name of 'Docklands', which was imposed on a number of very different areas, symbolizing the way in which official history overrides the grassroots and legitimates the marginalization of local populations.

The role of space and the relationship of spatial policy in the development of capitalism, however, is increasingly problematic. Space may contribute to the creation of disadvantage and social exclusion but is this still rational for capital? Spatial policy has helped mould a disciplined labour force that internalized its class position and which offered its loyalty in exchange for modest expectations. The renewal of these class relations has been the central feature in the organization of the city; town planning, for instance, distributed real income regressively (Reade 1987: 69–94) while using its monopolization of expertise to define the legitimacy of community groups and to circumscribe the limits of debate (Goodman 1974). In periods in which these relations are being restructured, the effects of the spatial organization of society and the state's role in that is less clear. While there is a loose relationship between economic crisis and neo-liberalism, the associated spatial patterns possess a large degree of contingency. Mass unemployment may not necessarily be most effectively translated into ghettos because if the poor are portrayed as 'the other', they are not necessarily regarded as a threat to jobs and their concentration does not therefore always discipline other workers. Creating immobile groups of the poor, for instance, merely adds to the soaring bill for housing benefit. The ghetto may instead pose a political threat, since its deviant subcultures may 'contaminate' surrounding areas (Goldsmith 1989). London may have one of the largest pools of unemployment and deprivation in Europe, but instead of ushering in the new phase of capital accumulation, it may demonstrate the anarchy of capitalist change and the futility of state intervention. The co-existence of unemployed resources and pressing needs may also demonstrate that anarchy to a wider population. When a fifth of all working couples cannot afford the cheapest one-bed flat (Brownill and Sharp 1992: 11), the only impact is economic disruption. On the other hand, this anarchy may be the only way that a backward and disorganized capitalism can restructure out of crisis.

The spatial distribution of the poor suggests that there is no simple relationship with social trends. If the emphasis on inner city deprivation has been used by neo-liberalism to lend support to the idea of the underclass, geographers should, at least, be aware of this, although, of course, no amount of work can unseat a powerful ideology. However, a majority of the poor live outside the inner areas. The multiplicity of processes that cause poverty are matched by a diversity of poor areas: overspill housing in Havant, Victorian terraces in the outer city of Haringey, modern estates in Glasgow, picturesque pockets of poverty in Cornwall and holiday resorts throughout the country. It must also be remembered that significant numbers of the poor are dispersed throughout rural Britain (Pacione 1995). As Cox (1976: 80) shows with regard to the Educational Priority Area programme of the 1960s, most disadvantaged children are not in disadvantaged areas, most of the children in disadvantaged areas are not disadvantaged, and positive discrimination policies often benefit those who least need them. The poor have no common interest or consciousness and move in and out of poverty and between core and peripheral labour markets over their life cycle and between generations.

Moving even further from visions of the inner city underclass, deprivation should not be seen as a characteristic of those at the bottom of society. Poverty has become a means for reducing the power of labour and enforcing class discipline. Changes originating in unskilled labour markets, such as the loss of employment rights through casualization, short-term contracts and part-time working, are spreading to the mainstream, to augment the steady weakening of the power of labour. New technology has eroded the privileges conferred by particular skills, which tends to create a more homogeneous and competitive workforce. With all jobs subject to similar pressures, the features thought to differentiate the core from the peripheral labour market, such as job security, unionization, skills and barriers to entry are now less distinct than they were. Even those who think of themselves as middle class are often only a salary cheque away from poverty, struggling to reproduce their status in society (Brown 1995) and are often unable to slip into the leisured materialism enjoyed by previous generations as they are increasingly incorporated into the working class.

In other words, deprivation and destitution are the tip of the iceberg and one should be looking not just at relative inequality, but at the conditions that face labour as a whole, particularly in the light of its global expansion (Callinicos 1989: 125). Poor inner city environments, for instance, are just one aspect of deteriorating environmental conditions for all. The enlarged sphere of accumulation, the growing mobility of capital and the sophistication of mechanisms of class control suggest the growing power of capital is causing increasing global immiseration. In some First World nations, nations such as Britain that Jordan (1982: 223) calls internal colonies for capitalist exploitation, absolute deprivation may be increasing for many more than just those in the bottom tenth of society.

Such an argument would be supported by Kuczynski's (1972: 11–36) work in which he takes a wide range of poverty indicators to assess what consumption patterns really mean for labour. He argues that additional nutritional intake, for instance, has to be linked to increased intensity of work, or that increased educational attainment is a benefit only if it is matched with increased freedom over what is learnt. If, as he suggests, a worker with a Ford is not necessarily better off than a worker without one, indicators that are used to judge living standards need to be reviewed. Certainly comparison with past generations suggests a permanent deterioration of the conditions facing entrants to the labour force. The failure of the mid-1980s boom to improve these conditions suggests that they are not cyclical but permanent changes in the balance of class forces.

CONSERVATIVE URBAN POLICY

Attitudes to deprivation depend on the politics of the enquirer. For the Right, the poor present problems of social stability and economic efficiency if they become too detached from the mainstream and require too much state support (*The Economist* 1995a). To the political centre, anti-poverty policy is essential for modernization because of the impact of class divisions on economic performance (Glyn and Milliband 1994, Oakeshott 1990). The centre is most strongly associated with disadvantage because of its sensitivity to what it labels as deserving groups. For the Left, poverty is symptomatic of the crisis-ridden nature of capitalism, but its concern is with developing class consciousness, challenging class relations and the international division of labour.

Social-democratic anti-poverty policy was concerned with stabilization, 'the transformation of the poor into a stable working class population of truck drivers and mail carriers' (Palmer 1972: 38), a task that was not too demanding in a period of growth. Policy for the poor is now balancing a number of pressures: restraint over labour in deflationary times, expanding the sphere of accumulation in contentious sectors such as in the inner city and in basic services, preventing the politicization of social problems around the state and the selective abandonment of the state's commitment to reproducing labour. The difficulties faced by policy reflects these different dimensions. Creating mass unemployment may undermine the work ethos; weakening the unions and welfare state destroys two superb instruments for labour's own self-discipline; commodifying essential services such as housing may lead to a breakdown in the reproduction of the labour force; physical renewal of inner areas may push social problems elsewhere; and concentrating the poor into small areas demonstrates their deviancy, yet hinders their integration into mainstream labour markets which benefits capital. By examining the Conservative's urban policies in the following paragraphs, we may see why alternative strategies that call for an enhancement of democracy have their attractions.

Early approaches to local economic regeneration assumed that economic development would generate substantial trickle-down effects. But support to growth sectors such as tourism or business services failed because competition between places and within these sectors is so intense that they could not bear the extra costs of a social programme (Cox 1993). Successful places are under pressure to retain their lead which dilutes concern for the disadvantaged. There have been many 'miracles' such as the transformation of derelict areas into thriving economies and pleasant environments. But as Donnison concludes (1993: 293), these have rarely benefited the most excluded groups since opportunities have often been taken by the skilled and the mobile. They have often been property-based exercises in gentrification that redistribute areas of wealth and poverty. Neither Birmingham's success in the conference business nor Sheffield's development of a sporting infrastructure, for example, have provided gains for the poor. Indeed, inner city policy has often been a means of exploring the possibilities for state and market to recapitalize volatile economies and to manage the social and political effects of re-establishing the land market there.

Urban policy is also strongly oriented towards the creation of a local consensus in order to promote growth opportunities, in particular those involving infrastructure and property development. Partnerships and growth coalitions help create the political conditions for large-scale, high-risk, physical and economic restructuring (Harding 1991). These coalitions are supported because they promise to convert that growth to local benefit, packaging the risk through their ability to organize a consensus. Local authorities and the poor therefore agree to prioritize growth in exchange for these benefits, as a precondition for central government finance.

One consequence of creating a consensus for growth is that welfare is increasingly conditional upon the local economy and subject to local discretion, thereby sidestepping the political force of national organization and undermining ideas of universal rights. Many local initiatives for the disadvantaged are discretionary and linked to the health of the locality, particularly if financed by business or central government. By subordinating welfare to growth, people are led to believe that the economy, not the state, will help the poor. There have been a number of attempts to make this link between welfare and growth at the local level. Partnerships between public, private and voluntary sector tie social and economic projects together in one package. City Challenge, similarly, gives central government leverage over local authorities by making funds available for social programmes in exchange for a commitment to promoting local growth. This undermines their autonomy and weakens their policies to safeguard communities from development pressure.

Urban policy stabilizes populations and creates political and social institutions appropriate to new rounds of accumulation; the Enterprise Zones, for instance, held out the carrot of jobs in exchange for the New Realism of Labour local authorities. Mainstream local economic initiatives attempt to integrate marginal populations into the labour market but they tend to increase competition and worsen conditions for all. Training for the unemployed is often in low-quality skills and reinforces notions of labour flexibility, easing their entry into competitive parts of the labour market. Education Compacts support those groups who would have difficulty entering the labour market with their schooling, while community enterprise gives them work experience and reinforces the work ethos. Schemes such as the Enterprise Allowance or the promotion of co-operatives may be seen as a sophisticated form of workfare that socializes marginal groups into the enterprise culture.

Urban policy often demonstrates attitudes to poverty, work, welfare and self-help in order to change expectations about the support that the state can give. Policies stress individualism: self-reliance in people and entrepreneurialism in cities. In particular, they aim at helping the poor not with the aid of collective values but by equipping them with the means of operating in the market. Self-help legitimates the disengagement of the state from welfare, pointing towards the informal economy as a replacement. Community enterprise, for instance, may replace state provision of goods and services on a quasi-market basis: welfare-through-enterprise. Again, this complements the neo-liberal agenda since self-help is feminized, practical, local and, above all, apolitical (Campbell 1989).

Business has been attracted by the anti-state sentiments that can be read into community politics. Its involvement in community affairs helps reinforce the consensus around the partnership approach to inner city development. In many instances there is a strong vein of enlightened self-interest as it tries to deal with skill shortages, poor education and other elements of the social infrastructure (Eisenschitz 1993). The policies that it supports, such as workshops for small business, particularly ethnic minority business, low-rent housing, training for the disadvantaged and environmental improvement, convey a strong political message. Emphasis upon personal development and self-help suggests that capitalism, not welfare, can solve the problems of inner cities. In the case of blacks the message is even more uncompromising: rather than capitalism causing racism through competition for scarce resources, the problem is the reverse: not enough capitalism. Business in the Community, for instance, cross-subsidizes community uses out of the profit made on development projects (Jacobs 1992: 242). While this gives business a 'can do' image compared to local authority bureaucracy, these projects generally exclude the really disadvantaged.

Improving co-ordination has also continued to be an important theme of anti-poverty policy. In the 1960s and 1970s, integrated local authority departments were created, a move that was followed by partnerships between the two arms of government. The third stage expands the idea of partnership to include business and the voluntary sector; these are often initiated by central government, as in the case of City Challenge and those financed by the Single Regeneration Budget, although business has played a part in developing them. They try to show that they can combine economic growth with policies to alleviate disadvantage, something that has eluded all national policy. They co-ordinate responses to poverty and bind welfare to local growth; they re-legitimate the idea of trickle-down and create consensus, reducing opposition to redevelopment by incorporating both local authorities and poor communities. Significantly, they tend to bypass the poor because organized labour is not represented, because community interests are swamped by representatives from state and private sector and because they are market-driven (Gaffikin and Morrissey 1994, Geddes and Erskine 1994).

Urban policy, then, is more to do with managing the tensions thrown up by the economic crisis rather than concern for the poor. To the political centre, urban policy has failed to integrate social aims with economic development. The Conservatives have, however, successfully redefined the parameters of social intervention, in particular by emphasizing the need for local intervention, by linking welfare to growth and by shifting the consensus over intervention to the Right.

GRASSROOTS OPPOSITION

What opposition has there been to the dominant urban policy? Following their election in 1979, policy polarized sharply as local authorities directly opposed Conservative central government, developing local anti-poverty initiatives in order to compensate for increasing poverty. The most radical were those of the Bennite Left, attempting a local socialism that was spearheaded by the GLC elected in 1981, the West Midlands County Council and Sheffield City Council. They developed a range of strategies aimed not just at poverty but at a comprehensive restructuring for labour that would resist the market-led restructuring for capital. Aspects of these strategies were copied on a lesser scale by many Labour authorities, particularly in London. Local socialist strategies had two elements: a productionist and a libertarian one. A strategy to restore the strength of the local economy was coupled with the development of a radical democracy in order that labour could benefit from the restructuring of capital (Eisenschitz and North 1985) and ultimately take power from capital and the state. There was also a comprehensive range of anti-poverty programmes that served both elements, including such policies as the development of combined heat and power to combat fuel poverty, transport subsidies and welfare rights (Alcock 1994).

Under a relentless flow of legislation including abolition of the Metropolitan counties, rate capping, the centralization of political power, the growth of the unaccountable quango, compulsory competitive tendering and the commodification of many basic services, the scope for opposition has been severely constrained (Cochrane 1993). Local authorities have been compelled to bid competitively for central government resources that are made available for agreed projects while the range of funding over which they have discretion has declined. The scope for a local anti-poverty policy has been steadily reduced to participation in partnerships where policies such as training, small firm formation and sectoral stimulation are targeted towards the disadvantaged. Targeting, however, has proved to be unsuccessful as benefits tend to leak to other areas and to better-off groups. Often the most able of the 'bottom third' gain the most, leaving even more intractable problems.

The political centre, however, has collaborated with neo-liberalism by adopting the manners of post-modernism (Taylor-Gooby 1994: 399). Anti-poverty policy has taken on a new flavour, replacing concern with injustice, disadvantage and communality with the notion of difference and the lived experience of groups (Miller and Bryant 1990: 324). A mass of competing, decentralized organizations, campaigning for particular interests and drawing on the experiences of ordinary people rather than experts, has replaced the idea of the welfare state and the rights enshrined in it. Opposition ceases to be guided by universal principles but is pragmatic and organized locally. This has generated conflict between groups that obscures their common interests. The concern of the poverty lobby is the survival of the poor as individuals in the competitive arena (Room 1995: 105), a fragmented approach that focuses on specific areas and groups. While it may change the faces of the poor, the places where they live and the forms that poverty takes, it cannot make the category of disadvantage redundant. It does not even point out the institutional and economic foundations of poverty, but presents deprivation in terms of the poor confronting the mainstream (Beresford and Croft 1995: 13), which gives the Conservatives the support they need in treating 'them' as a threat.

This fragmented approach, nonetheless, has two progressive features. Policies not only address more dimensions of deprivation with their concern for race, gender and environment but they are more firmly anchored in a bottom-up ethos. Disadvantaged communities are increasingly aware of the failure of attempts to help them and consequently are adopting bottom-up policies aimed at their empowerment which broaden the canvas on which they operate. In the final sections we examine some of these strategies that attempt to resist neo-liberalism by introducing greater elements of democracy.

Enhancing democracy

Grassroots opposition to neo-liberalism is increasingly responding to the recognition that social divisions and poverty have grown under welfare, corporatist and neo-liberal regimes and that three decades of local initiatives which have seen poor areas subjected to the 'alphabet soup' of acronyms have achieved little for disadvantaged local inhabitants. Movements for radical democracy have also been stimulated by the failure of the weak forms of democracy promoted by the political centre, such as greater participation in decision-making, decentralization of services or even tenant self-management. Such initiatives rarely extend autonomy to local populations, particularly as the local authority mediates relations between groups rather than making it possible for them to relate to each other directly. Rather than returning to the politics of collective consumption, more and more local responses are calling for radically extended forms of democracy. At the most fundamental level there is concern that democracy itself is at risk if there is no sense of a common interest in society (Arblaster 1989: 78), while the broader critique of the undemocratic nature of Britain has spread, significantly, to sections of capital itself (*The Economist* 1995b). Furthermore, it is commonly thought by ordinary people that the labour movement and the local authorities are too centralized, too remote and too hierarchical to represent the grassroots. The labour movement in particular has failed to get involved with issues outside the workplace with its narrow constituency and lack of interest in production. Grassroots movements attempt to renew the libertarian impulse that both these institutions crushed, opening choices to the disadvantaged by extending the democratic framework and removing sources of economic and political power. These are the new social movements that have risen in the wake of the death of socialism, the assumed disappearance of the working class and the emasculation of the local authorities.

Stronger forms of democracy, however, encompass a range of political positions, from pluralist stances to varieties of socialism as well as Right and Left versions of libertarian populism, which aim to extend the scope of democracy and autonomy into such areas as the provision of social infrastructure (anything from planning and housing to the police), the environment and production. Where they differ is in the extent of democratic control and in their ultimate social and political goals. While policies such as co-ops, for instance, may be shared between variants of radical democracy, what is expected from them differs sharply between them.

The GLC's Popular Planning Unit, for instance, supported groups, such as workers, community groups or tenants, to develop their own plans with the aim of rooting social change in people rather than the state (GLC 1983). As part of a socialist strategy, it was aimed at developing class consciousness in order that labour is strengthened to such an extent that, ultimately, it is able to take control over what is produced, the types of economic units in which production occurs, the means of distribution and production including land, finance and economic knowledge, the organization of the labour process, and the design of the capital goods and the technology used in production.

The politics of the GLC, however, can also be interpreted as one of associational democracy, in combining 'pluralism with a cooperative mutualism' (Martell 1992: 168). This was seen in its support for alternative forms of economic enterprise such as the co-operative or community business, which promised to introduce an element of economic democracy. This approach meets concentrations of power with networks of institutions located between the state and the market, which attempt to 'give voice' to disadvantaged regions, communities and groups by extending democracy throughout the state and economy (Amin and Thrift 1995). New economic and political institutions such as regional assemblies, partnerships and worker directors are to be complemented by democratic networks in civil society, such as credit unions and neighbourhood associations, and by a new public sphere in which the excluded have an economic stake.

All the policies for increasing democracy are open to conflicting political interpretations. Socially useful production – production for need rather than for exchange – is an important aspect of socialist, associationalist and green politics. The pioneering work of the swords-into-ploughshares proposals of the Lucas Combine (Wainwright and Elliott 1982) not only gives labour greater control over what is produced and how to produce it, but it can address other dimensions of poverty, for instance by developing energy-saving products or by enhancing choice for the disadvantaged. But, however many kidney machines and road-rail buses are produced, their use depends on finding new means of distribution which, in turn, requires significant changes in the concept of private property that lies at the heart of capitalist political relations. It is questionable if pluralist approaches to democracy could achieve such change.

Another extension of radical democracy concerns control over the knowledge relevant to production, which under capitalism is organized so as to weaken labour and devalue human skills. Market processes of de- and re-skilling tend to restructure the labour process in a way that renews existing class relations. The democratization of knowledge, therefore, is a means of politically transforming labour. Workers' plans and such experiments as the GLC's human-centred lathe built on the experience of labour to develop products that met human need and used technology for the improvement of living standards rather than for cutting costs and increasing control over labour (Murray 1985). Networks of co-ops, for instance, may create spaces that allow for a reversal of the market's tendency to increase the division of labour, thereby allowing greater control over the knowledge and skills to strengthen labour. Information networks allow knowledge to be shared internationally and let workers combine against multinationals (Wainwright 1994: 143–84). Particularly weak groups, such as homeworkers, may co-ordinate themselves against employers, while within firms, diffusing information downwards could empower rather than fragment workers. The reaction by employers to workers' plans, epitomized by the rejection of the Lucas initiative, confirms the importance of control over knowledge.

Green politics shares the associationalist approach of the Third Way, but by subscribing to a universal eco-imperative it often tends to an authoritarian politics (Saward 1993). It lends itself to incremental and apolitical strategies, reducing societal problems to problems of the city of or industrialism (Pepper 1993). It also fails to specify a link between justice for nature and social justice; yet ending the exploitation of nature may be consistent with continued class exploitation. These qualifications blunt the thrust of its democratic intentions; green policies tend to be distributionally regressive (Luke 1993). Local Exchange Trading Systems, for instance, intensify the inequalities of the formal economy since they cannot be used by those who have most to gain, the unemployed, as they would lose benefit (Williams 1995). Sustainability, too, will tend to reify existing class relations unless it is accompanied by radical politics; its appeal, however, is precisely its compatability with capitalist social relations.

It is, therefore, the political context within which policy is implemented that matters. Community enterprise, for instance, may demonstrate to people that their needs can be met in a society organized around co-operative rather than competitive social relations; that same policy may, however, introduce the ideas of entrepreneurialism so as to prepare marginal groups for capital's return to abandoned areas. Indeed, to the Conservatives, the voluntary sector is a stepping stone to the privatization of welfare, but it may, nonetheless, be able to provide the basis for alternative values to those of state or market (Rustin 1995: 31). Similarly, opposition to bureaucracy may legitimate a weakening of the state, but it may also allow for a radical decentralization of power. These alternatives then summarize the ambivalence of grassroots movements, so aptly outlined by Sivanadan (1990: 56), between the inward looking, identity politics that issues from the self, and the organic politics that emerges from collectivities, from need rather than from choice, and which extends from civil society to state in order to create new communities of resistance and a political culture in order to take on capital.

Productivist socialism: combining growth and welfare

Local socialism illustrates some of the ambiguity over the politics of democracy. The productivist socialist strategies of the early 1980s linked growth and welfare locally, but rather than letting that link be organized by business or central government, they relied on agreements being reached between the local authority, the firm and the labour force. Their rationale was that once the state had gained some leverage by providing finance, management skills or knowledge, it would be able to intervene in the firm and even the sector, so as to restore profitability, operating from the standpoint of collective capital. While that may often be the most rational approach for a firm to take, it is inhibited by a distrust of the state and by interfirm rivalry. For instance, a rational approach for London's furniture manufacturers, faced with the overwhelming power of the retailers, was to move upmarket, but individual firms were unable to organize such a strategy (GLC 1985b: 97–116). The extra profits generated by restructuring are then shared between capital and labour – to improve conditions, to enhance its involvement in management or to introduce positive discrimination in recruitment, training and promotion.

There was, however, more. The longer-term intention was that by extending democracy in production and by demonstrating what the state could achieve, there would be popular agitation for the socialization of more and more of the economy and for new forms of economic and political organization. However, the demands of a competitive market limited the extent to which social programmes could be financed out of profitability. The democratic elements and the links with groups outside the workplace were curtailed in deference to measures to improve profitability. Nonetheless this experiment demonstrated that to reach the disadvantaged the institutional structures that reproduce social stratification need to be changed through bottom-up intervention in production. It remains an important reminder of the power of the state in the accumulation process and its potential to use that power to modify social relationships.

Empowering the community

The drive behind initiatives for community empowerment is the attempt to reclaim the idea of community and to find alternatives to the limited gains achieved by the disadvantaged in participation in development partnerships and growth coalitions. It is to compensate for the devaluation of the idea of community that has been employed to oversee the outflow of productive resources and state finance from poor areas, and which has had the effect of renewing their dependent status (Nevin and Shiner 1995: 310). Attempts to empower communities are confused because there are so many different pressures on policy makers, many of which have little to do with the alleviation of poverty. As we have seen, anti-poverty and community policies take on many of these pressures. Thus policies for neighbourhood regeneration are concerned with informal welfare and the reproduction of labour (see Eisenschitz and Gough 1993: 140–70). Because of its importance, these policies are subject to many initiatives from numerous institutions – local and central state, the quangos, the private and voluntary sectors plus the locals themselves – and with every shade of bottom-up, top-down orientation.

Empowerment, however, is a term that allows much flexibility in interpretation. What is meant by it and how will it be achieved? Clearly degrees of empowerment exist, but one has always to look at the long-term effect over society as a whole and the degree of control that it is thought individuals should have over their lives. Empowerment should be defined as policies that change people's consciousness, that enhance mutuality and solidarity rather than those that reproduce self-help and competition, since it is not enough for policy to be bottom-up, to involve locals, to provide for community uses or to lever the poor into jobs. Many partnerships, for instance, use the language of community development and of associational democracy, yet their operation and their outcomes are often prejudicial to community interests. What matters are the institutional structures of power that channel the flow of economic resources; influencing them requires a redistribution of that power. Policy should aim at the enhancement of democracy in the social, political and economic spheres in order to grasp that power so as to create an environment in which new social arrangements can be formed. Limited forms of democracy, as the experience

Plate 6.1 An attempt at creating community in Covent Garden, London – a temporary community garden that was eventually displaced by development

of forms of democracy as the experience of urban policy has shown, will not change the relative position of the disadvantaged. Credit unions, for instance, have no power over the banking system and therefore have little future as a form of financial self-help, because people are unable to save the 50p or £1 a week that is needed (Miller and Bryant 1990: 319). However, they could be used to politicize labour against a financial system that drains resources from poor areas.

The evaluation of initiatives, then, is not straightforward. Removing the profit motive, for instance, does not necessarily signal an alternative to the market. There are several thousand non-profit Community Development Corporations in the USA, often described as democratic firms, which have built a lot of low-rent housing yet they simply reproduce patterns of stratification (Berndt 1977). Non-profit activities may either be part of a market-led strategy for renewal that helps individuals better themselves, or they may extend the collectivist values of 'sharing and looking after' (Callaghan 1992: 32). American blacks, for example, rejected black capitalism in favour of economic activities that favoured solidarity, arguing that they could not advance by competing with whites under a system that had produced their disadvantage (Tabb 1972). Self-help initiatives, however, are now pushing minorities into forms of 'separatist capitalism' which, paradoxically, are built upon the social infrastructure that has been laid down by non-profit activities such as enterprise training.

Plate 6.2 Grassroots response to the developers' image of the future on the Isle of Dogs in London's Docklands

If policy can be implemented in very different ways then the only way to achieve a particular outcome is to politicize the policy-making process – yet the current climate of local intervention is one of consensus, as epitomized by the proliferation of partnerships in disadvantaged areas. Yet very few represent local populations, making it virtually impossible for communities to assume power and engage in negotiations with the private and public sectors as an equal: premature partnership, according to Medoff and Sklar (1994: 276), is one of the central causes of failure. But communities still require greater powers that only the state can give them, such as a veto over development (Nevin and Shiner 1995: 318).

The Dudley Street initiative in Boston illustrates how regeneration through the values of mutuality can be achieved, which is in marked difference from most attempts at bottom-up renewal. This is an area of 12,000 people that was one of the most deprived in the city (Medoff and Sklar 1994). The strength of the community was its ability to merge the varied experience of its different groups into a politically effective organization that was run collectively. It used local campaigns to foster a community consciousness that let it form a partnership with the City Council in which it took on the responsibility for major redevelopment and gained control over the land market by getting the Boston development corporation to grant it ownership in perpetuity of the derelict land in the area. Compulsory purchase of the privately owned section was financed with a Ford Foundation loan and that was supplemented by the gift of the land in the hands of the public sector. This story is all the more remarkable because the Eminent Domain legislation that made this possible had been framed for the benefit of developers. Finally, the organization took a holistic approach that integrated physical renewal with a strategy for community empowerment that took in policies for human development, education, economic renewal, the social infrastructure and the environment.

A weak analogy is provided by London's Coin Street development on the south bank of the Thames. Public ownership of part of the site led to public inquiries and legal battles in which the community group's technical competence and campaigning abilities won through (Tuckett 1988). A key difference, however, was that Coin Street was a prime site and its development potential brought the group into conflict with central government and the property industry. However, both these examples demonstrate that an essential condition for successful community policies is state ownership of land or its willingness to assume powers over private interests. Similar attempts to control land in Tolmers Square in the early 1970s (Wates 1976) had floundered on the failure of central government to give loan sanction for compulsory purchase and the absence of legislation to allow the community rather than local government to own the land aquired in that way.

There are grounds for pessimism about bottom-up approaches. Conditions are much less favourable than they were because poor communities from the weakest groups in the population have less hold over the increasingly corporate local authorities. Those that are sympathetic to their disadvantaged populations have been so emasculated that they can do very little for them. Bottom-up approaches developed in a market environment tend to increase fragmentation and inequality, with rivalry between disadvantaged groups reducing solidarity among the poor (Craig *et al.* 1990: 289). In general, changes such as the increasing integration of land with the global financial system (Coakley 1994) reduces the chances of developing non-market solutions to poverty.

Community approaches reflect the ambiguity around democracy (Cater and Jones 1989) that are expressed by the tension between self-help and mutuality, or, as Medoff and Sklar put it, between policies 'based in communities' rather than being 'community based' (Medoff and Sklar 1994: 260). Community empowerment will require state support initially in radical legislation (Nevin and Shiner 1995); however, communities will need to distance themselves from the state in order to avoid becoming controlled by it. The state's role in the development of these democratic initiatives is a central issue for discussion.

Cultural strategies

Democratic cultural strategies are a response to the emergence of prestige cultural and heritage investments as cities compete for inward investment. These projects, however, have demonstrated the enormity of the gulf between the dominant culture and the needs of the locally disadvantaged. While one aim of cultural marketing has been to demonstrate to potential investors that the conflict-prone industrial past has been replaced by consensus, it is also creating new conflicts, not least by the creation of low-quality employment. The distorted history employed by the commercial heritage sector symbolizes acquiescence by the deprived to new rounds of accumulation.

In response, radical democratic strategies around local culture aim to enhance democracy, to empower local populations, to be a means of resistance and an avenue for self-development and for alternatives to the hegemonic culture that stress the creation of the social citizen. They are often concerned with the promotion of a civic culture that expresses local diversity to contrast with the standardization and individualization fostered by the market. Cultural production is to be reclaimed so as to develop community identities and increase choice for the disadvantaged. Since these products are often collective (swimming, cinema, sport, festivals), the city plays a large part in their provision (GLC 1985a). The renewal of a civic culture rests on the citizen's right to public space as an arena of democratic political debate. That renewal is in response to the way the market is shaping public space as urban theatre (Crilley 1993: 153) and limiting those rights (Davis 1992). As Fisher (1991: 6) argues, culture is democratized when the bus station has as much attention devoted to it as the local theatre. Policy therefore must root everyday life in the city by increasing the accessibility of the centre, breaking up its aesthetic monotony, returning the freedom of the streets and providing spaces for cultural activities.

Such a strategy promises to overcome social divisions (Hajer 1993) and combat alienation and social disintegration by recovering the identity destroyed by urban renewal, de-industrialization and the inflow of the service, retail and entertainment sectors. It could provide a more open city, a collective physical identity and sense of local ownership of central city space through night-time economies and intervention in the informal economy. In areas where old ways of life disappeared with industrial decline, cultural activities demonstrate alternatives to the dominant programme of reindustrialization and the enterprise culture. Consett's Making Music Work programme, for example, attempts to locate young people in working-class values that had been expressed in oppositional local musical traditions; it provides continuity with the past and shows that renewal requires neither competitive values nor outsiders (Hudson 1995).

Linking the city and the arts, however, promises even more. If culture becomes a part of everyday life and if support to local culture opens alternative values to people, then the arts may once more take on their role of assessing and challenging mass culture (Edgar 1991: 26). Its values, however, are increasingly in conflict with the values of community (Merrifield 1993); during Glasgow's Year of Culture, for instance, its indigenous socialist culture was systematically downgraded (McLay 1990).

A democratic cultural strategy may be an apolitical irrelevancy or a challenge to the status quo. Its attractions to that status quo are not hard to see: it can be non-confontational and may be reached by anything from community festivals to pedestrianization or the stimulation of the indigeneous music industry (Bianchini *et al.* 1988: 40). But unless these policies are securely 'community based' they will, however, run the risk of being appropriated by boosterist local agencies.

CONCLUSION

What then is the potential of urban policy in alleviating poverty? While space may be important in the construction of disadvantage, the effects of its policies are limited. Many centrist policies still reflect 'the triumph of hope over experience': that pressure-group politics will lead to incremental improvement; that help to one group will not be balanced by adverse movements elsewhere or that integrating the poor into mainstream labour markets will overcome their social exclusion. The failure of those policies underpins the movement towards greater democracy. But implemented in the current political climate, these policies tend to reinforce neo-liberalism. One has to ask what empowering the disadvantaged means and what conditions would be needed for that to occur. Geographers need to be aware of how their own practice, how their ideas, policies and indeed language can contribute to the reproduction of disadvantage. The very idea of culture, for instance, can mystify the material practices occurring in cities (Mitchell 1995), while the growth of *local* responses to poverty, for instance, often demonstrates a failure to appreciate the relationship between disadvantage and capitalism. The poor are hardly surplus to capital as Geddes and Erskine (1994: 296) argue; this view obscures how they are produced by the 'therapy' of deflation as capitalism restructures out of crisis. The possibilities for alleviating poverty depend on recapitalization; policy has to get involved with what is an unpredictable and political process and decide whether to reform it, work with it or mobilize opposition to it.

Rather than imposing the agenda of the poverty lobby, one has to understand the restructuring process and thus the logic of neo-liberalism. To the centre, neo-liberalism is an unacceptable variant of capitalism and one of a number of possibilities (Therborn 1989). Such a view, however, obscures its logic as a solution to crisis and overplays the ability of the political centre to resist it. One cannot choose which variant of capitalism one wants: Keynesianism and neo-liberalism are symbiotically linked, a conclusion that creates dilemmas for activists (see Tickell 1995). Language is an element of that restructuring process. Chaotic concepts such as inner city or community often reinforce what Mills (1970: 87–112) called the problem-solving practicality of the political centre. The term 'locally disadvantaged', for example, denies the significance of class in poverty work, while area and race often act as proxies for it, providing illusions of solutions.

The ambiguity of the strategies discussed suggests that a form of opposition is to work within the tensions thrown up by neo-liberalism. Conservative policies define areas for opposition. Compulsory competitive tendering, for instance, draws attention to outputs and performance indicators, inviting agitation over the need for nationally agreed standards. The notion of the enterprise culture defines new spaces for popular opposition, the worker's plan and popular entrepreneurialism (Cooke 1984). Similarly, the commodification of mainstream culture has prompted questions of 'whose city?' and pressure for a democratized culture. The line between class and community responses in anti-poverty policy is a fine one (Berry 1988: 271); broadening the idea of deprivation may lead either to mundane community work or to class action. Socialist productionism, similarly, may save weaker firms in a recession or may herald a socialized economy. Local policies may increase consciousness (act local, think global), although local economic initiatives have, in practice, contained reformist pressures (Gough and Eisenschitz 1996). What will happen if the disadvantaged cannot gain control over their local economies (Hudson 1995: 471)?

With the collapse of socialism, resistance is increasingly being channelled into initiatives to enhance democracy. Movements toward democracy, however, can be consistent with neo-liberalism and Marxism, with containment, reform or revolution. Advanced forms of capital attempting to overcome the problems of hierarchy and disciplinary labour relations are also experimenting with greater democracy; while it may be tempting for labour to enter into alliances with the more progressive forms of capital, history should provide a degree of caution over the benefits that such a politics can achieve. Can associationalism challenge entrenched power in the advanced capitalist countries? Can new forms of common life be developed 'out of the wreckage left by the Conservatives' (Sayers 1995: 4)? Or does democratic self-determination require an advance in social relations that cannot be accommodated under capitalism (Wood 1991: 176)?

The closure of social and political alternatives may also point to a failure of our imagination to reorder our existence in the face of the hegemony of the market. As Thompson (1967: 68) so cogently puts it, we need to 're-learn some of the arts of living' and renew our capacity for experience. For instance, since the 1930s the idea that social progress involves the reduction of work rather than its creation has been progressively abandoned (Hunnicutt 1989: 118). Thompson's argument is increasingly relevant as the political and consumer culture converge.

GUIDE TO FURTHER READING

For an account of life in a deprived area, set within an analysis of structural change in London, that vividly contrasts market and 'community' values, David Widgery's (1993) book on East London is unsurpassed. The experience of the GLC is as relevant today as it ever was. Its flavour can be gauged from its London Industrial Strategy (GLC 1985), particularly the introduction. An excellent discussion of contemporary means of opposition to the market is given by Wainwright (1994). Medoff and Sklar (1994) provide a detailed account of what may prove to be a significant experiment in community regeneration in Boston, while Arblaster (1989) gives a masterly exposition of approaches to democracy that gives a basis for analysing much contemporary urban policy. Amin and Thrift (1995) give an excellent account of the politics of the 'Third Way' between state and market that is so important in grassroots politics, while Callinicos' tract on post-modernism provides an element of caution to some of the new political initiatives.

REFERENCES

Alcock, P. (1994) 'Welfare rights and wrongs: the limits of local anti-poverty strategies' *Local Economy*, 9(2): 134–52.

Ambrose, P. (1994) *Urban Process and Power*, London: Routledge.

Amin, A. and Thrift, N. (1995) 'Institutional issues for the European regions: from markets and plans to socioeconomics and powers of association', *Economy and Society*, 24(1): 41–66.

Arblaster, A. (1989) *Democracy*, Milton Keynes: Open University Press.

Barclay, P. (1995) *Inquiry into Income and Wealth* vol. 1, York: Joseph Rowntree Foundation.

Beresford, P. and Croft, S. (1995) 'Time for a new approach to anti-poverty campaigning?' *Poverty*, 90: 12–14.

Berndt, H. (1977) *New Rulers in the Ghetto*, Westport CO: Greenwood Press.

Berry, L. (1988) 'The rhetoric of consumerism', *Community Development Journal*, 23 (4): 266–72.

Beynon, H., Hudson, R., Lewis, J., Sadler, D. and Townsend, A. (1989) ' "It's all falling apart here": coming to terms with the future in Teeside' in P. Cooke *Localities*, London: Unwin Hyman: 267–95

Bianchini, F., Fisher, M., Montgomery, J. and Worpole, K. (1988) *City Centres, City Cultures*, Manchester: CLES.

Bodington, S., George, M. and Michaelson, J. (1986) *Developing the Socially Useful Economy*, Basingstoke: Macmillan.

Boyer, M. (1993) 'The city of illusion: New York's public places' in P. Knox *The Restless Urban Landscape*, New Jersey: Prentice Hall: 111–26

Brown, P. (1995) 'Cultural capital and social exclusion: some observations on recent trends in education, employment and the labour market', *Work, Employment and Society*, 9 (1): 29–51.

Brownill, S. and Sharp, C. (1992) 'London's housing crisis' in A. Thornley *The Crisis of London*, London: Routledge; 10–24.

Callaghan, G. (1992) 'Locality and localism', *Youth and Policy*, 39: 23–33.

Callinicos, A. (1989) *Against Postmodernism*, Cambridge: Polity Press.

Campbell, B. (1989) 'New times towns' in S. Hall and M. Jacques *New Times: The Changing Face of Politics in the 1990s*, London: Lawrence & Wishart: 279–99.

Campling, J. (1994) 'Social policy digest', *Journal of Social Policy*, 23(3): 405–23.

Cater, J. and Jones, T. (1989) *Social Geography – An Introduction to Contemporary Issues*, London: Arnold.
Coakley, J. (1994) 'The integration of property and financial markets', *Environment and Planning A* 26: 687–713.
Cochrane, A. (1993) *Whatever Happened to Local Government?*, Milton Keynes: Open University Press.
Cooke, P. (1984) 'Workers' plans: an alternative to entrepreneuralism', *International Journal of Urban and Regional Research*, 8(3): 421–37.
Cox, H. (1976) *Cities: The Public Dimension*, Harmondsworth: Penguin.
Cox, K. (1993) 'The local and the global in the new urban politics: a critical view', *Environment and Planning D: Society and Space*, 11: 433–48.
Craig, G., Mayo, M. and Taylor, M. (1990) 'Empowerment: a continuing role for community development', *Community Development Journal*, 25(4): 286–91.
Crilley, D. (1993) 'Megastructures and urban change: aesthetics, ideology and design' in P. Knox *The Restless Urban Landscape*, New Jersey: Prentice Hall: 127–64.
Davis, M. (1992) *City of Quartz: Excavating the Future in Los Angeles*, London: Vintage.
Donnison, D. (1975) *An Approach to Social Policy*, Dublin: National Economic and Social Council.
——(1993) 'The challenge of urban regeneration for community development', *Community Development Journal*, 28(4): 293–8.
Edgar, D. (1991) 'From Metroland to the Medicis: the cultural politics of the city state' in M. Fisher and U. Owen *Whose Cities?*, Harmondsworth: Penguin: 19–31.
Eisenschitz, A. (1993) 'Business involvement in the community' in D. Fasenfest *Community Economic Development*, London: Macmillan: 141–56.
Eisenschitz, A. and Gough, J. (1993) *The Politics of Local Economic Policy: The Problems and Possibilities of Local Initiative*, Basingstoke: Macmillan.
Eisenschitz, A. and North, D. (1985) 'The London Industrial Strategy: socialist transformation or modernising capitalism?', *International Journal of Urban and Regional Research*, 10(3): 419–40.
Fisher, M. (1991) 'Introduction' in M. Fisher and U. Owen *Whose Cities?*, Harmondsworth: Penguin: 1–7
Gaffikin, F. and Morrissey, M. (1994) 'In pursuit of the holy grail: combating local poverty in an unequal society', *Local Economy*, 9(2): 100–16.
Geddes, M. and Erskine. A. (1994) 'Poverty, the local economy and the scope for local initiative', *Local Economy*, 9(2): 192–206.
Glyn, A. and Milliband, D. (1994) *Paying for Inequality: The Economic Cost of Social Injustice*, London: Rivers Oram Press.
Glyn, A. and Sutcliffe, B. (1972) *British Capitalism, Workers and the Profits Squeeze*, Harmondsworth: Penguin.
Goldsmith, W. (1989) 'Poverty, isolation and urban politics', *Review of Radical Political Economics*, 21(3): 91–8.
Goldthorpe, J. (1987) *Social Mobility and Class Structure in Modern Britain* 2nd edn, Oxford: Clarendon.
Goodman, P. (1974) *After the Planners*, Harmondsworth: Penguin.
Gordon, D. (1981) 'Capitalist development and the history of American cities' in W. Tabb and L. Sawyers *Marxism and the Metropolis*, New York: Harper & Row: 25–63.
Gough, J. and Eisenschitz, A. (1996) 'The construction of mainstream local economic initiatives: mobility, socialisation and class relations', *Economic Geography*, 72(2): 178–95.
Greater London Council (1983) *The People's Plan for the Royal Docks*, London: GLC.
——(1985a) *The State of the Art and the Art of the State*, London: GLC.
——(1985b) *The London Industrial Strategy*, London: GLC.

Hajer, M. (1993) 'Rotterdam: re-designing the public domain' in F. Bianchini and M. Parkinson *Cultural Policy and Urban Regeneration*, Manchester: Manchester University Press: 48–72.
Hamnett, C. (1994) 'Social polarisation in global cities: theory and evidence', *Urban Studies*, 31(3): 401–24.
Hanlon, J. (1995) 'Cold comfort', *Red Pepper*, 10: 10–11.
Harding, A. (1991) 'The rise of urban growth coalitions, UK-style?', *Environment and Planning C: Government and Policy*, 9: 295–317.
Harloe, M. and Fainstein, S. (1992) 'Conclusion: the divided cities' in M. Harloe and S. Fainstein, *Divided Cities*, Oxford: Blackwell: 236–68.
Holman, B. (1991) 'It's no accident', *Poverty*, 80: 6–8.
Hudson, R. (1995) 'Making music work? Alternative regeneration strategies in a deindustrialized locality: the case of Derwentside', *Transactions of the Institute of British Geographers*, 20(1): 461–73.
Hunnicutt, B. (1989) 'Why has there been no significant reduction in working hours for over fifty years in the United States?' *LSA Conference Papers*, no. 34: 108–19.
Jacobs, B. (1992) *Fractured Cities*, London: Routledge.
Jameson, F. (1991) *Postmodernism: The Cultural Logic of Late Capitalism*, London: Verso.
Jordan, B. (1982) *Mass Unemployment and the Future of Britain*, Oxford: Blackwell.
Knox, P. (1993) 'Capital, material culture and socio-spatial differentiation' in P. Knox *The Restless Urban Landscape*, New Jersey: Prentice Hall: 207–36.
Kuczynski, J. (1972) *A Short History of Labour Conditions under Industrial Capitalism in Great Britain and the Empire 1750–1944*, London: Muller.
Loney, M. (1983) *Community against Government: The British Community Development Project 1968–78*, London: Heinemann.
Luke, T. (1993) 'Green hustlers: a critique of eco-opportunism', *Telos*, 97: 141–54.
Mandel, E. (1980) *The Second Slump*, London: Verso.
Marcuse. H. (1992) 'Ecology and the critique of modern society', *Capitalism, Nature and Socialism*, 3(3): 29–38.
Martell, L. (1992) 'New ideas on socialism', *Economy and Society*, 21(2): 152–72.
Massey, D. (1984) *Spatial Divisions of Labour*, Basingstoke: Macmillan.
McLay, F. (1990) *The Reckoning*, Glasgow: Workers City.
Medoff, P. and Sklar, H. (1994) *Streets of Hope: The Fall and Rise of an Urban Neighbourhood*, Boston: South End Press.
Merrifield, A. (1993) 'The Canary Wharf debacle', *Environment and Planning A*, 25: 1247–65.
Miller, A. (1993) 'Social science, social policy, and the heritage of African-American families' in M. Katz *The 'Underclass' Debate: Views from History*, Princeton NJ: Princeton University Press: 254–92.
Miller, C. and Bryant, R. (1990) 'Community work in the UK: reflections on the 1980s', *Community Development Journal*, 25(4): 316–25.
Mills, C.W. (1970) *The Sociological Imagination*, Harmondsworth: Penguin.
Mitchell, D. (1995) 'There's no such thing as culture: towards the reconceptualisation of the idea of culture in geography' *Transactions of the Institute of British Geographers*, 20 (1): 102–16.
Murray, R. (1985) 'Benetton Britain: the new economic order', *Marxism Today*, 29: 28–32.
Nevin, B. and Shiner, P. (1995) 'Community regeneration and empowerment: a new approach to partnership', *Local Economy*, 9(4): 308–22.
Oakeshott, R. (1990) *The Case for Workers' Co-ops*, 2nd ed., Basingstoke: Macmillan.
Pacione, M. (1995) 'The geography of deprivation in rural Scotland', *Transactions of the Institute of British Geographers*, 20(2): 173–92.

Palmer, J. (1972) 'Introduction to the British Edition' in R. Goodman *After the Planners*, Harmondsworth: Penguin: 9–50.
Pepper, D. (1993) *Eco-socialism*, London: Routledge.
Poverty (1994) *Poverty*, 89: 17.
Reade, E. (1987) *British Town and Country Planning*, Milton Keynes: Open University Press.
Room, G. (1995) 'Poverty in Europe: competing paradigms of analysis', *Policy and Politics*, 23(2): 103–13.
Rustin, M. (1995) 'The idea of community and the non-profit sector', *Renewal*, 3 (2): 25–34.
Saward, M. (1993) 'Green democracy?' in A. Dobson and P. Lucardie *The Politics of Nature*, London: Routledge: 63–80.
Sayers, S. (1995) 'The value of community', *Radical Philosophy*, 69: 2–4.
Sivanandan, A. (1990) 'All that melts into air is solid: The hokum of New Times' in A. Sivanandan *Communities of Resistance*, London: Verso: 15–59.
Stedman Jones, G. (1971) *Outcast London*, Oxford: Clarendon Press.
Tabb, W. (1972) 'Viewing minority: economic development as a problem in political economy' in American Economic Association *The Second Crisis of Economic Theory*, New Jersey: AEA: 31–8.
Taylor-Gooby, P. (1994) 'Postmodernism and social policy: a great leap backwards?', *Journal of Social Policy*, 23(3): 385–404.
The Economist (1995a) 'Always with us?' 334(7901): 36–7.
——(1995b) 'Britain's constitution: the case for reform' 337(7936): 23–6.
Therborn, G. (1989) 'The two-thirds, one-third society' in S. Hall and M. Jacques *New Times: The Changing Face of Politics in the 1990s*, London: Lawrence & Wishart: 103–15.
Thompson, E.P. (1967) 'Time, work discipline and industrial capitalism', *Past and Present*, 38: 56–97.
——(1980) *The Making of the English Working Class*, Harmondsworth: Penguin.
Tickell, A. (1995) 'Reflections on "Activism and the academy"', *Environment and Planning D: Society and Space*, 13(2): 235–7.
Tuckett, I. (1988) 'Coin Street: there *is* another way . . .', *Community Development Journal*, 23(4): 249–57.
van Rees, W. (1993) 'Neighbourhoods, the state and collective action', *Community Development Journal*, 26(2): 96–102.
Wainwright, H. (1994) *Arguments for a New Left: Answering the Free Market Right*, Oxford: Blackwell.
Wainwright, H. and Elliott, D. (1982) *The Lucas Plan*, London: Allison & Busby.
Wates, N. (1976) *The Battle for Tolmers Square*, London: RKP.
Widgery, D. (1993) *Some Lives*, London: Simon & Schuster.
Williams, C. (1995) 'Informal networks as a means of local economic development: the case of local exchange trading systems (LETS)' *IBG Conference*, University of Northumbria at Newcastle.
Wood, E. (1991) *The Pristine Culture of Capitalism*, London: Verso.
Worpole, K. (1991) 'The age of leisure' in J. Corner and S. Harvey *Enterprise and Heritage*, London: Routledge: 137–50.

Part II
PLANES OF DIVISION

7

INCOME AND WEALTH

Anne Green

INTRODUCTION

In a recent guide 'to finding the best places to live in the UK' (Focas *et al.* 1995), a composite index of economic prosperity is derived from the economic activity rate, the unemployment rate, gross domestic product (taken as a measure of salaries, company income and job quality) and the average income of housebuyers. Richmond upon Thames, Kingston upon Thames, Harrow, Berkshire, Hillingdon, Surrey, Sutton, Buckinghamshire and Bromley appear at the top of the rankings, while the Isle of Wight, Liverpool, Mid Glamorgan, Knowsley, Sunderland, Sheffield, Manchester and South Tyneside are ranked lowest. The 'bottom 10' is summed up as 'a litany of the most economically depressed parts of the country. Inner city areas and overcrowded estates; high unemployment and low earnings'(Focas *et al.* 1995: 54). By contrast, at the other end of the rankings, in terms of highest levels of prosperity 'We look naturally to the suburbs of our big cities, and London especially. People earning big salaries "in town" go home to spend them in the commuter heartland'. Significantly, while Richmond upon Thames and Kingston upon Thames head the rankings, 'the ultra-wealthy Westminsters and Kensingtons' are not ranked so high; here there are '*some very wealthy people,* but *some very poor ones too*' (Focas *et al.* 1995: 54, emphasis added) – illustrating the existence of marked intra-urban variations in prosperity in some areas.

This chapter explores some of the main dimensions of variation in income and wealth at the local scale in more detail. The first section is concerned with a review of data availability at national, regional and local scales, and in the absence of 'direct' income measures at the local scale highlights some of the most appropriate 'proxy' indicators available. In particular, the importance of the unemployment rate as one such indicator is highlighted. In the second section the relevance of a labour market focus – in which the unemployment rate is the most widely used indicator in local scale analyses – is rehearsed. The key features of economic change are outlined, and the main 'losers' and 'winners' in the face of economic restructuring and ongoing labour market trends are identified. Empirical evidence on the geography of

Table 7.1 Variations in income by region, 1993

Region	*GDP per head (UK=100)*	*Personal income per head UK=100)*	*Disposable income per head (UK=100)*	*Consumers' expenditure per head (UK=100)*
South-east	116.0	113.1	109.6	115.5
Greater London	124.7	121.2	117.7	122.7
Rest of south-east	110.5	108.0	104.4	110.9
East Anglia	101.7	100.9	101.4	98.1
Scotland	98.4	99.9	101.5	94.8
South-west	96.6	99.2	100.3	98.8
East Midlands	95.8	93.5	94.1	92.7
West Midlands	93.1	94.9	96.0	88.7
Yorkshire and Humberside	91.2	92.1	93.6	93.0
North-west	90.8	92.4	93.9	95.0
Northern	89.4	89.5	91.2	90.4
Wales	84.7	87.5	90.5	90.2
Northern Ireland	81.9	87.1	93.3	85.0

Source: CSO Regional Trends 1995

economic (in)activity is presented in the third section. Key dimensions of variation on selected measures of economic disadvantage are presented, and the main patterns emerging are contrasted with those of economic advantage. Some remarks on likely future prospects and policy responses are presented in the final section.

The most commonly used indicator for measuring variations in income is GDP. GDP measures the value of goods and services produced in a country/region – it is the total sum of incomes earned from productive activity. Table 7.1 outlines the main patterns of regional variation in GDP in the UK, together with key indicators of personal income and expenditure. GDP per head relative to the UK average remains higher in the south-east than in other regions, despite a relative decline over the period from 1989. East Anglia is the only other region to record a higher level of GDP per head than the UK average. The lowest levels of GDP per head are recorded in Northern Ireland, Wales and the northern region. These same three regions also display personal income per head more than 10 per cent below the UK average. By contrast, the south-east displays the highest values of any region on income and expenditure indicators. Two major macro-geographical divides, highlighted by Dunford (1995) in a study of socio-spatial inequality in Britain from a long-term perspective, are well illustrated by these data: the first between the Greater London metropolitan area and the rest of Britain, and the second between south and north.

Data on GDP are also available at the county scale. In nineteen counties/ Scottish regions, GDP per head was higher than the UK average in 1991.

Table 7.2 Variations in gross average weekly full-time earnings (£ per week) by county, April 1994

County	Highest Males	Highest Females	County	Lowest Males	Lowest Females
Greater London	467.3	336.5	Powys	288.1	n/a
Berkshire	439.9	288.3	Isle of Wight	291.4	n/a
Surrey	423.3	291.0	Cornwall	292.2	221.4
Buckinghamshire	393.9	276.5	Dumfries & Galloway	296.3	n/a
Hertfordshire	391.7	276.0	Dyfed	304.0	n/a
Grampian	382.1	250.3	Lincolnshire	307.5	221.9
Bedfordshire	379.0	261.2	Gwent	307.8	222.7
West Sussex	376.2	274.0	Mid Glamorgan	308.3	244.1
Cheshire	374.7	248.7	Shropshire	308.4	221.5
Wiltshire	372.5	251.3	Gwynedd	315.3	n/a

Source: CSO Regional Trends 1995

In Greater London, Grampian, Berkshire, Buckinghamshire, Cumbria, South Glamorgan, Lothian and Wiltshire, GDP per head exceeded the UK average by 10 per cent, while in Mid Glamorgan, the Isle of Wight, Cornwall, Merseyside, Northumberland, Gwynedd, East Sussex and Durham, GDP per head was at least 20 per cent lower than the UK average.

Income from employment accounted for about two-thirds of GDP in all regions in 1993. Hence, earnings data are important variables for economic analysis and occupy a central place in models of the labour market. However, there is a dearth of spatially disaggregated data at the local (particularly the sub-county) scale. The prime source of data on wages in Great Britain is the New Earnings Survey (NES), which yields a vast amount of information on earnings by industry, occupation and age group. The weakest element of the survey is its limited spatial disaggregation, so that only average wage rates for full-time adult male and female workers are provided by county; it is only at the regional scale that some (limited) disaggregation by industry and occupation is available. Table 7.2 shows the ten counties with the highest and lowest average earnings in 1994. The majority of areas with highest average earnings are located in the south-east, while those with lowest average earnings are predominantly more remote rural areas.

No NES data are published at the sub-county scale. However, it is possible to make some provisional estimates of income at the sub-county scale using estimates derived from NES data. For example, Gordon and Forrest (1995) present estimates of income at the local authority district scale in England by multiplying the number of men and women in each of seventy-seven occupation categories by the average full-time weekly earnings of that occupation as recorded in the 1991 New Earnings Survey, adjusting for numbers in part-time work, those on government schemes and the unemployed, before

Table 7.3 Estimated mean earnings (£ per week) by local authority districts in England, 1991

Highest district	*Earnings*	*Lowest district*	*Earnings*
City of London	300	Knowsley	175
Richmond upon Thames	278	Liverpool	182
Elmbridge	271	Kingston upon Hull	187
Wokingham	268	Middlesborough	189
Hart	266	Scunthorpe	191
Surrey Heath	266	Wansbeck	191
St Albans	265	Hartlepool	191
Mole Valley	264	Easington	191
Windsor and Maidenhead	263	Sunderland	192
Kensington and Chelsea	263	Great Grimsby	193

Source: Gordon and Forrest 1995

dividing by the economically active population in order to give an average estimated income from earnings. Unsurprisingly, Table 7.3 shows that the districts with the highest mean earnings are concentrated overwhelmingly in London and the south-east, while large cities and declining manufacturing areas dominate the list of districts with lowest earnings.

In the absence of comprehensive direct income measures at the local scale it is necessary to make use of 'proxy' indicators. In a study of changing patterns of income and wealth in Oxford and Oldham, Noble *et al.* (1994) used households in receipt of income support as a direct measure of low income. However, there is no single central source of such spatially disaggregated benefit data, and in any case since poverty is a multi-dimensional phenomenon comprising a whole set of experiences there are some drawbacks in basing analyses on a single indicator. The main source providing comprehensive local data is the Census of Population. Such data is widely used in studies of deprivation (see Simpson 1993, Hirschfield 1994, Robson *et al.* 1995, Pacione 1995), with deprivation often being taken as a surrogate for the existence of poverty (Goodwin 1995). There has been a long debate over the causes, definition and measurement of poverty (are 'absolute' or 'relative' measures appropriate?), and about the operationalization of composite indices based on such indicators.

Recent research by Davies *et al.* (1993) has attempted to examine the validity of commonly used Census of Population indicators as predictors of income using Family Expenditure Survey and General Household Survey data sets in which income is also available. This research confirms lack of a job (i.e. unemployment and economic inactivity) as a key predictor of low income – alongside lack of access to a car, and living in rented accommodation. Indeed, research on the changing composition and incidence of the poor

Figure 7.1 Proportion of unemployed and pensioners in bottom decile of income distribution
Source: adapted from Goodman and Webb 1994

over recent decades reveals a sharp increase in the proportions who are unemployed. Classifying the bottom decile group of the national income distribution by economic status, Goodman and Webb (1994) highlight the emergence of mass unemployment as a key cause of low income in the United Kingdom – particularly during the 1980s, to the extent that by 1991 the unemployed accounted for a higher proportion of the poorest decile group than did pensioners. This represented a significant turnaround over the previous thirty years (see Figure 7.1).

KEY FEATURES OF LABOUR MARKET CHANGE

Since lack of a job is a key indicator of poverty, and for most adults of working age the main source of (potential) income is earnings from paid work, a labour market focus would appear fundamental to a study of (lack of) income and wealth at the local scale. Clearly, participation in the labour market, with the object of gaining employment, is crucial for most people. However, it is important to consider both the *quantity* and *quality* of such participation by population sub-group and by area. During the 1980s there was an increase in female participation rates in virtually all industries, occupations and areas (Green 1994a). This rise was associated particularly with the expansion of part-time employment opportunities for married women. At the same time, economic activity rates amongst working-age males declined. While in the youngest age group higher levels of participation in full-time education was a factor in the decline, it was in the older age groups that the decline was most marked. Indeed, non-participation rather than unemployment has been identified by Forsythe (1995) as accounting for the major proportion of the increase in the male jobless total.

Two key features of labour market restructuring help to explain the decline in male participation rates and the rise in such rates for females. The first is the changing industrial structure: employment in the primary and manufacturing sectors (where male employees easily outnumber females) has continued to decline since the 1970s, while throughout much of the service sector (where the vast majority of women's employment is concentrated) the number of jobs increased over the same period. Second, in tandem with these changes in industrial structure there has been a decline in traditional, full-time, permanent employment relationships (nearly three-quarters of male employees fall into this category, compared with just under half of female employees) and a growth in non-traditional employment relationships – notably part-time working (nearly 90 per cent of part-time employees are females), but also self-employment and other forms of flexible working.

There have also been important changes in occupational structure in the 1980s and 1990s. In part, such changes are explained by changes in the industrial structure: white collar employment in the service sector has expanded at the expense of manual jobs in the primary and manufacturing sectors. Technological and organizational changes within industries have also tended to favour higher skilled, white collar, non-manual jobs (notably managerial and administrative, professional and associate professional occupations) at the expense of blue collar, low skilled manual ones (Institute for Employment Research 1994). Hence, long-term changes in the structure of the labour market in the UK have led to an increase in the salience of educational qualifications as a determinant of employment prospects, and a decline in the opportunities available to those without skills or qualifications. The implications of these changes for income generation potential through earnings are that, in general, high level, non-manual occupations and those working full-time command higher rates of pay than those in less skilled occupations and/or working part-time, while the increase in non-permanent employment relationships tends to undermine the security/stability of income flows through earnings.

In Table 7.4 an attempt is made to identify those population sub-groups who are most likely to be disadvantaged/advantaged by the uneven impacts of labour market restructuring, and also the locations in which these 'losers'/winners' are most likely to be concentrated (see Green 1994b, Henley Centre 1994 for more detailed background information). It should be noted that this does not mean that all the disadvantaged/advantaged will be from these sub-groups, or that all people in the locations identified will be 'losers'/winners'.

Several commentators (see Casey and McRae 1990, Goodman and Webb 1994, Boorah and Hart 1995) have suggested that the main labour market trends outlined above have led to a *polarization* of life experience. Social polarization is most generally taken to mean the widening of the gap between specific groups of people in terms of their economic or social circumstances and opportunities (Woodward 1995). The experience of polarization and social division will be different in different places, according to local economic, social, political and cultural factors. Indeed, there is growing concern that divisions in society are increasingly manifest in spatial terms, as revealed in the literature on 'dual' or 'quartered' cities (Sassen 1991, Fainstein *et al.* 1992, Marcuse 1993). The effects of spatial concentration of disadvantaged groups in specific geographical areas have been implicated in the growth of an underclass in some parts of the urban and regional system, notably in inner cities (Boorah and Hart 1995, Smith 1992, Wilson 1987). According to the thesis highlighting the significance of ecological effects, if non-employment is concentrated in 'pockets' of cumulative disadvantage, where the chances of obtaining employment may be compounded by localized job losses, these areas may come to be seen as undesirable places in which to live. As those able to migrate out of such areas do so, there is a danger that these neighbourhoods will become even more 'trapped in disadvantage', such that location becomes a determinant not only of present position but also of future prospects.

In the following section, empirical evidence is presented on what key indicators of local economic change reveal about polarization and the rise of an underclass.

SELECTED ASPECTS OF THE GEOGRAPHY OF ECONOMIC (IN)ACTIVITY

The geography of unemployment

The unemployment rate is the most widely used indicator of disadvantage – at regional, local and intra-urban scales. At the regional scale, by the end of the 1970s it had come to be regarded as accepted wisdom that unemployment rates were lower in the south and Midlands than in the north. However, the early 1980s recession took a particular toll on manufacturing industry, and the Midlands regions – characterized by a stronger reliance on manufacturing employment than the rest of Britain – suffered larger than average increases in unemployment. In the period of economic expansion from the mid to late 1980s unemployment rates fell in *all* regions. Then, in the early 1990s, local areas in the south were the first to experience increasing unemployment rates (Green *et al.* 1994), and there was a convergence in

Table 7.4 The 'losers' and 'winners' – the effect of labour market restructuring

The 'losers'	*The 'winners'*
Who are they?	
• those currently/previously employed in *manufacturing and primary industries* in long-term decline;	• those employed in *growth industries* – notably business services;
• those currently/previously employed in *semi-skilled and unskilled manual occupations*;	• those in *high level non-manual occupations* – despite talk of a 'white-collar recession' in the early 1990s, professional and managerial groups remain less vulnerable to unemployment than those in most other occupations;
• certain groups of *full-time employees* and certain groups of *part-time employees* – particularly those in low paid jobs with poor career prospect, and those in *precarious employment* with limited opportunities for any (except the most limited job-specific) training, or any kind of advancement;	• *core workers* in non-precarious employment, with ample opportunities for general and specific training;
• those with *limited journey-to-work horizons*;	• those with *extensive journey-to-work horizons*;
• those with *no or few* formal educational or vocational *qualifications*;	• those with *higher level* educational and vocational *qualifications* – despite relatively high levels of unemployment amongst recent graduates;
• the *long-term unemployed* – they tend to occupy the last positions in the queue for employment opportunities	• those with *no or limited 'breaks' from employment* – those with a (near) continuous work history tend to fare better than those with unstable work histories; similarly 'returners' who fare best tend to be those who return to work with the same employer after only a relatively short 'break'.

The 'losers'	*The 'winners'*
Where are they?	
• inner parts of the largest metropolitan areas;	• areas on the fringes of the 'Greater South East' region extending from London to the south coast, westwards along the M4 corridor, along the M40 through Oxfordshire to Warwickshire, and north-eastwards to Cambridge: these areas are most likely to benefit from a self-sustaining combination of factors which act as crucial 'drivers' of economic prosperity: – a highly skilled and professional workforce, – a strong entrepreneurial base, – relatively good infrastructure links with the rest of the UK and Europe;
• areas (formerly) characterized by a relatively narrow industrial base – notably coal-mining, heavy manufacturing and port-related activities – in long-term decline;	
• some resort and retirement areas – which are vulnerable to restructuring of welfare services, and where unemployment rates are often higher than average (particularly in winter); • remoter rural areas – where the quantity and variety of employment opportunities are limited – particularly those without private transport;	• mixed urban/rural and accessible rural areas;
• those in the least desirable public rented housing – the operations of housing and labour markets interact so as to produce spatial concentrations of people with similar characteristics: employment vulnerability reduces choice of residence, so that those who are disadvantaged in labour market terms tend to be found in the least desirable housing – the 'residualization' of council housing has become more intense in the 1980s.	• areas with a favourable *image* and offering a good '*quality of life*'.

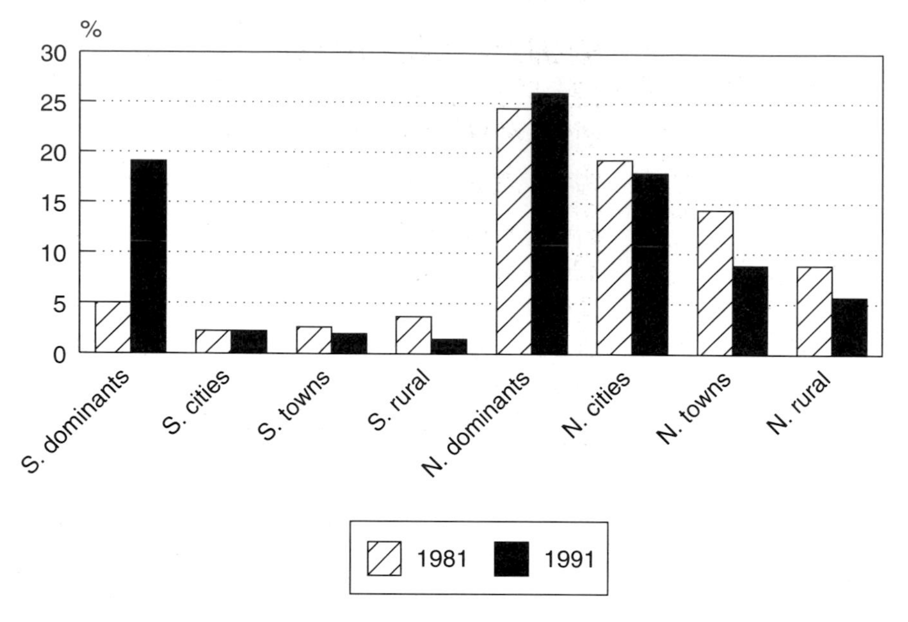

Figure 7.2 The geography of unemployment – percentage of wards in 'top' decile group
Source: 1991 Census of Population

regional unemployment rates over the period 1991–3. Nevertheless, over this period the incidence of unemployment in the south-east, East Anglia and the south-west remained lower than the UK average.

In addition to these regional differentials, intra-regional variations in unemployment rates have become more marked over the 1980s: there are south–south divides and north–north divides alongside a north–south divide. Using a simple 'north–south' distinction, the proportion of wards in the south ranked in the 'top' decile group of all wards in Great Britain (i.e. that tenth of all wards in Great Britain with the highest unemployment rates) rose from 3 per cent in 1981 to nearly 6 per cent in 1991, while the comparative share in the north fell from over 16 per cent in 1981 to 14 per cent in 1991. Figure 7.2 shows that the increase in the representation of 'top' decile group wards in the south is attributable solely to the increase in such wards in the southern dominants (i.e. the largest cities in the south) (further investigation reveals that the bulk of the increase in localized high unemployment neighbourhoods was in Inner London). It is evident also from Figure 7.2 that there was a higher proportion of 'top' decile group wards in the northern dominants (i.e. the largest cities in the north) in 1991 than in 1981, whereas in other urban size categories in the north there were decreases in the share of 'top' decile wards. The greatest concentration of 'pockets' of high unem-

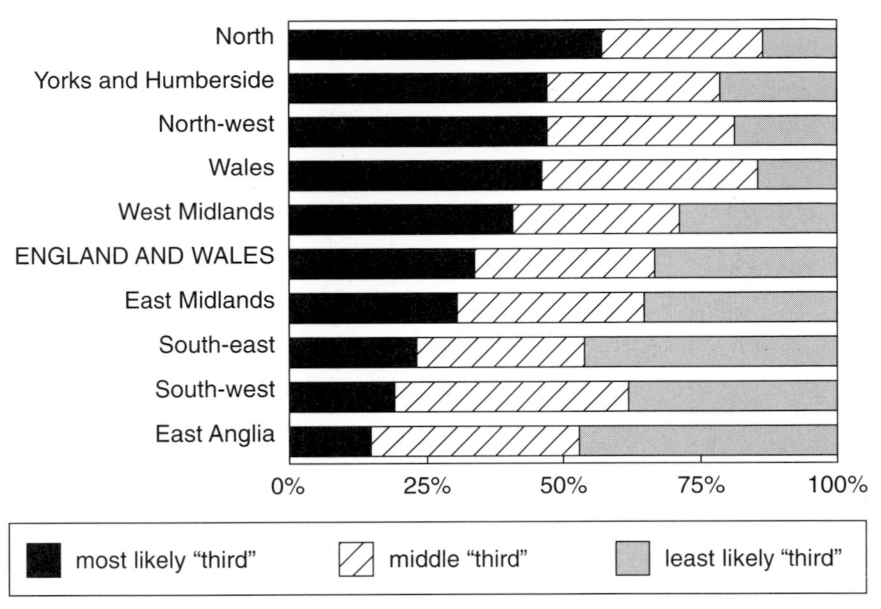

Figure 7.3 The geography of 'idleness' – likelihood of long-term unemployment
Source: Dorling and Tomaney 1995

ployment in both 1981 and 1991 was in the largest metropolitan districts outside London, and this concentration increased over the decade so that by 1991 over half of the wards in such areas fell into the 'top' decile group.

While evidence has been presented for a narrowing of regional differentials in unemployment (Martin 1994), the evidence regarding *neighbourhoods* with the highest unemployment rates suggests a continuing north–south variation and an increasingly important urban–rural variation – with pockets of extremely high unemployment becoming increasingly concentrated in the largest cities. Analyses focusing on long duration of unemployment also reflect a continuing north–south variation. Dorling and Tomaney (1995) have mapped the 'geography of idleness', dividing wards in England and Wales into thirds on the basis of long-term unemployment rates. Figure 7.3 shows the representation of each of these thirds by region, and highlights the concentration of long-term unemployment in northern England (the northern region, Yorkshire and Humberside and the north-west) and Wales. Problems of 'hard core' long duration unemployment remain concentrated in northern Britain.

The geography of non-employment

Increasingly, traditional concerns about unemployment have extended to cover non-employment (i.e. all of those of working age without employment:

Plate 7.1 Environments of disadvantage (a) the Wood End district of Coventry provides some support for a hypothesized correlation between multiple deprivation and the size of the canine population, while (b) in the Castlemilk estate in Glasgow the proliferation of satellite dishes reflects another possible palliative for life in areas of urban disadvantage

Source: (for 7.1a) M. Gould, University of Warwick Photographic Service

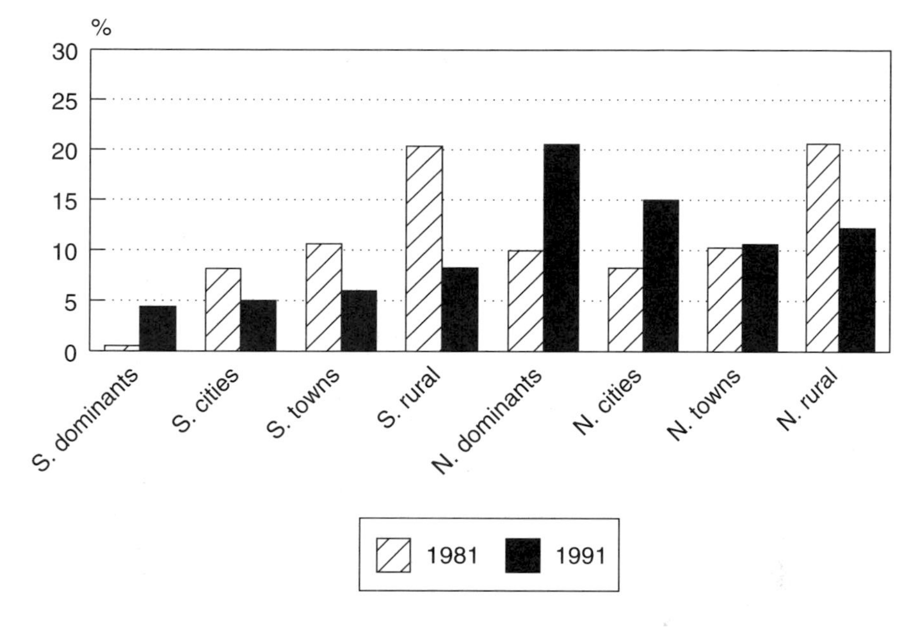

Figure 7.4 The geography of non-employment – percentage of wards in 'top' decile group
Source: 1991 Census of Population

the unemployed, those on government training schemes and the inactive). Indeed, McCormick and Philo (1995) suggest that beyond the unemployment rate, the most telling indicator of what is going on is that of non-employment. In 1981, 8.5 per cent of wards in the south were ranked in the 'top' decile group of all wards in Great Britain on the non-employment rate, while in the north the proportion was 11.4 per cent. Over the decade to 1991 the proportion of 'top' decile group wards in the south decreased to 5.5 per cent, while in the north the share rose to 14.1 per cent. Hence, the north–south divide in non-employment became more pronounced. Even more marked, however, is the increase in the share of all wards suffering 'extreme' non-employment rates in the largest cities (i.e. 'dominant' local labour market areas) from a mere 4.9 per cent in 1981 to 12.1 per cent in 1991. Over the same period there was a substantial decrease in the proportion of wards in rural areas categorized in the 'top' (i.e. worst) decile group, with the non-employment rate falling from over 20 per cent in 1981 to 11 per cent in 1991. Using a combined urban size/regional location categorization, Figure 7.4 confirms that this increase in 'extreme' non-employment rates in the dominant areas is evident in both southern and northern Britain, as is the decrease in 'top' decile group wards in rural areas. In 1991 the largest concentration of wards with high non-employment rates was in

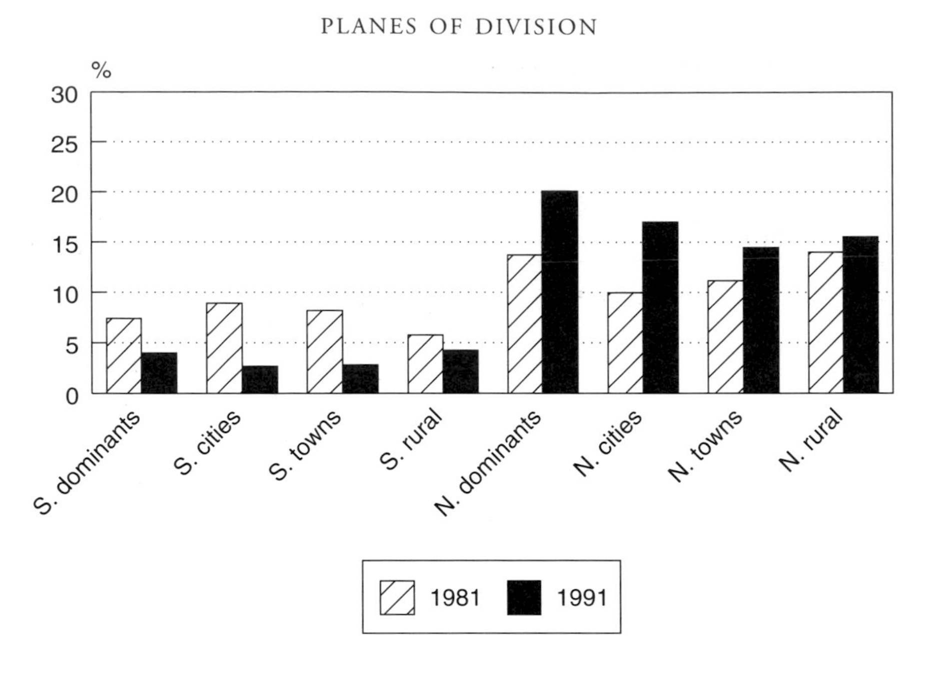

Figure 7.5 The geography of male inactivity – percentage of wards in 'top' decile group
Source: 1991 Census of Population

the northern dominants (21 per cent of all wards in this category fell in the 'worst' decile group for non-employment).

A major 'driver' of this pattern of change is the impact of increasing employment opportunities for females – with the highest relative increases in rural areas, where participation rates were lowest at the start of the decade (Green 1994a). However, another important factor is the *increase in male inactivity*, which was particularly pronounced in the larger urban areas of the north (see Figure 7.5). Indeed, in all four urban size categories identified in the north there was an increase in the share of wards falling in the 'top' decile group of the indicator of inactivity amongst working age males.

Economic inactivity, like unemployment, is highest amongst the sector of the population with the lowest levels of educational attainment and learning power (Wadsworth 1994). Figure 7.6 charts the changing unemployment, inactivity and employment fortunes of low skilled workers in the period 1984–9 when the British economy underwent recovery. Unemployment fell fastest, employment grew most and inactivity rates rose least in those local authority districts with the best aggregate performance (as measured by lowest unemployment rates). By contrast, in the more disadvantaged districts (those with the highest unemployment rates) the decrease in unemployment was

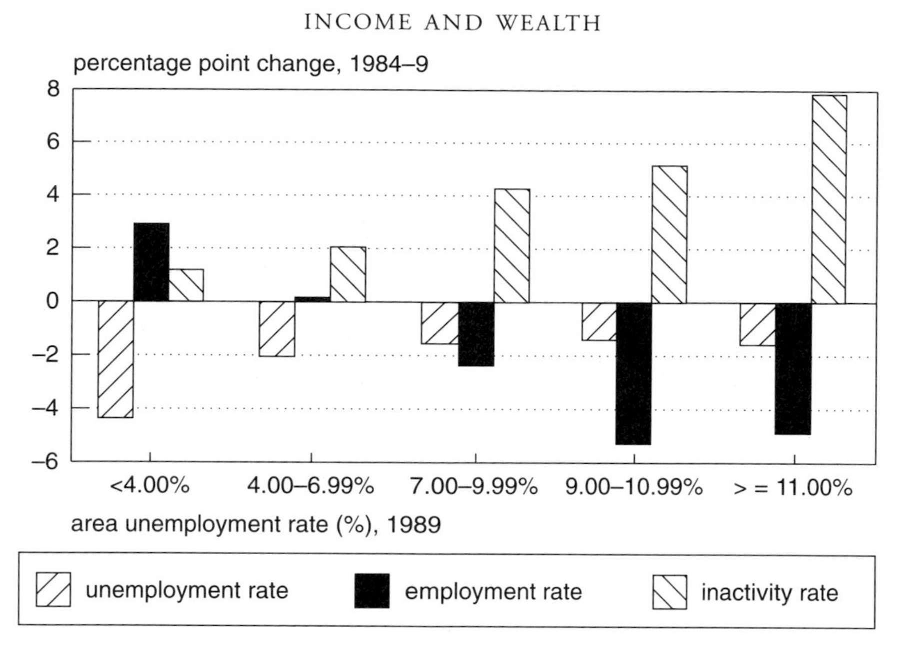

Figure 7.6 Change in unemployment, employment and inactivity rates of low-skilled workers
Source: Wadsworth 1994

less marked, the inactivity rate increase was greatest and the employment rate declined. Similarly, using a range of measures of segregation at the neighbourhood level, Green (1995a) (in an analysis encompassing local labour market areas across the whole of Great Britain) and Noble *et al.* (1994) (in a more detailed comparative study of Oxford and Oldham) have revealed that the distribution of unemployment became more uneven (i.e. the unemployed became more spatially segregated from those who were not unemployed) over the period 1981–91.

Changing intra-urban variations: the example of Oldham

It is informative to examine the geography of non-employment and changing patterns of disadvantage in more detail in one particular urban area. In this section, drawing on the detailed study of Oldham presented in Noble *et al.* (1994), trends on selected indicators of disadvantage are presented for three contrasting neighbourhoods: Holts (a post-war peripheral council estate), Westwood (an inner city area and the centre of Oldham's Bangladeshi community) and Grasscroft (a village on the outskirts of metropolitan Oldham, consisting largely of substantial detached houses with large gardens). Table 7.5 presents scores on selected indicators of disadvantage for each of

Table 7.5 Indicators of disadvantage for selected neighbourhoods in Oldham

Indicator	*Holts (%)*	*Westwood (%)*	*Grasscroft (%)*	*Oldham (%)*
Unemployment rate				
1981 (%)	19.84	17.29	1.93	9.89
1991 (%)	33.52	27.33	3.53	10.74
1981–91 change	13.68	10.04	1.60	0.85
Households with no car				
1981 (%)	74.30	73.75	17.66	50.09
1991 (%)	73.51	69.72	10.02	42.62
1981–91 change	–0.79	–4.03	–7.64	–7.47
People in families on Housing Benefit/Council Tax Benefit as % of 1991 residents	64.24	53.36	4.96	26.60
People in families on Income Support as % of 1991 residents	42.78	40.91	1.71	14.77

Source: adapted from Noble *et al.* 1994
Note: the data on benefits relate to 1993

these neighbourhoods. Holts and Westwood not only display the highest initial incidence of unemployment in 1981, but also show massive increases in unemployment rates over the decade to 1991, compared with a modest increase for Oldham as a whole. All neighbourhoods identified in Table 7.5 shared in the reduction in the proportion of households without access to a car, but it is notable that the decrease was more marked in Grasscroft than elsewhere, while in Holts the reduction was very modest; in 1991 approximately three in every four households in Holts had no access to a car compared with only one in ten in Grasscroft. Holts and Westwood also had considerably higher proportions of people in receipt of benefit than the Oldham average.

Looking at the position of particular neighbourhoods, in Oldham there is a consistent picture of 'richer' neighbourhoods improving their position, and the 'poorer' neighbourhoods deteriorating. In terms of types of area, council estate areas revealed greatest deterioration; in such areas there tended to be an increase in already high levels of unemployment, a greater than average decline in economic activity for males and a decline in economic activity for females (contrasting with the general trend for rising female activity rates) (see Table 7.6). The increasing polarization of income and wealth in Oldham in particular applies to people living in households headed by a person born in the New Commonwealth – revealing an ethnic dimension underlying other patterns of labour market disadvantage.

Table 7.6 Economic status by type of area in Oldham, 1981–91

Indicator	*Settled owner-occupied (%)*	*Mixed status owner-occupied (%)*	*Mixed tenure areas (%)*	*Council estates (%)*
Male unemployment rate				
1981 (%)	6.2	13.6	4.2	22.1
1991 (%)	6.6	14.9	14.3	28.7
Female unemployment rate				
1981 (%)	4.4	8.0	8.2	12.2
1991 (%)	4.6	9.7	10.0	18.1
Male economic activity rate				
1981 (%)	82.3	78.1	77.2	76.9
1991 (%)	77.7	73.1	70.6	67.2
Female economic activity rate				
1981 (%)	56.4	47.2	48.4	45.4
1991 (%)	59.6	48.2	47.6	40.3

Source: adapted from Noble *et al.* 1994

No earner and multi-earner households

Gregg and Wadsworth (1994) have highlighted a further example of social polarization in the simultaneous growth in no earner and dual earner households over the 1980s. Part-time jobs are particularly attractive to women married to men who have full-time jobs. These same part-time jobs are relatively unattractive to women whose husbands are unemployed (due to the system of benefit regulations – whereby income from part-time work would mean a reduction in income from benefits, leaving the household no better off). Hence, available employment is becoming increasingly concentrated in fewer households.

In 1991, the Census of Population included for the first time a question on the numbers of economically active adults and the number of earners per household. The local authority districts recording the highest proportions of *households* containing economically active residents but *no earners* are listed in Table 7.7. In four inner London boroughs (Hackney, Tower Hamlets, Newham and Southwark) and four large metropolitan districts from northern Britain (Knowsley, Liverpool, Manchester and Glasgow) over one in ten households containing one or more economically active residents had no earners. In some of the most disadvantaged wards within these districts more than one in five wards were so categorized.

By contrast, the 'Greater south-east' has the greatest concentration of households with two or more earners (Green 1994b, 1995b). More suburban/accessible rural districts close to the major urban centres of the East and West Midlands (for example, in Leicestershire, Warwickshire and

Table 7.7 Local authority districts with the highest proportions of households with one or more economically active residents but no earners, 1991

Local authority district	*Households* (%)	*Wards in 'top' decile group* (%)	*Average across 3 highest wards* (%)
Hackney	12.83	100.00	15.86
Tower Hamlets	12.37	94.74	16.15
Knowsley	12.17	78.57	18.51
Liverpool	11.51	81.82	21.60
Manchester	10.38	78.79	21.00
Glasgow City	10.32	55.28	21.99
Newham	12.27	91.67	13.00
Southwark	10.25	84.00	15.25
Haringey	9.82	73.91	12.72
Islington	9.80	95.00	12.22
Lambeth	9.78	72.73	13.65
Middlesbrough	9.37	64.00	15.81
Kingston upon Hull	8.68	52.38	14.91
Newcastle upon Tyne	8.50	50.00	18.84

Source: 1991 Census of Population

Hereford and Worcester) are also well represented. However, within large metropolitan areas there are also marked differences in experience. Figure 7.7 shows variations in the proportion of multiple earner households in London, expressed as a proportion of all economically active households. A clear pattern emerges of more multiple earner households in Outer London (at least 55 per cent in Harrow and Hillingdon) and far fewer in the Central and Inner London boroughs that saw some of the most marked increases in unemployment during the 1980s and early 1990s.

The informal economy

Unable to gain a foothold in the formal economy, what are the prospects for the unemployed to gain income through participation in the informal economy? For many years the assumption was that 'black market' work was concentrated amongst deprived populations and in deprived localities. However, empirical studies show that the character and magnitude of black market work is heavily dependent upon the employment situation in a particular area: research in Britain (Bunker and Dewberry 1984), France and the Netherlands (van Geuns *et al.* 1987) has shown that the higher the level of unemployment, the *less* black market work takes place. Moreover, studies have shown that multiple earner households are more likely to be active in black market work than no earner households (Warde 1990). Hence, not only is there a tendency for more black market work to be undertaken in

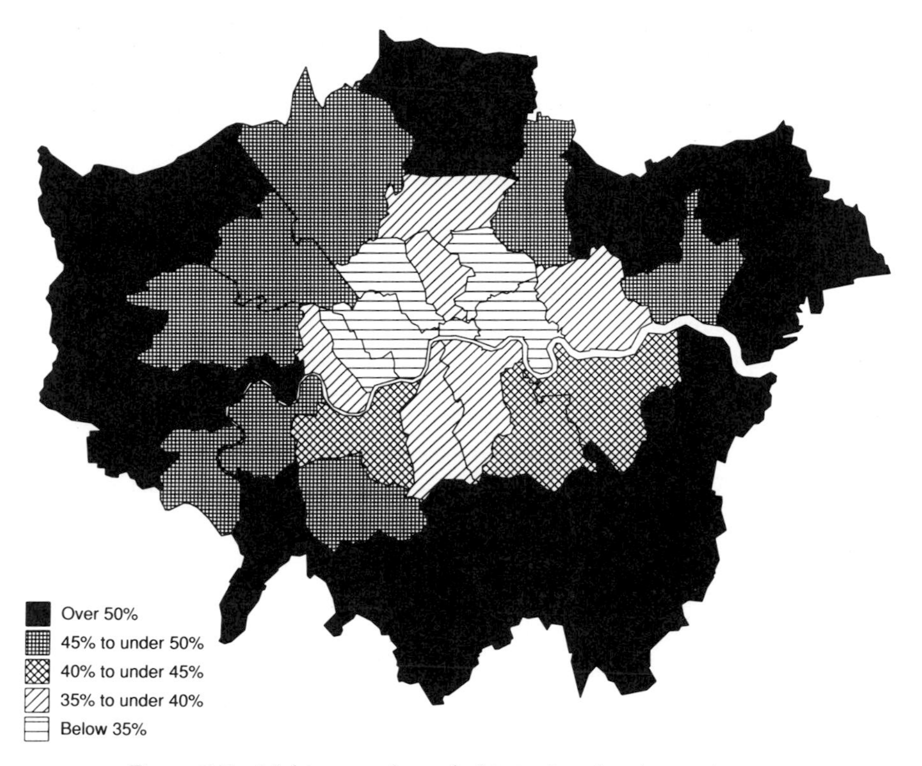

Figure 7.7 Multi-earner households by London borough, 1991
Source: 1991 Census of Population

affluent localities by more affluent people, but such work is usually of a more rewarding and less exploitative kind than that in poorer areas undertaken by poorer people (Williams and Windebank 1995). Specific forms of skill, work experience and social networks determine, to a significant extent, access to informal work and tend to further disadvantage those already unemployed, particularly in the long term (Pahl 1984). In a study of 'fiddly work' (i.e. working 'undeclared' [usually in irregular, low-paid and temporary jobs] whilst in receipt of unemployment benefits) in Cleveland in north-east England, MacDonald (1994) also highlighted how such work tended to be restricted to such groups. White, working-class males in their twenties and thirties, in particular neighbourhoods, with particular skills and work experience, with particular sorts of cultural lives and social networks, were the most likely to partake in fiddly work. The more common experience of unemployment (and social isolation) excluded many from the networks through which such work is distributed. Hence, black market work represents a 'survival strategy' for a small minority only.

POLICY PERSPECTIVES AND FUTURE PROSPECTS

As outlined above, there is some evidence for the increasing spatial segregation of the non-employed within cities (particularly the largest cities) and within regions. As a result, the concept of space becomes central to non-employment, as those without jobs are increasingly insulated from those with jobs, and from job opportunities, in other parts of cities; Gaffikin and Morrisey (1994) use the phrase 'social containment' to describe this condition. The Joseph Rowntree Foundation *Inquiry into Income and Wealth* (Barclay 1995) highlighted the fact that the already substantial differences between deprived and affluent neighbourhoods grew further over the 1980s, and suggested that ongoing support for marginalized areas through a co-ordinated package of measures – involving not only the labour market, but also education and training, housing, social security and benefits – is imperative if such areas are not to fall further from the socio-economic mainstream.

Recent policy initiatives – at national, supra-national and local scales – have recognized the need for a broadly based policy approach highlighting principles of:

- *multidimensionality*: tackling disadvantage through a range of activities at different area, individual and institutional scales – in order to address different dimensions of disadvantage simultaneously;
- *partnership*: co-ordinated, shared activity, involving a range of public and private sector bodies working together in designing programmes to tackle disadvantage;
- *participation*: active involvement by disadvantaged/excluded groups in the planning and conduct of programmes/activities to tackle disadvantage.

At the local scale, it is increasingly recognized that the different constellations of circumstances call for different policy measures, integrated and adapted to address local conditions, in order to redress processes of labour market disadvantage and poverty. Combining the three principles identified above, the recent emphasis of local policy has been on building partnerships at the local level and tailoring existing policies, and generating new bespoke initiatives, to meet local needs.

Geddes and Erskine (1994) have identified the following five main approaches to local economic development and regeneration policy, designed to address the issue of poverty and poor areas:

- the local labour market;
- small business enterprise;
- community enterprise;
- property development and inward investment;
- welfare rights and income support.

They argue that some combination of these policy approaches may bring some economic improvement to some localities. It is also recognized that

Table 7.8 Contradictions, tensions and issues in local development programmes

Issue	*Comment – possible solution*
Should programmes attempt to promote jobs in the local area where they are difficult to sustain or in the broader labour market where many disadvantaged fail to get them?	Could intervene to stimulate the broader labour market but look at transport access, discrimination, etc.
Problem of skills 'mismatch' even if more jobs were to become available	Could customize training to meet employer demands
Non-work culture amongst long-term unemployed	Training must involve more than skills enhancement; some post-recruitment follow-up may be appropriate
Degree of local linkage of large/ prestigious projects	Local-labour clauses – rewards for exceeding targets, penalties for falling below; improving the information base about local contractors, etc.
Poorer than average survival chances of small enterprises in deprived areas	Provision of better support services utilizing economies of scale in provision
Job quality: jobs easiest to create tend to be low paid and tend not to go to the unemployed	Concentration on higher order skills
Upward social mobility is usually accompanied by spatial mobility	Diversifying housing tenure mix – to increase degree of housing choice
Difficulties of spatial targeting: leakage, blighting other areas, use of inappropriate spatial areas	Enhance flexibility of spatial targeting; complement spatial targeting by sectoral and individual targeting

Source: Adapted from Gaffikin and Morrisey 1994

there are contradictions and tensions inherent in many local development programmes: for example, when new employment is generated in a poor neighbourhood, factors such as skills 'mismatch' may militate against unemployed residents gaining access to the jobs, or the jobs may be of poor quality and provide such low pay that problems of low income remain (see Table 7.8 for further details).

Even if solutions can be found to circumvent some of these contradictions and tensions, Geddes and Erskine (1994) argue that if the wider economic context is one of austerity policies and a depressed labour market, local employment initiatives will have very limited success or will achieve it at the expense of other localities. Clearly, local policies have their limits: critical policy decisions, institutions and powers are national, European or global; the lack of controls on the movement of capital affect the prospects for stable employment creation – and in this context local initiatives can have only a marginal material impact. Moreover, since many people in poverty live outside deprived areas, policies concentrated only on the most disadvantaged areas do not tackle the position of many poor people. While it is important to be alert to the limits of local policies, numerous commentators have related the success of particular localities to the effectiveness of particular regional/local economic policies and institutional structures and relationships (Murray 1991). Hence, the most appropriate way forward appears to be one involving a variety of different policies at a variety of different scales, involving a variety of different organizations and institutions, and tackling a variety of different dimensions of disadvantage.

GUIDE TO FURTHER READING

For an overview of recent evidence on the distribution of income and wealth in the UK, drawing together the findings of a research programme commissioned by the Joseph Rowntree Foundation, see the two volumes of the summary report launched by the Joseph Rowntree Foundation in February 1995:

Barclay, P. (1995) *Inquiry into Income and Wealth*, vol. 1, York: Joseph Rowntree Foundation.
Hills, J. (1995) *Inquiry into Income and Wealth*, vol. 2, York: Joseph Rowntree Foundation.

Indicators of deprivation from the Census of Population are used in many studies to highlight differences in economic and social well-being at the intra-urban scale. For an overview of the issues associated with the selection and use of such indicators see:

Hirschfield, A. (1994) 'Using the 1991 Population Census to study deprivation', *Planning Practice and Research*, 9: 43–54.

For an introduction to debates on the underclass see an edited collection of contributions stemming from a conference held at the Policy Studies Institute:

Smith, D.J. (1992) *Understanding the Underclass*, London: PSI.

A useful review of the extent and role of the informal economy in local economies throughout Europe is contained in:

Williams, C.C. and Windebank, J. (1995) 'Black market work in the European community: peripheral work for peripheral localities', *International Journal of Urban and Regional Research*, 19: 23–39.

An overview of anti-poverty initiatives and policies at the local scale is provided in:

Geddes, M. and Erskine, A. (1994) 'Poverty, the local economy and the scope for local initiative', *Local Economy*, 9: 192–206.

REFERENCES

Barclay, P. (1995) *Inquiry into Income and Wealth,* vol. 1. York: Joseph Rowntree Foundation.
Boorah, V.K. and Hart, M. (1995) 'Labour market outcomes and economic exclusion', *Regional Studies*, 29: 433–8.

Bunker, N. and Dewberry, C. (1984) 'Unemployment behind closed doors', *Journal of Community Education*, 2: 31–43.
Casey, B. and McRae, S. (1990) 'A more polarised labour market?', *Policy Studies*, 11: 31–9.
CSO (1995) *Regional Trends* 30, London: CSO.
Davies, H., Joshi, H. and Clarke, L. (1993) 'Is it cash the deprived are short of?', Paper presented to the 'Research on the 1991 Census' Conference, University of Newcastle upon Tyne, September.
Dorling, D. and Tomaney, J. (1995) 'Poverty in the old industrial regions: a comparative view', in C. Philo (ed.) *Off the Map: The Social Geography of Poverty in the UK*, London: Child Poverty Action Group: 103–22.
Dunford, M. (1995) 'Metropolitan polarization, the North–South divide and socio-spatial inequality in Britain', *European Urban and Regional Studies*, 2: 145–70.
Fainstein, S.S., Gordon, I. and Harloe, M. (1992) *Divided Cities*, Oxford: Blackwell.
Focas, C., Genty, P. and Murphy, P. (1995) *Top Towns: The Guinness Guide to Finding the Top Places to Live in the UK*, Enfield: Guinness Publishing.
Forsythe, F.P. (1995) 'Male joblessness and job search: regional perspectives in the UK', *Regional Studies*, 29: 453–63.
Gaffikin, F. and Morrisey, M. (1994) 'In pursuit of the holy grail: combating local poverty in an unequal society', *Local Economy*, 9, 100–16.
Geddes, M. and Erskine, A. (1994) 'Poverty, the local economy and the scope for local initiative', *Local Economy* 9: 192–206.
Goodman, A. and Webb, S. (1994) *For Richer, For Poorer*, London: IFS.
Goodwin, M. (1995) 'Poverty in the city: you can raise your voice, but who is listening?' in C. Philo (ed.) *Off the Map: The Social Geography of Poverty in the UK*, London: Child Poverty Action Group: 65–82.
Gordon, D. and Forrest, R. (1995) *People and Places 2: Social and Economic Distinctions in England*, Bristol: SAUS.
Green, A.E. (1994a) 'The geography of changing female economic activity rates: issues and implications for policy and methodology', *Regional Studies*, 28: 633–9.
——(1994b) *The Geography of Poverty and Wealth*, Coventry: IER, University of Warwick.
——(1995a) 'The changing structure, distribution and spatial segregation of the unemployed and economically inactive in Great Britain', *Geoforum*, 26: 373–94.
——(1995b) 'The geography of dual career households: a research agenda and selected evidence from secondary data sources for Britain', *International Journal of Population Geography*, 1: 29–50.
Green, A.E., Owen, D.W. and Winnett, C.M. (1994) 'The changing geography of recession: analyses of local unemployment time series', *Transactions of the Institute of British Geographers*, NS 19: 142–62.
Gregg, P. and Wadsworth, J. (1994) 'More work in fewer households', *NIESR Discussion Paper*, 72, London: NIESR.
Henley Centre (1994) *Local Futures*, London: Henley Centre for Forecasting.
Hirschfield, A. (1994) 'Using the 1991 Population Census to study deprivation', *Planning Practice and Research*, 9: 43–54.
Institute for Employment Research (1994) *Review of the Economy and Employment: Occupational Assessment*, Coventry: IER, University of Warwick.
McCormick, J. and Philo, C. (1995) 'Where is poverty? The hidden geography of poverty in the United Kingdom' in C. Philo (ed.) *Off the Map: The Social Geography of Poverty in the UK*, London: Child Poverty Action Group: 1–22.
MacDonald, R. (1994) 'Fiddly jobs, undeclared working and the something for nothing society', *Work, Employment and Society*, 8: 507–30.

Marcuse, P. (1993) 'What's so new about divided cities?', *International Journal of Urban and Regional Research*, 17: 355–65.
Martin, R. (1994) 'Regional inequalities in income and poverty: the North–South divide lingers on' in S. Hardy, G. Lloyd and I. Cundell (eds) *Tackling Unemployment and Social Exclusion: Problems for Regions, Solutions for People*, London: Regional Studies Association: 3–11.
Murray, R. (1991) *Local Space: Europe and the New Regionalism*, Manchester: Centre for Local Economic Strategies.
Noble, M., Smith, G., Avenell, D., Smith, T. and Sharland, E. (1994) *Changing Patterns of Income and Wealth in Oxford and Oldham*, Oxford: Department of Applied Social Studies and Social Research, University of Oxford.
Pacione, M. (1995) 'The geography of deprivation in rural Scotland', *Transactions of the Institute of British Geographers Transactions*, NS 20: 173–92.
Pahl, R. (1984) *Divisions of Labour*, Oxford: Blackwell.
Robson, B., Bradford, M. and Tye, R. (1995) 'A matrix of deprivation in English authorities' in Department of the Environment (ed.) *1991 Deprivation Index: A Review of Approaches and a Matrix of Results*, London: HMSO: 73–163.
Sassen, S. (1991) *The Global City*, Princeton NJ: Princeton University Press.
Simpson, S. (1993) *Census Indicators of Local Poverty and Deprivation*, Newcastle upon Tyne: LARIA.
Smith, D.J. (1992) *Understanding the Underclass*, London: PSI.
van Geuns, R., Mevissen, J. and Renooy, P. (1987) 'The spatial and sectoral diversity of the informal economy', *Tijdschrift voor Economische en Sociale Geographie*, 78: 389–98.
Wadsworth, J. (1994) 'Job search, male unemployment and joblessness in the regions', in S. Hardy, G. Lloyd and I. Cundell (eds) *Tackling Unemployment and Social Exclusion: Problems for Regions, Solutions for People*, London: Regional Studies Association: 75–8.
Warde, A. (1990) 'Household work strategies and forms of labour: conceptual and empirical issues', *Work, Employment and Society*, 4: 495–515.
Williams, C.C. and Windebank, J. (1995) 'Black market work in the European community: peripheral work for peripheral localities', *International Journal of Urban and Regional Research*, 9: 23–39.
Wilson, W.J. (1987) *The Truly Disadvantaged: The Inner City, the Underclass and Public Policy*, Chicago IL: University of Chicago Press.
Woodward, R. (1995) 'Approaches towards the study of social polarization in the UK', *Progress in Human Geography*, 19: 75–89.

8

HOUSING

Mark Goodwin

INTRODUCTION

Housing is one of the key planes of division in contemporary British society. Differential access to accommodation, and one's subsequent experience of it, is crucial in many aspects of social and economic life. Most obviously perhaps, the nature of the dwelling that one lives in – its condition, location, size and price, for instance – structures the elementary experience of shelter and of everyday space. In this sense, one's accommodation is literally *the* place where we live. Its impacts on the way we live, however, are much broader. Through the operation of the housing market, our accommodation governs access to material resources and social rewards. Indeed, at various points in the housing booms of the 1980s, the dwelling became more important for many people as a source of value than as a source of shelter, and decisions on where to live and when to move were made accordingly. The house is also a key cultural object that at once expresses and depends on all kinds of values and beliefs. In many ways our houses, and the objects within them, are the most visible forms of our own cultural identities. The location and type of house one lives in can help to structure an individual's life trajectory, from school onwards. Recognition of this has led numerous parents to move house in an effort to gain access to the catchment area of a 'better' school. Later in life, the house becomes the base from which people look for work, and its location fixes the labour market in which one is able to work. It can help shape the way one feels about oneself and others, as well as provide security for this generation and the next. Put simply, housing plays a key role in any person's social, cultural, economic and political experience.

The opportunities that arise from this for opening up and sustaining divisions, both between individuals and between social groups, are numerous. In order to make sense of such complexity, this chapter will move beyond the traditional notion of housing tenure, which is commonly used as a basic conceptual shorthand in housing studies. The problem with such shorthand, however, is that it is often falsely 'assumed that taxonomic collectives like "owner-occupation" or "social housing" correspond with significant break-points in concrete real world categories like housing quality, social status, or financing mechanisms (Barlow and Duncan 1988: 219). This assumed correspondence has grown more problematic in recent years as council housing has been sold to tenants, as owner-occupiers have been increasingly subject to mortgage default and repossession, as schemes for part-ownership have become more extensive, and as housing associations have been given the primary responsibility for providing social housing. These changes, along with many others, have made it increasingly difficult to analyse housing divisions in terms of tenure. There are disadvantaged groups in each tenure, and in order to highlight these the chapter will explore housing divisions through looking at homelessness, the housing market and social housing. It will look at the way in which these divides have deepened in recent years, and at the policies that have helped to cause them. It will also consider critiques of these policies, and examine alternatives recently put forward from a variety of political perspectives.

HOUSING DIVISIONS

Homelessness

The most obvious divide in contemporary urban housing is between those who have accommodation and those who do not. The 1977 Housing (Homeless Persons) Act placed a duty on local housing authorities to ensure that homeless people who are categorized as being in 'priority need' and who are not 'intentionally homeless' are provided with permanent accommodation. This duty is now contained in Part III of the 1985 Housing Act (Table 8.1). The records that these authorities keep provide the most comprehensive homelessness data. Since the legislation came into force, the number of households accepted as homeless has nearly trebled, from 53,100 in 1978 to over 146,000 in 1994 (DoE homelessness statistics). The latter year is not exceptional – indeed, three years previously over 149,000 families had been officially accepted as homeless, representing some 425,000 people. The crucial point about these figures is that they represent only those officially accepted as homeless, which on average amounts to only 35 per cent of all those who apply (O'Dwyer 1995). Those turned away are either not classified as homeless, or are not felt to be in priority need. The latter definition includes the majority of single people and childless couples who approach local authorities for help under the homeless legislation. Even though they may well be homeless, or living in hostels, or sleeping on a friend's floor, they are not entitled to housing if they are not classified as being 'in priority need'. And if they are not entitled to housing they do not appear in the statistics – even though they are technically without a home.

Thus the homelessness problem is much greater than the official figures suggest. These are by definition only partial, but estimates of the true numbers of homeless are notoriously difficult to generate. We do know how many

Table 8.1 Major housing legislation since 1980

Year	*Title*	*Main focus*
1980	Housing Act	Right to Buy, Tenant's Charter, New subsidy system
1982	Social Security and Housing Benefits Act	Housing Benefit introduced
1985	Housing Act	Consolidation of legislation
1986	Housing and Planning Act	Extended Right to Buy: Power to sell whole local authority estates
1988	Housing Act	New private regime for Housing Associations: Deregulated Private Renting; Tenants' Choice
1989	Local Government and Housing Act	New financial regime for local authorities

Source: adapted from Williams 1992

applicants local authorities turn away, but we do not know about those who never approach the council in the first place. A recent report on single homelessness estimated that there were 43,000 single homeless people in London alone, none of whom had been accepted as being in priority need and therefore officially classified as homeless. These were people in squats, bed and breakfasts or sleeping rough. The latter category, those literally on the streets, was estimated to number over 1,000 (SHiL 1995). Parliamentary records suggest that in England and Wales as a whole over 8,000 people were sleeping rough in 1991, with a further 50,000 in squats. Moulding all this incomplete and geographically specific evidence together is a complex task, but we can estimate that at any time in the 1990s, well over half a million people in England and Wales lacked a permanent roof over their heads. In some years in the early 1990s the total was probably closer to 750,000.

The crisis is undoubtedly an urban phenomenon – in both 1993 and 1994, the London boroughs accepted almost a quarter of all officially homeless households in England. A walk around the centre of any large city serves to confirm the scale of the problem. The proliferation of people begging, lying in the streets, curled up in shop doorways or underneath park benches is testimony to the inexorable growth in homelessness – and to the inability of government to solve the homelessness crisis. As suggested, these 'visible homeless', those who make their presence felt, are only the tip of the iceberg. The vast majority of homeless do not beg, or sit by the side of the street, or in the entrance to shops and tube stations. They are not immediately noticeable and in many respects they do not differ significantly from the majority of the general population. The one big difference is, of course, that they consistently have to cope with the day-to-day chore of living in temporary

Plate 8.1 Frozen out of the system: the tents of the homeless in Lincoln's Inn Fields, one of the most prestigious squares in cental London, during the cold winter of 1991
Source: David Hoffman

accommodation, in hostels and bed and breakfasts, or in cramped conditions with friends and relatives. They are eking out an existence that often involves highly insecure accommodation, and it can mean whole families trying to live in a single room. They have few 'possessions' and little or no furniture. They can never settle into the local community, into its schools and doctors and libraries, because they are often moved on into the next 'stage' of accommodation. For over half a million people in urban Britain today, these appalling conditions represent home.

The unequal housing market

Those with the resources to gain access to what they like to think of as 'permanent' housing are infinitely better off than those who are homeless. But here too we find vast differences and divisions. The Conservative governments of the 1980s and 1990s have largely succeeded in selling the idea that their housing policies were about promoting choice and opportunity. Sales of council housing to tenants and housing associations alike have been promoted as 'Tenants' Choice', the rapid expansion of owner-occupation is presented in terms of providing opportunities for home ownership, and the 1995 housing White Paper, setting out future objectives, was called *Our Future Homes: Opportunity, Choice and Responsibility.* Yet the language belies

Table 8.2 Categories of housing tenure and stock of dwellings (in millions) in Great Britain, 1914–91

Date	*Rented from local authority or new town*		*Privately rented*		*Owner-occupied*		*Total number of dwellings*
	No.	*%*	*No.*	*%*	*No.*	*%*	
1914	0.1	1.8	7.5	88.2	0.9	10.0	8.5
1944	1.6	12.4	8.0	62.0	3.3	25.6	12.9
1951	2.5	17.9	7.3	52.5	4.1	29.6	13.9
1961	4.4	26.8	5.0	30.5	7.0	42.7	16.4
1971	5.8	30.6	3.6	18.9	9.6	50.5	19.0
1981	6.6	31.0	2.7	12.6	11.9	56.4	21.2
1991	5.7	24.8	1.7	7.4	15.6	67.7	23.1

Source: Carter 1995

the real situation. In practice Britain's housing market is amongst the most restricted in Europe. Since nearly 70 per cent of British homes are now owner-occupied (Table 8.2), the choice in many areas is simply between buying and buying – in other words there is no real choice. Those that cannot afford to buy are being forced to rely on an ever dwindling, socially rented sector, and a privately rented sector that has less housing than any other European country. Elsewhere in Europe, households can choose between an array of housing types in order to meet their particular circumstances and objectives. Private renting is more widespread, but so is co-operative renting, self-build, self-development and social renting (see Barlow and Duncan 1994). Households can choose the type of housing they require, according to family size, labour market position, or age for instance. In Britain, by comparison, the choice is severely limited, and speculative forms of owner-occupation predominate in every locality.

This situation would not in itself matter so much, if Britain's form of owner-occupation was more successful in building and providing homes. Over the past decade or so, however, our dominant tenure has overwhelmingly failed to meet people's housing needs. The overarching context, as Hutton (1995) has pointed out, is a national economy in which full-time permanent jobs are getting scarcer, thus making it harder for families to take on and service long-term mortgage payments. The interaction between extremely volatile house prices and the new insecurity of employment 'has created a vicious inequality of gains and penalties as well as an unprecedented level of personal financial crisis (for people) unable to meet their commitments' (Hutton 1995: 205). Four hundred thousand mortgages are now in arrears, and more than 3 million households are unable to move because they are in a position of negative or insufficient equity. For the 1.5 million in negative equity, the value of their mortgage still to be paid exceeds the

Plate 8.2 Repossession of homes became a common occurrence in south-east Britain in the early 1990s as borrowers defaulted on mortgage repayments. In many instances resale realized only half of the equity
Source: Philip Wolmuth

market value of their home (*Guardian* 6.3.1996). For those 1.7 million with insufficient equity, the value of the dwelling exceeds their mortgage by less than £5,000, which is not enough to finance a move (*Roof* 1995). For these people the freedom and choice promised by owner-occupation has become 'an intolerable burden, a financial trap' (Hutton 1995: 209). For thousands of others, their dream home has actually been taken away, as mortgage lenders have repossessed the house, afraid that the borrowers would never be able to make up the gap between what they owed on their mortgage and what they could afford to pay. Repossessions rocketed from the small figure of 3,000 in 1980 to over 75,000 in 1991. Over 250,000 houses were repossessed in the four years from 1991 to 1994, meaning that over 600,000 people lost their accommodation. For these families, the property owning democracy has become a slogan with very little substance.

A particular group of urban households who have found themselves in considerable difficulties have been those who decided to buy their former council dwelling. Since the 1980 Housing Act, which made it compulsory for local authorities to sell to those tenants who wished to buy and which introduced a scheme of discounted prices for council property, almost 1.5 million council houses have been sold – roughly a quarter of the entire council stock. Whilst the majority of those properties sold have been desirable suburban houses with gardens, a not insignificant number of inner urban flats and maisonettes were bought with the lure of substantial discounts and cheap mortgages. Since these dwellings were often on estates with considerable maintenance problems, the former tenants who have bought these properties have subsequently faced huge service and repair bills. Such bills were previously paid as part of their council rent, and although they continued to be met by the local authority for a fixed time after purchase, once this time limit had passed, households found it difficult to cope with a sudden rise in their housing costs. The inevitable outcome has again often been repossession. Even when the new owners can meet the charges, the accommodation can be very difficult to sell, especially as building societies are increasingly reluctant to advance mortgages on such properties.

For those who lack the regular income or the social profile required to gain access to owner-occupation, and who are unable to obtain social housing, the private rented sector is the only alternative. Renting from private landlords has been in steady decline since the end of the First World War, when it was the source of nine out of every ten homes in Britain. It was still the majority tenure in the early 1950s, but by 1991 it accounted for only 7 per cent of the nation's housing (Table 8.2). It is, however, an important sector of the housing market in Britain's cities. Indeed, in most large cities there are many inner urban wards where private renting accounts for up to a third of all housing stock. Despite its dramatic overall decline, it is still an important form of urban housing, especially for the young, the single and those forming a household for the first time.

The government have attempted to stem the decline of private renting by deregulating private tenancies, thus making the sector more attractive to potential landlords. Rent levels are no longer controlled, and the ability for tenants to register a fair rent has been removed. This has inevitably pushed rents up. Indeed, Department of Environment statistics suggest that, on average, weekly rents for all types of tenancies have doubled since 1988 when the Housing Act introduced the changes (Table 8.1). This has inevitably made it difficult for those whose income lies just above the threshold for housing benefit. Even for those who do qualify, housing benefits are no longer guaranteed to cover the full rent, again increasing the likelihood of eviction for the poorest and most disadvantaged tenants. Protection from eviction generally has been lessened as part of the deregulation of the sector, and tenants now have far fewer rights than they did before the 1988 Act was introduced. Latest data in the Department of Environment's 1995 Housing and Construction Statistics suggest that the private rented sector is expanding again, partly as a result of the 1988 deregulation and partly due to the recession in the owner-occupied market. However, this has done nothing to help the most disadvantaged in the sector, who are having to cope with the combined effects of higher rents, lower benefits and more insecure tenancies. Many of these disadvantaged tenants will be found in the traditional heartlands of the private rented sector, in the run-down Victorian terraces of our inner cities.

Social housing

At one stage in the not too distant past, it looked as if council housing would provide a safe and secure alternative to private renting for Britain's poorer urban residents. By 1981, local authorities and new towns had come to provide almost a third of all the country's housing (Table 8.2). The vast bulk of this state housing was concentrated in Britain's urban areas, both on sprawling suburban estates and in inner urban redevelopment schemes. The municipal estate was an established and accepted feature in every British city. The 1980s saw a massive shake-up in public housing provision, and the system of providing and managing council housing, which had been gradually established since 1919, was effectively dismantled.

Government investment in public housing was cut severely from £7.2 billion in 1978/9 to £1.6 billion in 1989/90 (Blake and Dwelly 1994). The number of council houses built fell even more dramatically, from 79,000 in 1979 to just 1,200 in 1992/3. This reflects government policy that housing associations, who housed only 3 per cent of Britain's population in 1988, should be the main providers of social housing. Local authorities no longer have a role in providing housing – they simply manage and maintain their existing stock, which has been reduced by 1.5 million since 1980 through council house sales. These sales have inevitably been concentrated in the best areas of council stock, leaving local authorities to manage a dwindling supply of increasingly run-down housing. Despite the dramatic assault on public housing by the Conservatives, it still numbers around 20 per cent of the nation's housing stock. In most large cities the figure is considerably higher – a legacy from the days when council housing dominated their housing provision. For tenants in this stock, as well as for those hoping to gain access to it, the decline in government investment has also been keenly felt. Council rents have been deliberately forced up by Conservative administrations that felt they had been kept artificially low. Between 1979 and 1993 average council rents rose from £6.41 to £33.50 a week.

For many urban residents a council house had once seemed an almost natural choice. Large urban authorities in London and the other conurbations were running housing stocks of up to 100,000, and waiting lists for a council house regularly stood at over 10,000. This route is now closed to all but a tiny handful of families. Those wanting social housing, as it has been officially renamed, must now rely on housing associations. These are semi-charitable bodies who have now been given the primary role for providing new homes for those in need. But the system of provision is not the same as the old local authority one. It is not simply a case of exchanging one social landlord for another. The rules of the game have been considerably changed. The funding method involves the associations in raising 40 per cent of the money they need for every new home from private sources. Whereas local authorities could borrow at favourable rates to finance housing investment, housing associations are forced to look to the money market. This has pushed up the cost of housing association schemes and, inevitably, of the rents charged. In just three years, from 1992 to 1995, public investment in housing associations via the Housing Corporation development programme has been more than halved – necessitating in the near future even higher levels of private borrowing and resulting in even higher levels of rent. This will also mean that housing associations are unlikely to be able to afford to make up the shortfall in social housing provision that has accumulated over the past 15 years through a combination of low new building and high council house sales. Many housing pressure groups argue that around 100,000 affordable homes need to be built each year to meet existing need. Housing association completions are currently running at around half this figure.

HOUSING DISADVANTAGE AND EXCLUSION – A SUMMARY

As the previous sections suggest, there is currently a wide range of housing disadvantage in British cities. At its most obvious, for those who have to live on the streets, such disadvantage means being excluded from all forms of housing. But beyond this most visible form of exclusion, we find many other forms of disadvantage. There are hundreds of thousands of people who are forced to share overcrowded accommodation with friends and relatives, or live in insecure and run-down squats, hostels and bed and breakfasts. For some of those accepted as officially homeless, a permanent dwelling will be gained eventually. For others, especially those not in priority need, living in these conditions becomes a way of life and can easily last for several years. A route out of such conditions was once offered by council housing, but now this avenue is no longer available. New social housing is restricted to housing associations, whose financial regimes increasingly mean higher and more market-oriented rents – again excluding those who cannot afford to pay.

Even for those who have secured council accommodation, there is often division and disadvantage. Management policies and council house sales have combined to create a tiered system of council properties, with some urban estates being notorious for their drug dealing, high crime levels and street violence. In Liverpool, for instance, housing workers try to confine their visits to certain estates to between 9.00 and 11.00 a.m. – in theory before the drug dealing begins (Birch 1995). Fear of reprisal means that once such dealing is established it is extremely difficult for other tenants to stop it. Those who can move out do so, whilst the properties that fall vacant become increasingly difficult to let, thus reinforcing the feeling of decline and despair. These problems are not unique to Liverpool, and indeed are probably worse in Manchester and certain parts of Inner London, where the American phenomenon of drug-related drive-by shootings has recently been imported.

There are also huge divides within private housing. Renting from private landlords spans a wide spectrum from those occupying penthouse apartments in the latest waterside development to those managing to survive in the most run-down bedsit in the inner city. The point to be made is that the plight of those occupying the poorer end of this market has worsened in recent years as rents have increased and security of tenure has decreased. Private renting at the lower end of the market is overwhelmingly an urban phenomenon, and behind the peeling facades that still predominate in some neighbourhoods there exists a twilight world of houses in disrepair that lack adequate safety precautions. A recent 'unlawful killing' verdict in respect of a baby girl who died after fire swept through a house in multiple occupation in Scarborough has highlighted the problem (Campbell 1994), but it is one that still has to be faced on a daily basis by thousands of inner city tenants. Fire deaths in such properties still number 150 a year.

Owning a home may protect you from such dangers, and from having to cope with nuisance and harassment on the worst council estates, but it is no longer a guarantee of trouble-free housing. The recent slump in the home ownership market, bringing a huge rise in repossessions and in negative and insufficient equity, has shown that an Englishman's home is no longer his protected castle. Even in this most privileged sector of the housing market, millions of people are suffering from housing disadvantage, and in the case of repossession, from housing exclusion. The huge price rises that accompanied the housing boom of the 1980s in London and the rest of the south-east has meant that towns and cities in this region have been especially hard hit by the housing recession of the 1990s.

London provides the clearest example of the ways in which these housing divides manifest themselves in a specific urban context. (For a more detailed look at the London housing situation see Brownill and Sharp 1992.) The divisions here are sharper and more fragmented than in any other city. London is the site of the largest concentration of homeless people in the country, and the place where house prices and repossessions are at their peak. As stated earlier, in both 1993 and 1994 the London Boroughs accepted over a quarter of the nation's officially homeless. At the end of the 1980s the cost of providing bed and breakfast accommodation in the capital for these people was estimated at £113 million. These official homeless are by definition the most vulnerable in society, those classified as being in priority need. Such bed and breakfast accommodation in London housed around 8,000 children in 1990, often in situations where whole families were occupying single hotel rooms with limited kitchen and bathroom facilities on different floors or in different parts of the building. In addition to these official homeless, SHiL (1995) estimates that a further 43,000 single people (not classified as in priority need, and therefore not officially homeless) were homeless in London, and the London Research Centre calculated that another 250,000 were living unwillingly as part of someone else's household.

If the scale of the homelessness crisis in London is shocking, the capital also leads the way in house prices, negative equity and repossessions. The three factors are all linked, of course. London saw the most dramatic rise in house prices in the 1980s, when average prices almost doubled between 1985 and 1989. Those who scrambled into the London owner-occupied market by their fingertips at the height of the boom in the late 1980s are now those who face negative equity or repossession in the mid-1990s. The capital contained 14 per cent of the nation's householders in negative equity in 1993, and saw over 27,000 repossessions in 1991 (Mohan 1995). Meanwhile, London's housing stock is in decline. New completions fell from 23,000 at the beginning of the 1980s to only 12,500 at the end. The high price of housing means that 50 per cent of households in London with a mortgage were spending between a third and a half of their incomes on housing costs, and 50 per cent of social tenants were paying up to 41 per cent of much lower incomes (Sharp *et al.* 1990). Brownill and Sharp conclude that there is a housing crisis in the capital 'that not only affects all Londoners but also permeates many other aspects of the condition of life in London and threatens its economic viability' (Brownill and Sharp 1992: 10). On the basis of the data that is available it is hard to disagree. High prices coincide with short supply, and increasing levels of homelessness coincide with falls in new building. The irony is that those who can afford to buy comfortably into the capital's owner-occupied stock are residing in some of the most expensive and desirable properties in the country.

HOUSING POLICIES

The current situation

There is little doubt that the housing policies pursued by the Conservative governments of the 1980s and 1990s have exacerbated housing disadvantage and divisions. For a start, although the formation of new households has continued to grow, the amount of houses being built has shown a steady decline. The stagnant owner-occupied market of the 1990s has ensured that the private sector is unable to make up for the dramatic decline in public sector provision. As the rise in homelessness would suggest, there are simply not enough new, affordable houses being built to meet housing need. Government policy has been to aid the consumers of housing rather than the suppliers. Subsidies have gone to home owners through mortgage tax relief, rather than into bricks and mortar. This gives large benefits to those already owning a home, but in a depressed housing market this does little to boost housing output. At its peak in 1991, support to owner-occupiers via mortgage tax relief reached almost £8,000 million. Although subsequent drops in interest rates and in the rate of tax relief have caused a decline in this support, over £6,000 million in subsidy was still given to home owners in 1993/4, via tax relief and income support payments on mortgage interest charges. In the 15 years from 1980, almost £70 billion has been given to owner-occupiers by the government in this way (Joseph Rowntree Foundation 1995). This is money that could have been spent elsewhere in the housing system – and just a fraction of it spent on new social housing would have made a considerable dent in the homelessness statistics.

Government policy has also been directly responsible for the decline in council housing. The 1980 Housing Act made the right to buy compulsory in all housing authorities, and introduced the system of discounted prices that made these sales so attractive to so many sitting tenants. But by not giving local authorities the finance to provide replacement housing, the government was ensuring that overall levels of affordable rented property would fall. This situation was exacerbated by the deregulation of the private sector contained in the 1988 Housing Act. This led to increased rents in the private sector, and also introduced less secure forms of tenancy. The 1988 legislation also introduced the new financial regime for housing associations, at the same time as making them responsible for providing social housing. The new regime introduced a situation where 'fair' rents set by the Rent Officer were replaced by new market-oriented rent setting procedures – a situation exacerbated by the association's increased dependency on private finance. The end result of these procedures has been to make housing association dwellings more costly to both produce and consume, which in turn has led to doubts over the sector's long-term capacity to replace local authorities as the main arm of affordable social housing. The Conservative's 1995 Housing White Paper (Department of Environment 1995), which presaged legislation in 1996, does little to change their overall policy direction. Indeed, the main thrust of the White Paper is to continue and extend existing policies. Thus the right to buy will be extended to housing association tenants; those accepted as homeless will not be offered public sector tenancies but will instead be referred to the private sector; local housing companies will be set up to take control of large urban estates from local authorities; private developers will be able to bid for development money from the Housing Corporation, blurring the distinction between housing associations and the private sector still further, and continued attempts will be made to entice private financiers to invest in the private rented sector.

Alternative approaches

In the face of Britain's continuing housing crisis, a range of alternative policies have been put forward by political parties and housing agencies. None draws on the 1995 White Paper, but several do accept the continuing popularity of the right to buy provisions and acknowledge the political difficulty of tampering with financial support for owner-occupation. Major differences arise, however, in terms of using receipts from council house sales, and in the balance of housing provision between the private and social sectors. Almost all alternative proposals begin from the premise that the output of social housing needs to be boosted, and that allowing authorities to spend their income from right to buy sales is a major way of funding this. Under the Conservatives, local authorities have been forced to use sales receipts to reduce their borrowing levels. Indeed, aside from a very brief post-election period in 1992/3, they have been unable to finance any new building from sales. This would change under alternative policies put forward by both Labour and the Liberal Democrats, and the latter party has set a target of providing 100,000 units of new social housing each year for five years.

Both opposition parties accept that housing associations have an increased role to play in providing social housing, but both also stress that the sector needs to work in partnership with local authorities, rather than instead of them. In the field of homelessness, both parties favour an emergency building programme to tackle the current crisis, but it is not clear how this would be differentiated from the general increase promised in social housing. Both also favour some kind of modifications to the allocation system, reflecting the moral panic of current political discourse which accuses the homeless in general, and single mothers in particular, of making themselves deliberately homeless in order to jump the housing queue. In fact only 10 per cent of single mothers are living in social housing six months after the birth of their child.

This, and the continued reluctance significantly to alter the financial support given to home owners, demonstrates how the mainstream political parties are hampered in their housing proposals by their desire not to lose the votes of the home-owning majority. In the light of this, the most innovative alternatives are coming from housing campaigners, pressure groups. One alternative favoured on the Continent is self-build, where the eventual occupiers buy a plot and organize construction. This accounts for 80 per cent of detached houses in Austria and 60 per cent in Germany (Bayley 1995), whilst in France such self-promotion accounts for over 50 per cent of all new buildings (Barlow and Duncan 1994). Schemes in Britain account for around 18,000 homes a year, but community self-build would appear to meet many of the problems that currently top the political agenda. It gives people a stake in their own homes, and can be used to give skills to the unemployed. Labour costs are cheaper, and houses can be designed to suit individual needs. Ownership can vary again according to need, and schemes can easily encompass owner-occupation, shared ownership and rented properties.

Another alternative, co-operative housing, can give tenants the ability to buy and sell the occupancy rights – thus allowing movement – but because the dwelling itself is rented, and maintenance costs shared, the overall costs are affordable. This form of housing accounts for around a quarter of all new completions in Sweden (Barlow and Duncan 1994), but is virtually non-existent in Britain. Housing finance is also handled differently on the Continent, where state intervention in a variety of guises is used to control either construction or consumption costs. In some countries, such as Sweden, the price of development land for new housing is also controlled. The end result in each case is cheaper and more affordable housing.

In Britain none of these alternatives has reached the discussion stage, in spite of the fact that they have had proven success elsewhere. The potential votes of owner-occupiers, the continued drive to control public sector finance, and the power of landowners and developers have all combined to restrict practical alternatives to current mechanisms of housing provision. At best all that is on offer are a few minor adjustments around the margins of accepted practice, whilst at worst current policies will be continued and even extended.

However, as charted above, these policies are directly responsible for much of the current housing crisis. If this is to be halted, let alone reversed, and if housing disadvantage and division in Britain's cities are to be effectively tackled, then the current housing system needs root and branch reform rather than the minor modifications that presently pass for housing policy.

GUIDE TO FURTHER READING

Shelter's magazine *Roof,* which is published six times a year, is the best up-to-date guide to contemporary housing issues. It also contains fact sheets, legal updates, policy discussions and book reviews.

For the most comprehensive guide to the development and current management of council housing see Malpass, P. and Murie, A. (1995) *Housing Policy and Practice* (4th ed.), London: Macmillan.

Those interested in comparing Britain's housing provision with that of the rest of Europe should consult Barlow, J. and Duncan, S. (1994) *Success and Failure in Housing Provision: European Systems Compared,* Oxford: Pergamon. This concentrates on housing in growth areas, but does contain an interesting discussion of the respective roles of states and markets in housing provision.

The best guide to the current housing policies of the three major political parties is Blake, J. and Dwelly, T. (1994) *Home Front,* London: Shelter. This contains a review of policy plus a transcript of an extended debate between the parties' housing spokesmen.

For the best collection of general data on housing finance, provision and consumption see the Joseph Rowntree Foundation's *Housing Finance Review.* This is now into its third annual edition, and contains regional data and analysis, plus articles on a range of contemporary housing issues.

For those concerned about the future, Holman, A. (1995) *Housing Demand and Need In England, 1991–2011,* York: Joseph Rowntree Foundation, is an authoritative guide. It makes rather bleak reading, estimating a shortfall of some 430,000 dwellings in 1991 – a backlog that will take two decades to eradicate even if we build 110,000 new homes a year.

REFERENCES

Barlow, J. and Duncan, S. (1988) 'The use and abuse of housing tenure', *Housing Studies*, 3(4): 219–31.
——(1994) *Success and Failure in Housing Provision: European Systems Compared*, Oxford: Pergamon.
Bayley, R. (1995) 'Self-Assured?', *Roof*, 20(5): 38–9.
Blake, J. and Dwelly, T. (1994) *Home Front*, London: Shelter.
Birch, J. (1995) 'Not out of place in Palermo', *Roof*, 20(4): 20–1.
Brownill, S. and Sharp, C. (1992) 'London's housing crisis' in A. Thornley *The Crisis of London*, London: Routledge.
Campbell, R. (1994) 'After the Scarborough affair', *Roof*, 19(6): 17.
Carter, H. (1995) *The Study of Urban Geography*, London: Arnold.
Department of Environment (1995) *Our Future Homes: Opportunity, Choice and Responsibility*, London: HMSO.
Hutton, W. (1995) *The State We're In*, London: Jonathan Cape.
London Research Centre (1990) *The Demand for Social Rented Housing in London*, London: LRC.
Mohan, J. (1995) 'Missing the boat: poverty, debt and unemployment in the south-east' in C. Philo *Off the Map: The Social Geography of Poverty in the UK*, London: Child Poverty Action Group: 133–52.
Joseph Rowntree Foundation (1995) *Housing Finance Review 1994/5*, York: JRF.
O'Dwyer, C. (1995) 'Top 40 councils refusing homeless applications in 1994', *Roof*, 20(2): 17.
Roof (1995) 'Homeowners unable to move in 1994 – fast facts', *Roof*, 21: 16.
Sharp, C., Jones, C., Brownill, S. and Meirrett, S. (1990) *What Londoners Pay for Their Housing*, London: The London Housing Aid Centre.
Single Homeless in London (1995) *Time to Move On*, London: SHiL.
Williams, P. (1992) 'Housing' in P. Cloke *Policy and Change in Thatcher's Britain*, Oxford: Pergamon Press: 159–98.

9

HEALTH

Ian Jones and Sarah Curtis

> The manner in which the great multitude of the poor is treated by society today is revolting. They are drawn into large cities where they breathe a poorer atmosphere than in the country; they are regulated to districts which, by reason of the method of construction, are worse ventilated than others; they are deprived of all means of cleanliness, of water itself, since pipes are laid only when paid for, and the rivers so polluted that they are useless for such purposes; they are obliged to throw all offal and garbage, all dirty water, often all disgusting drainage and excrement into the streets, being without other means of disposing of them; they are thus compelled to infect the region of their own dwellings.
>
> (Engels 1844: 128)

> All the environmental sources of illness are stronger in poorer areas, and even more so in the inner city: smoking is more prevalent especially amongst women; diets contain more sugar, starch and fats because foods high in protein, minerals and vitamins are more expensive; damp and cold give rise to chest complaints and rheumatism; nervous tensions, air pollution, moulds from damp, all contribute to allergies, asthma, eczema; low birth weights are common because of poor maternal diet and smoking during pregnancy; there are more accidents in the home and on the street, from the use of paraffin heaters, lack of safe toys, absence of safe play facilities; injuries from family or street violence, and from work, are more common; there are health risks arising from poor refuse-disposal systems, shared toilets, ducted air heating; exhaustion and overwork and the effort of coping with stress leads to reduced resistance to other diseases. The inner city incubates illness, and attracts to itself those whose earning capacity is limited by long-standing illness.
>
> (Harrison 1983: 255–6)

INTRODUCTION

Almost 150 years separate the work of Engels and Harrison but they share a concern for the effect of material deprivation on the health of the poor of the inner cities. When discussing health and medicine in Western societies, a distinction is often made between communicable diseases, as the dominant health problem of the nineteenth century, and chronic and disabling diseases, the dominant health problem of the twentieth century. This distinction is apparent when we contrast Engels' description of health among the working classes, which focuses on infectious diseases such as typhus and scarlet fever, with Harrison's emphasis on the importance of long-standing illness which is more likely to arise from degenerative, non-infectious disease. However, both authors give systematic accounts of higher adult mortality rates, higher rates of infant mortality and higher rates of childhood accidents among the city's poor. Equally they speak of the risks to the health of people in the city (Engels emphasizing alcoholism and Harrison talking of smoking and poor diet). When their accounts are placed side by side, one is struck by their common concern with the health disadvantages of poor populations in the inner city. Both in the past and today, influential sections of public opinion have also been concerned about the risks to the health of more privileged urban dwellers. Parallels can be drawn between nineteenth-century perceptions of threats to middle-class city areas from the risk of contagious disease spreading from poorer areas (Hobsbawm 1962) and the twentieth-century accounts stressing the perceived threat from contemporary risks (Beck 1992). Despite this, it would not be wholly accurate to argue that as cities move into the twenty-first century they are experiencing nineteenth-century divisions. Although problems of poverty, poor housing and poor working conditions still prevail today, the nature of the risks involved are often different, as are the responses to these risks. For example, the threat to health of motorized transport in terms of air and noise pollution and the risk of traffic accidents is a problem that is becoming especially severe at the end of the twentieth century.

The responses to urban health hazards, such as the nineteenth-century sanitary movement and the twentieth-century ecological movement, share some common themes and strategies in relation to public health, but in some respects the twentieth-century 'new public health' movement is genuinely new (Ashton 1992). Public health movements in the late twentieth century are looking for new ways to promote joint action to improve health. In the nineteenth century the main strategy was to mobilize middle-class opinion and action to help bring about change. Today, there is also emphasis on strategies that seek to empower disadvantaged communities to take action themselves and produce grass-roots pressure for change. Furthermore, urban life is a more complex phenomenon today than it was in the nineteenth century. The resulting health problems are more diverse and need to be seen

Figure 9.1 Boundary map of London boroughs and, in bold, East London and City Health Authority (ELCHA)

both as local issues that often affect particular communities in our cities and also as reflections of global processes influencing life in world cities.

This chapter focuses on these questions with particular reference to London and discusses ways in which health inequalities are influenced by factors associated with the socio-economic profile of the population, housing conditions, ethnic diversity, environmental factors, health service organization and local effects of national health policy.

Cities are not necessarily unhealthy places to live in. Indeed, this chapter will discuss some evidence of good health among Londoners. Initiatives that are specifically designed to enhance the well-being of individuals living in the inner city will also be considered. However, the main focus for this chapter will be health disadvantage in Britain's cities, so the emphasis is on the uneven distribution of factors beneficial or injurious to health of urban populations. We illustrate our arguments with examples from cities across the country but focus mainly on London and especially on the London boroughs of Tower Hamlets, Newham, Hackney and City of London, which together make up the East London and City Health Authority (ELCHA) in London's East End (Figure 9.1). A note of caution is necessary here for, although developments and trends in health in the capital are mirrored else-

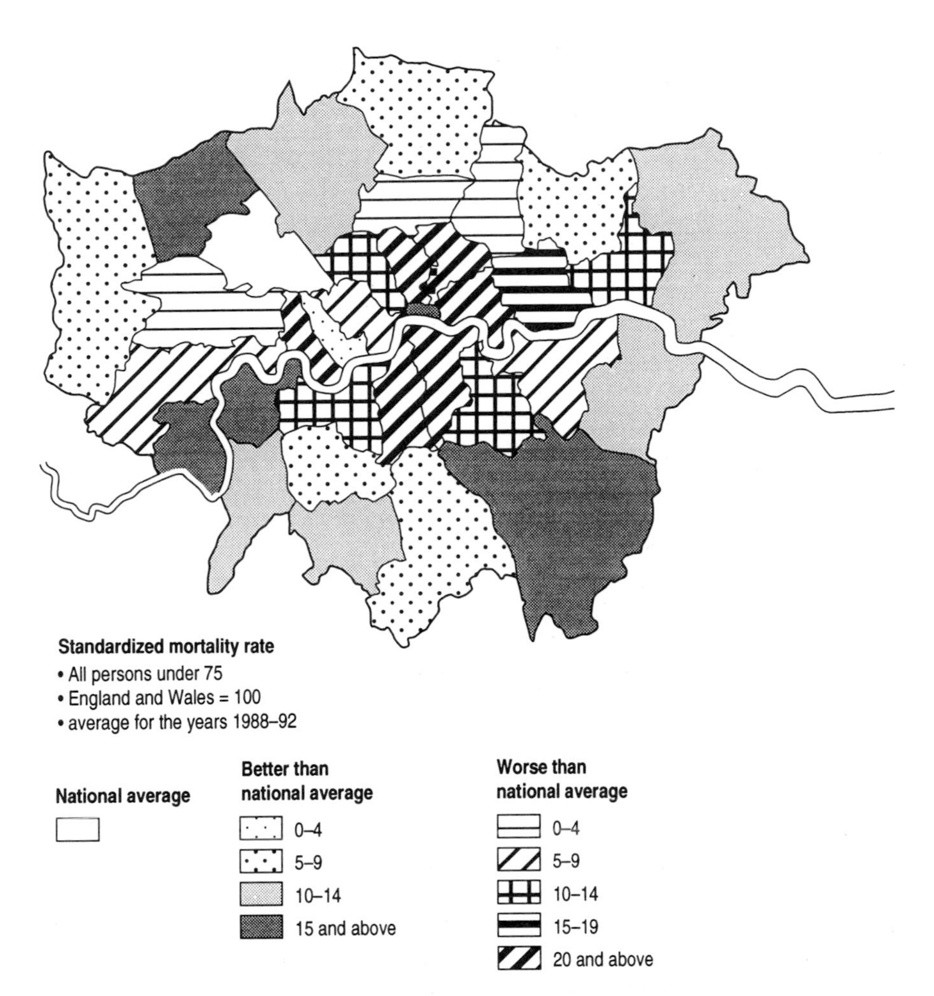

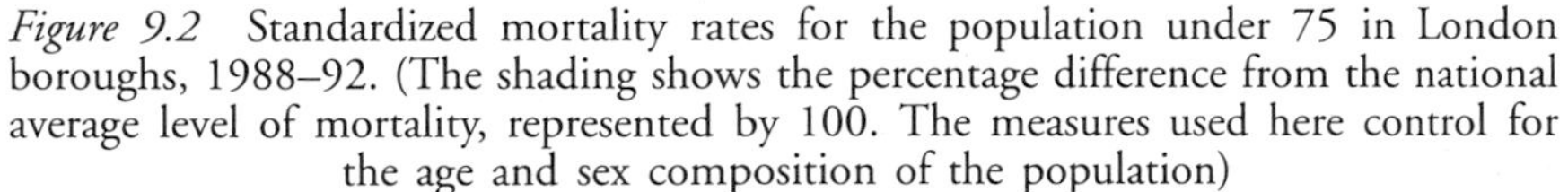

Figure 9.2 Standardized mortality rates for the population under 75 in London boroughs, 1988–92. (The shading shows the percentage difference from the national average level of mortality, represented by 100. The measures used here control for the age and sex composition of the population)

where, as we shall see, London has particular characteristics that may not be wholely indicative of trends in the rest of the country.

MORTALITY AND MORBIDITY IN LONDON

As we have noted, it would be misleading to depict the population of London as generally unhealthy relative to the rest of the nation. Indeed a comparative study of the health status of Londoners (Benzeval *et al.* 1992) concluded that Londoners appeared to be more likely to be healthy than their counterparts in other areas of the country. Nevertheless, poor areas of London and of other major cities in Britain stand out as having marked health disadvantage. Britton's (1990) analysis of mortality in urban government districts shows, for example, that London boroughs in Inner and East London, and inner city metropolitan districts in Manchester, Merseyside and Newcastle upon Tyne, figure among the worst twenty local government districts nationally with respect to mortality from several different causes of death. Figure 9.2 shows that standardized mortality rates for the population under 75 years are high in Inner London compared to the national average and can be contrasted with relatively low standardized rates in areas of Outer London. A similar pattern can be found when standardized long-term illness ratios are examined (Figure 9.3).

Geographic differences in health are related to the characteristics of the people living in different areas and to the characteristics of areas as places (Macintyre *et al.* 1993). This has particular importance for our understanding of the relationship between health and living in cities and for policy intended to improve the health of urban populations. The following discussion considers some of the factors that help to explain the relatively poor health of East London's population.

SOCIO-ECONOMIC FACTORS AND HEALTH

The relationship between material deprivation and ill-health has been well documented, at the scale of both individual people and small geographical areas (Townsend *et al.* 1988a). A considerable body of work developed during the 1980s looking at geographical differences in health and comparing these with indexes of material deprivation for use by health services (e.g. Townsend *et al.* 1988b, Morris and Carstairs 1991). Phillimore and Reading (1992) also found wider health inequalities in urban areas compared to rural areas. Eyles (1987) highlighted the shortcomings of these area-based analyses (primarily because of the ecological fallacy) and showed that deprivation could not be seen solely as an inner city problem. However, some studies of individuals also point to the particularly poor health experience of poor populations living in inner cities as compared with other types of area. For example, Blaxter (1990) found that the psycho-social health of manual occupational groups was generally worse than for non-manual groups. Furthermore, living in a deprived city area had an adverse effect on health in both manual and non-manual families. There is increasing evidence to show that socio-economic differences in health increased during the 1980s (Davey Smith and Morris 1994) and that these differences are tangible in the inner cities (McCarron *et al.* 1994).

The diverse and unequal nature of inner city populations is an important characteristic of modern urban life. Extremes of wealth and poverty can exist

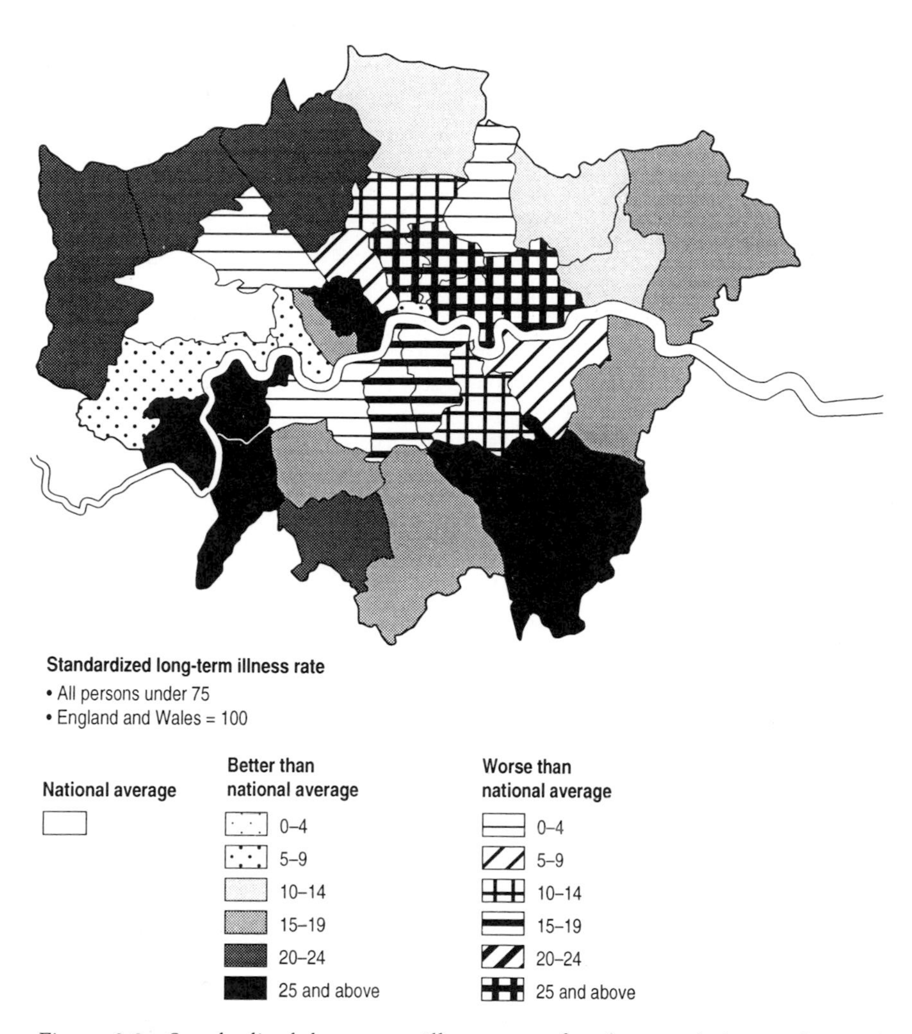

Figure 9.3 Standardized long-term illness rates for the population under 75 in London boroughs, 1991. (The shading shows the percentage difference from the national average level of long-term illness, represented by 100. The measures used here control for the age and sex composition of the population)

within a stone's throw of each other. Although middle classes remain in fashionable areas of long standing, apparently wealthy areas will have pockets of poverty whilst some deprived areas have been transformed as middle classes arrive into run-down areas on waves of gentrification. The imperatives of capital in the late twentieth century have transformed city landscapes rapidly and the impact this has had on city populations is profound. Communities

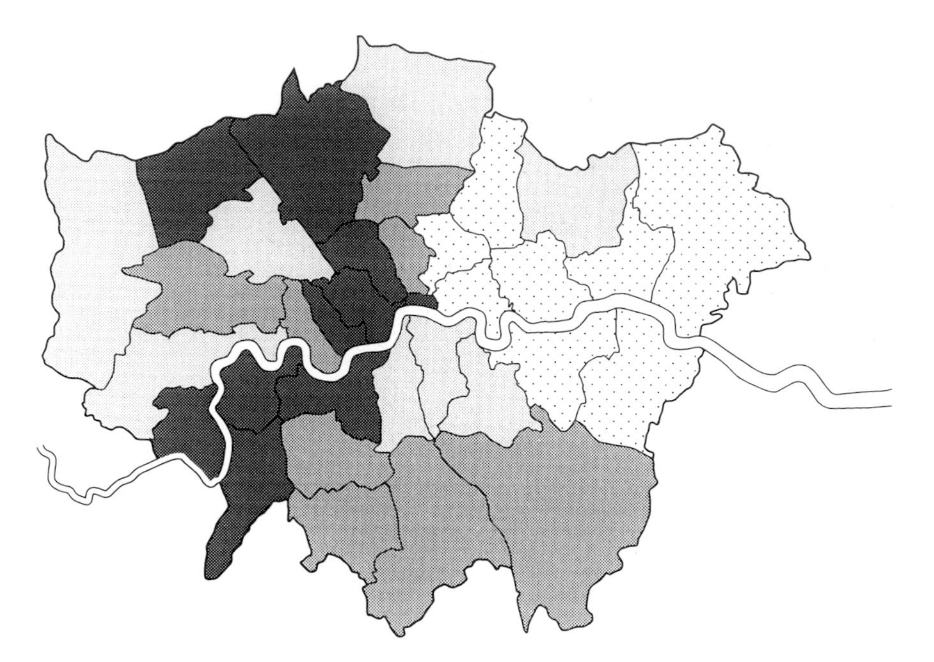

% households with head of household in social class I, 1991

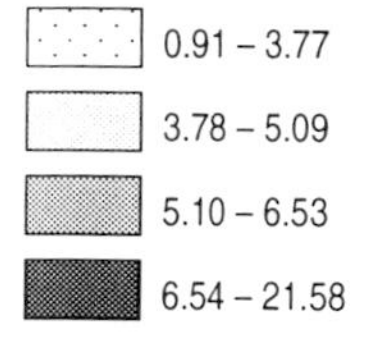

Figure 9.4a Distribution of households headed by professional workers in London boroughs, 1991

have been displaced and fragmented and the poor and underprivileged are often left isolated.

Figures 9.4a and 9.4b of the distribution of people in social class I (professionals) and in class V (unskilled manual workers) reveal the concentration of relatively disadvantaged groups in London's East End. Figure 9.4c, for enumeration districts in Tower Hamlets, also shows that, while this borough has a generally low proportion of professionals and a high proportion of unskilled workers, there are nevertheless local concentrations of wealthier social groups within the area. These are associated with variation in local housing markets. There are gentrified older properties in the north-east of the borough and there has been large-scale construction of new properties for sale in the southern part of the borough, which is in the area of the

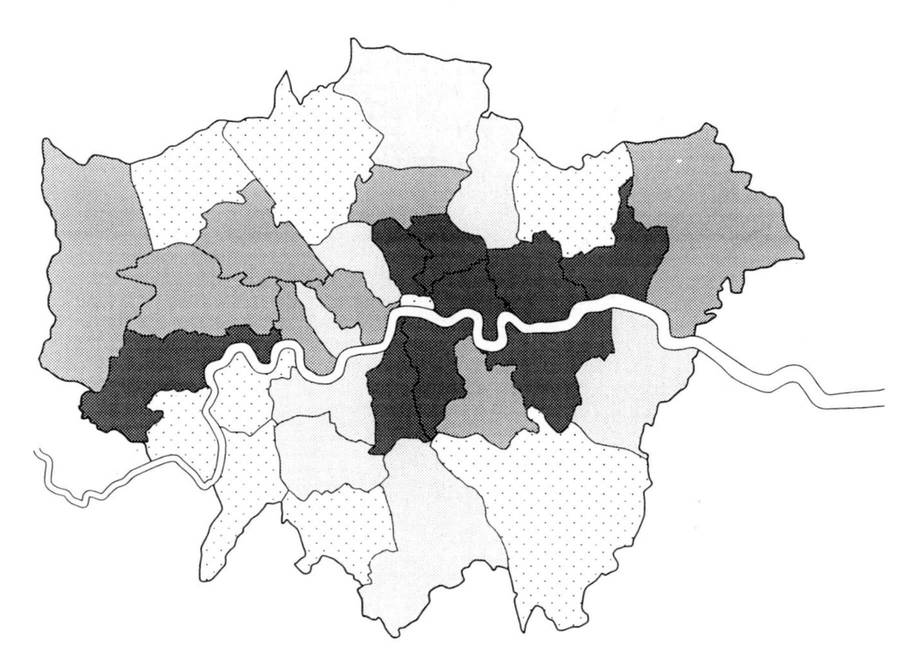

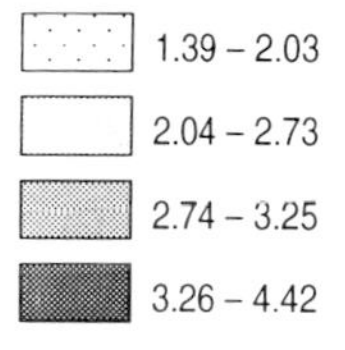

Figure 9.4b Distribution of households headed by unskilled manual workers in London boroughs, 1991

London Docklands Development Corporation and has seen a recent influx of young, professional people, sometimes accompanied by a displacement of original, poorer residents (Ogden 1992).

British cities have undergone selective suburbanization in recent years as wealthier population groups have moved to the suburbs more rapidly than others. However, this has not happened to the same degree as in American cities, and the notion of an inner city 'underclass' is perhaps more relevant to the American experience (Wilson 1994, Wallace 1990, Dear and Wolch 1987). Some have interpreted the British experience in terms of the creation in inner cities of an industrial 'reserve army' of relatively poorly paid workers in insecure employment, which matches the needs of a post-Fordist system of production for a flexible labour force (Byrne 1995). Whether or not this

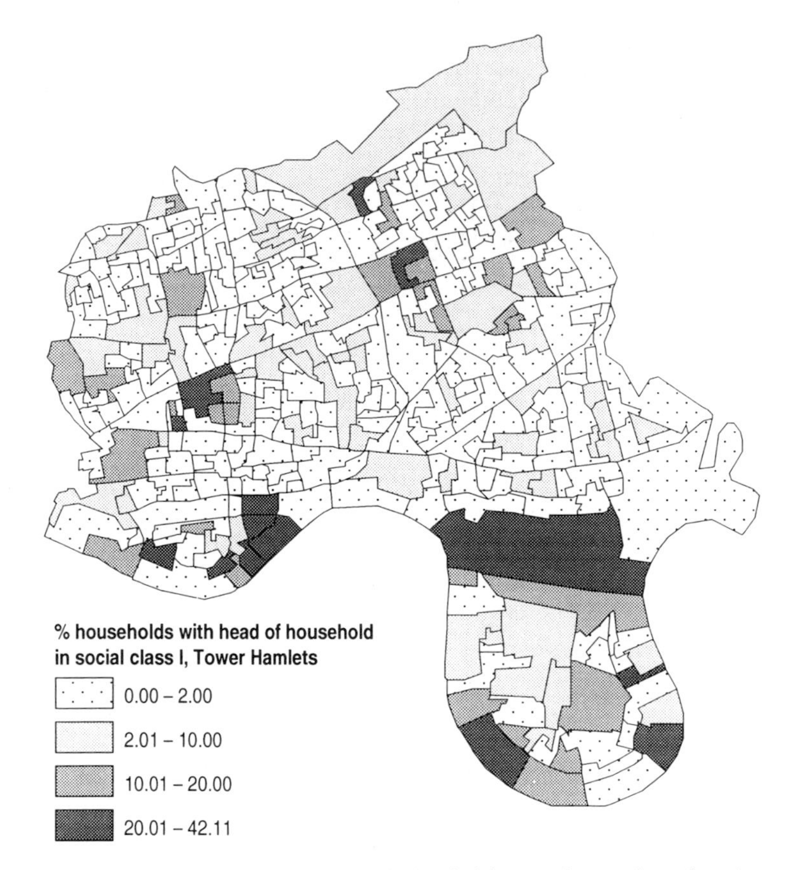

Figure 9.4c Distribution of households headed by professional workers in enumeration districts in Tower Hamlets, 1991

view is justified, the consequences for people's health are dramatic and often health professionals are left feeling angry and frustrated by the magnitude of health problems in the face of seemingly irresistible social forces (Widgery 1991).

OTHER IMPORTANT POPULATION ATTRIBUTES

In addition to the social class profile of East London, there are other features of the population that are relevant to health differences and health disadvantage. We consider here the effects of housing conditions, homelessness, high levels of mobility and ethnic diversity which are significant factors for health in this area.

Table 9.1 Estimates of homelessness in London, January 1993

Area	*Temporary accommodation*	*Squats*	*Hostel*	*Sleepout*	*Travellers*	*Total*	*Total population (%)*
East London	12,970	6,785	448	111	242	20,556	3.6
Inner London	59,224	13,146	3,035	1,106	943	77,454	2.8
All of Thames	128,572	14,921	3,403	1,623	8,543	157,062	1.1

Source: Access to Health 1993

Housing, homelessness and health

Poor housing conditions are a persistent feature of inner city areas, and the relationship between poor housing and ill-health has been documented extensively (Lowry 1989, SCOPH 1994). Cold, damp housing and overcrowding have been found to be associated with high levels of respiratory illness and general ill-health in East London (Hyndman 1990), whilst other research has associated cold housing with the risk of breathing difficulties, circulatory illnesses and hypothermia (Collins 1993).

Changes in the structure of social housing provision since the early 1980s (especially the 'residualization' of the local authority rented sector and changes in benefit entitlements) have been mirrored by homelessness becoming an increasing urban problem in the UK. This is manifested in the rise of the numbers of visible street homeless in our cities and in the numbers of people in housing need (Department of Environment 1994). Table 9.1 gives estimates for East London and compares these with estimates for Inner London and all Health Authority Areas within the Thames Regions. The figures show considerable differences in the relative number of homeless people and type of homelessness. Accurate estimates of the numbers of homeless people are difficult to achieve and some of these figures may represent only the tip of the iceberg in terms of absolute numbers of homeless people.

The large homeless populations in London give rise to a range of health-related problems. Common problems among single homeless people include mental illness, alcohol and drug problems, and a broad range of illnesses including TB, arthritis and skin diseases; all of which are compounded by the experience of living on or near the streets (Balazs 1993). High levels of depression, disturbed sleep, poor eating, over-activity, bed wetting and aggression have been found among homeless children living in bed and breakfast accommodation (Patterson and Roderick 1990). There is evidence of high rates of mental illness among homeless populations (Access to Health 1992) and there are high rates of acute mental illness, psychosis and schizophrenia among street homeless people. There is evidence that inner city hostels are filling up with psychiatric patients discharged from long-term psychiatric institutions as these close down (Marshall 1989). This has led some to argue that:

> Many psychiatrically very ill people are living like feral children in the forest of the city, scavenging for garbage and subsisting from charitable hand-outs. This neglect is constantly before us, and yet we seem to be inured to the desperate plight of these vulnerable individuals, busying ourselves instead with the closure of the psychiatric hospitals.
>
> (Weller *et al.* 1989: 1509)

This seems to add weight to descriptions of inner cities as 'sinks' that attract vulnerable groups. This has important implications for health in the inner cities for these vulnerable groups are often vulnerable because they are ill and ill because they are vulnerable.

In spite of their high levels of health care need, services are often ill adapted for homeless people. Homeless people may have difficulty in accessing health services, especially since a normal requirement is registration with a general practitioner, which is problematic for highly mobile populations (Hinton 1994). Partly as a result of this, homeless people seem relatively more likely than resident populations to make unplanned use of hospital services (Scheuer *et al.* 1991). Even where homeless people are able to make use of services, these services are not always equally appropriate for the diversity of patients. The experience of psychiatric services for homeless women may be different to that of homeless men. Homeless women with mental illness have been found to be more socially stable prior to admission than their male counterparts but less likely to co-operate with treatment and more likely to discharge themselves from hospital (Herzberg 1987). These differences have been linked to variations in the perception of and responses to disturbed behaviour by women on the streets.

The government response to homelessness in London has principally been in the form of the DoE's Rough Sleepers Initiatives, attempting to provide bed spaces for rough sleepers in the capital. In addition there have been cold weather shelters initiatives (1991) and a homeless mentally ill initiative (1991) providing outreach services and rehabilitation units for the homeless mentally ill. However, the impact of large homeless and mobile groups on the demand for health care is not clearly understood. Some attempts have been made to measure the health status of homeless groups and to make allowances for homeless populations in health care resource allocation formulas (Access to Health 1993). The new national formula for allocating resources to health authorities on a capitation basis does not, however, have a weighting for homeless populations, although the report on which this formula was based recommended that further research was required on the impact of homeless populations on the need for extra resources (NHS Executive 1994).

Plate 9.1 Health promotion workers in East London
Source: East London and City Health Authority Health Promotion Department

Ethnicity and health

Ethnic diversity is also an important feature of inner urban populations in most British cities, and especially in London. This arises from a long history of international migration flows to and from the capital, which continues today and which often includes migrants who are moving under particularly difficult circumstances. For example, in ELCHA, the population includes refugees from many parts of the world affected by political disturbance in recent decades: Ugandans, Vietnamese, Somalis, Kurds have all moved into this part of London, which is a traditional urban 'reception area' for poor immigrants. Estimates suggest that as many as one in ten people living in East London have come to London as refugees (ELCHA 1995). The ethnic minority communities of London also include people who were born in Britain and are not migrants, although they are still ethnically distinct from the majority population in Britain as a whole. Within particular neighbourhoods in East London, groups that might be considered ethnic 'minorities' in national terms are locally forming large, sometimes majority, proportions of the resident population.

The concentration of minority ethnic groups in urban areas means that health and health care in the city must be responsive to the health and health care needs of these populations. Figure 9.5 shows, for example, the

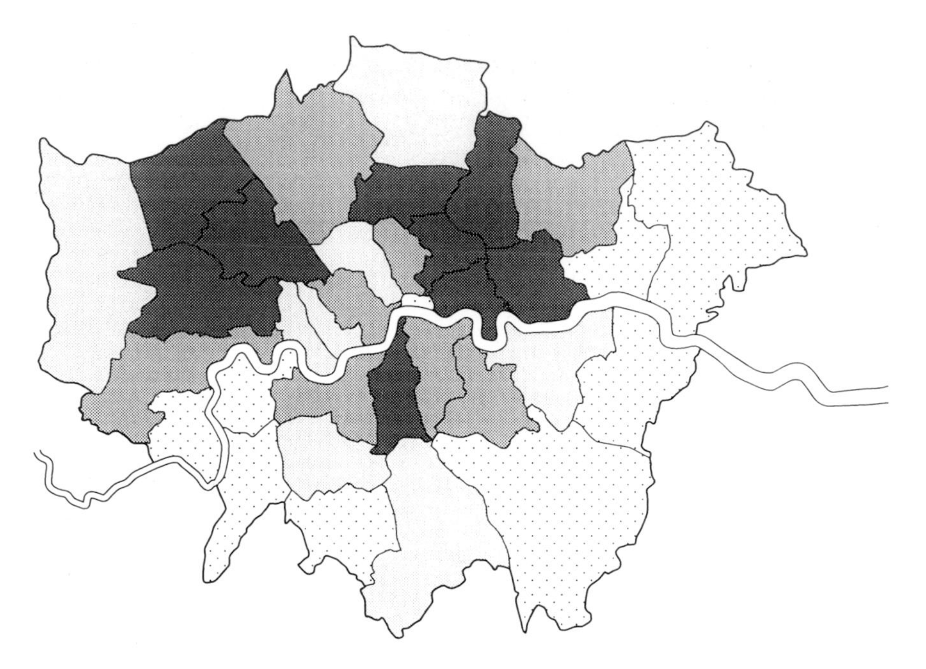

% population defined as "non-white", 1991

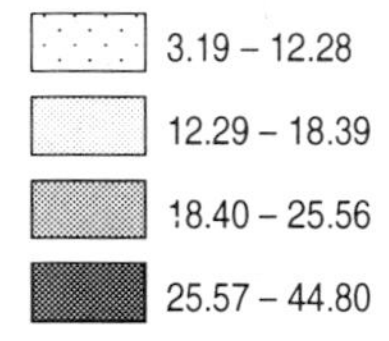

Figure 9.5 Proportion of borough populations belonging to 'non-white' ethnic minority groups in London, 1991

distribution among London boroughs of people who identify themselves as belonging to ethnic groups other than the white majority. This provides a rough indication of the uneven distribution of black and other ethnic minority populations in the city, and shows the relatively high concentration in the inner city, including boroughs in ELCHA.

The health experiences of minority ethnic groups are multifaceted and often conflicting. For example, compared with majority 'white' populations, higher rates of stroke and hypertension and low rates of heart disease have been found in Afro-Caribbean populations. South Asian populations, however, have high rates of heart disease and diabetes and low incidence of some cancers. The comparatively younger age structure of many minority ethnic groups has important consequences for planning health services in urban areas (Roderick *et al.* 1994). The contingent resource implications for health authorities with large minority ethnic populations (in terms of meeting demand from higher rates of illness and the need to provide translation and advocacy services that address the needs of these populations) are particularly acute in inner city areas like ELCHA.

Despite the evidence that the greatest burden of disease for minority ethnic groups is due to chronic disabling conditions that also commonly occur in the British population as a whole, much health care research has tended to concentrate on 'exotic' conditions such as TB, rickets and venereal disease (Smaje 1995, Donovan 1986). This tendency has the undesirable effect of focusing on cultural or individual factors producing ill health in groups defined in terms of 'racial' characteristics. In doing so it detracts attention from the important health effects of poverty and material disadvantage experienced by large proportions of people in minority ethnic groups. Such disadvantage results more from conditions in the wider society than from specific behaviour of ethnic groups themselves, as evidenced, for example, by research on tuberculosis by Bhatti *et al.* (1995) and Mangtani *et al.* (1995).

The impact of racism on the health of city populations is very important for the problems associated with the health experience of minority ethnic groups. High levels of racial harassment and multiple victimization (i.e. repeated attacks) have been documented on estates in East London (Foreman 1989, Sampson and Phillips 1992) and the immediate effects on the health of individuals range from the consequences of physical injury to fear, feelings of vulnerability and symptoms of stress. The fear of racial attack results in individuals placing constraints on their own activity (preventing children from playing outside being just one example). In parallel to the effects of racial harassment, the continued existence of institutional racism in the health service has important implications for those health care institutions that are based in inner city areas. The problems involved are not simple difficulties of language barriers, but include also insensitivity to cultural and religious customs and beliefs and failure to recruit health professionals from among the ethnic minorities that make up a large part of the local population. The drive towards ethnic monitoring in the health service, the development of equal opportunity policies and the adoption of culturally sensitive practices in health care settings represent important trends within the National Health Service. They underline the importance of the NHS in the capital, not only in providing for the needs of the population's health, but also as a major local employer and significant actor in the local economy.

Implications of mobility and diversity for health services

The practical impact of mobile, displaced and vulnerable populations is all too clear for local health authorities and health professionals who need to be able to identify and enumerate the nature and scale of the health service needs of urban populations (Warren 1993). The plurality of city populations, their ethnic, racial and social mix, the fluidity of inner city populations, the transient nature of the squatters, homeless and refugee populations, creates a very mixed socio-demographic profile that gives rise to a complex set of health care issues, several of which have been considered above.

There are also important implications for the allocation of National Health Service resources, since some of these populations do not figure in official statistics. Jacobson (1993) talks of the 'health of London's invisible populations'. The nature of inner city populations creates particular difficulties of measurement, for example for District Health Authorities (DHAs) in British cities. Rapid population changes and census underenumeration affects the accuracy of their census-based population estimates (OPCS 1994) and, because health authorities receive resources from central government on a capitation basis, health authority budgets are ultimately compromised. In addition, the calculation of rates of mortality and morbidity for inner city populations are distorted by this 'denominator' problem, making the planning of services more uncertain.

CITY ENVIRONMENTS AND HEALTH

In addition to the socio-economic factors already considered, the physical environment in London is important for the health of the population. Over two weeks in 1952, 4,000 people died as a result of the London smog. Four years later the Clean Air Act was passed and smogs from stationary pollution sources (factories and homes) became a thing of the past. In the 1990s, however, pollution in our cities has become a major concern. With increasing emissions from mobile sources, cities are experiencing summer smogs in contrast to the winter smogs of earlier periods. Diesel fumes are the major source of particulate in urban areas, and on still, sunny days WHO guidelines for peak ozone concentrations are regularly exceeded in UK cities. These levels are known to cause acute and long-term damage to lung function and there is growing evidence of the role of summer smog in cardiorespiratory disease (Godlee 1992). Photochemical smogs result from interaction between hydrocarbons, oxides of nitrogen and ozone in sunlight. As they have become an increasing characteristic of the urban environment there is concern that they are having a damaging effect on lung functions, particularly in vulnerable groups such as people with asthma, children and the elderly (Mathew 1992). Further evidence linking air pollution and asthma in children has been highlighted by Read (1991). Increasing trends in air pollution over the past forty years have coincided with increases in death rates from asthma and hospital admission for asthma and allergic diseases (POST 1994, Schwartz 1994, SEIPH *et al.* 1994). Populations in East London, including the ELCHA area, show relatively large proportions of the population affected by respiratory diseases including asthma, bronchitis and lung cancer (ELCHA 1994, 1995, Britton 1990).

Despite this, the evidence linking photochemical air pollution to asthma and cardiorespiratory disease is contested. A number of confounding factors are associated with the health data; most notable are smoking behaviour, exposure to other atmospheric pollutants in the workplace or at home, and changes in disease classifications and recording of cases. The contested nature of the evidence, together with the perception of pollution being outside the sphere of action of conventional medical services, appears to constrain public health responses to the problem (Phillimore and Moffatt 1994).

While air pollution is one of the most obvious examples of a topic of current concern for public health in our cities, there are many other aspects of the urban environment and the organization of city life that have important impacts on health, and have the potential to be especially damaging to the health of poor people. While space does not permit a full discussion here, we could mention as illustrations issues such as access to safe and affordable public spaces for children's play and adult leisure and recreation; access to retail outlets providing healthy food at reasonable prices; facilities for community meetings; opportunities for adult education and learning; freedom from noise pollution; efficient domestic and industrial waste disposal; and domestic and outdoor urban design to minimize risks of accidents. All of these depend on the social and economic organization of the city. The growing tendency to involve the free market to a greater extent in the provision and supervision of municipal services and to introduce charging for users at the point of consumption for some of them may act to disadvantage poor populations especially. Furthermore, as wealthier urban populations tend to move out of many city centre areas towards the suburbs, there is a tendency to decentralize key facilities with them. In run-down urban areas it may become progressively more difficult to maintain a reasonable urban environment for healthy living. Gentrification and urban renewal projects, while they may transform the inner urban environment, often have the effect of bringing in wealthier populations to displace existing disadvantaged residents, so that such urban improvements are of little direct benefit to the welfare of the urban poor (Ogden 1992, Brownill 1990).

THE STRUCTURE OF HEALTH SERVICES IN THE CITY

Modern urban space is continuously restructured (Harvey 1989) and the drive to restructure city spaces has deeply affected hospital provision in UK cities. Hospital provision in the UK has been historically concentrated in inner city areas (Rivett 1986) but in contrast to this the UK has experienced consistent population decentralization since the Second World War, a process that has persisted throughout the 1980s (Inner London being an exception, having experienced a small increase in its population between 1981 and 1991) (Breheny 1995). Table 9.2 compares age and sex standardized beds per 1,000

Table 9.2 Age/sex standardized beds used by the residents of major cities in England, 1992–3

Area	*Beds/1,000*	*Difference from national average (100) (%)*
England	2.62	100
Leeds	4.80	183
Newcastle	3.68	140
Manchester	3.52	134
Liverpool	3.00	114
Birmingham	2.94	112
Bristol	2.81	107
London	2.73	104

Source: Adapted from Edwards and Raftery 1995

population in English cities with the national average in 1992/3 and shows that there is considerable variation in bed supply levels, with London having relatively low levels of provision on this measure.

The fortunes of health care institutions are closely connected to the local economy and local labour markets. The desire to rationalize hospital services in the cities grew from a perception of fragmented services in old, outdated Victorian buildings, the need to respond to demographic changes (depopulation of urban areas) and the need to develop more efficient land use in the face of cost constraints. This had major consequences for health authorities based in London where a large number of hospitals targeted for rationalization were sited. At the same time as they were experiencing pressure to reduce resources from the centre they were having to meet the inflationary wage spiral associated with staff shortages in the capital (Mohan 1995). This was given an added dimension during the 1980s by a property boom that gave health authorities the opportunity to raise capital from land sales; but the property slump of the late 1980s and early 1990s had disastrous consequences for many of these capital plans. Nevertheless, rationalization was given momentum by central government, particularly in London where recommendations to reduce acute service levels in the city had considerable implications for bed numbers (Tomlinson 1992). The assumptions behind this drive for rationalization have been questioned (Jarman 1993) and there is some evidence to suggest that there are acute shortages of beds and care for elderly people in the capital (Edwards and Raftery 1995). Inner London health authorities such as ELCHA are particularly affected by these problems.

Private health care

Private sector provision has experienced considerable growth in recent years in the nursing home sector and in the acute sector. During the period

1979–89 the number of private beds in the UK increased by 58 per cent and the proportion of the population with private health insurance increased from 5 per cent to 13 per cent (Calnan *et al.* 1993). These developments were not evenly distributed across the country, as hospital developments have focused mainly in the large cities, especially around London, close to specialist and teaching hospital provision and in wealthier areas where the population with private insurance coverage is relatively large. The pattern of provision in the independent sector means that, for those who are able to pay, the choice of health services will be enhanced in some parts of the city. Since the introduction of an internal market into the NHS, the independent sector also influences provision of state-funded health care. Salter (1995) reveals the complexity of the relationship between the NHS and the private sector before suggesting that the trend will be towards the NHS purchasing health and social care from the private sector (particularly from private nursing homes for the elderly) and for an expansion in private insurance companies purchasing pay-bed facilities in NHS hospitals. The impact of such developments on the availability and accessibility of health services in cities will depend very much on the uneven geography of the private health care sector and of populations with insurance cover that enables them to use independent health facilities. We have already noted the relative concentration of people in less privileged social classes in East London, and these groups are least likely to be able to benefit from private health care provision, even if it is available locally. The national data from the General Household Survey in 1987 (OPCS 1989), showed, for example, that of those aged 45 to 64 in semi- or unskilled classes, only 2 per cent had private insurance cover, while among employers and managers the proportion was 17 per cent and for professionals 21 per cent.

Primary and community care

Primary care in the inner city areas such as ELCHA has suffered from long-term underinvestment (Leavy 1985, Jarman and Bosanquet 1992) which is manifested in poor primary care facilities in some practices. The area is part of the London Implementation Zone, within which extra resources are being allocated by the NHS centrally to develop primary care in Inner London. Some commentators have questioned whether this applies equally to all inner cities in Britain (Wood 1983) and it may be a peculiarity of the capital. The need to allocate resources to primary care in inner cities to improve service provision and to balance the drive towards rationalizing acute services is a persistent feature of the health care debate in the 1990s (Craft and Dillner 1994).

New legislation implemented in April 1993 has given local municipal authorities (as opposed to District Health Authorities) the lead responsibility for assessing the needs of elderly and disabled people and for buying packages of care on their behalf. This has placed considerable pressure on municipal authorities. In spite of the transfer of some funds from the NHS to local government in order to implement these reforms, the general budgets of local councils in inner cities have been strictly controlled, leaving little scope to develop community social care. In London the complexity of implementing community care in the face of inadequate resourcing has been clearly identified (Jowell 1992). In the inner cities the problems of providing community care for the disabled, elderly and mentally ill are compounded by isolation, poor housing, low incomes, poor transport facilities and alienation. The provision of community care in cities presents particular problems in terms of access to services and barriers to helping individuals maintain their independence. This is closely bound up with the contested notions of 'community' in urban areas (Moon 1990). Some commentators have suggested that various factors, including decisions of health professionals, rejection by wealthy communities and the behaviour of service users, can produce 'ghettos' in poor inner city areas where there are large concentrations of people needing services for chronic mental or physical illness but where service provision is particularly overstretched (Dear and Wolch 1987, Tudor-Hart 1971, Wallace 1990).

THE LOCAL EFFECTS OF NATIONAL HEALTH POLICY

Important strands of recent health policy in England include: the rationalization of hospital services (combined with attempts to shift from hospital to community care), the drive towards capitation-based funding in the NHS and the emphasis on national targets for improving the population's health. These are closely related, for capitation-based funding is producing a squeeze on the resources of inner cities which are experiencing population decline whilst they are simultaneously having to cope with high per capita needs in the population and the costs of service reorganization within their areas. As in many countries, national health policy in Britain is trying to move towards setting objectives for population health gain, not only for providing health services. The setting of national targets for improvement of the health of the population has the advantage of focusing attention on how to improve health, but the 'Health of the Nation' policy (Department of Health 1992) has been rather insensitive to the sort of local conditions that prevail in inner city areas. In order to reach national targets for improvement of health in terms of cardiovascular disease, cancer, accidents, sexual and mental health, many poor inner city areas will have much further to go than more advantaged and healthier parts of the city (Bryce *et al.* 1994, McCloone and Boddy 1994). Furthermore, the national 'priorities' do not always fully match those for health improvements of poor urban areas. In East London, for example, tuberculosis, diabetes, asthma, learning difficulties and physical and sensory difficulties have also been identified as important health problems for the local population, reflecting the particular conditions prevailing in the area (ELCHA 1995).

Healthy cities

Some government responses to the health problems of UK cities have been alluded to already. Many responses concern areas that are outside the conventional activities of the National Health Service (initiatives relating to homelessness and urban regeneration programmes being two examples). Some local agencies concerned with urban health are adopting new strategies to try to cope with these pressures and to address local health problems in appropriate ways. The healthy cities movement grew out of a project instigated by the World Health Organisation Regional Office for Europe in 1986, and it has been a feature of public health responses to health problems in many cities around the world. The healthy cities movement is underpinned by WHO 'Health for All' principles, which emphasize health promotion and illness prevention using strategies that are broad based, multisectoral and action oriented. Healthy cities programmes aim to stimulate social change for better health and are part of the 'New Public Health' agenda that operates both outside and within mainstream public health institutions (Ashton 1992). The principal objective is to change priorities by developing public health decision-making that is based on strategies of prevention and health promotion. In this sense it has a strong political dimension and promotes equity through community participation and the development of new healthy alliances between various statutory and non-statutory agencies (WHO 1992). The plurality of the work that comes under the healthy cities banner has prompted some to describe it as a 'post-modern' solution to the health and health care problems of late twentieth-century society (Kelly *et al.* 1993). Healthy cities projects have been undertaken, for example, in Glasgow, where the Healthy Cities Women's Health Working Group has worked towards developing a women's health centre and the Drumchapel healthy cities project has co-ordinated health action focusing on a deprived housing estate on the edge of Glasgow. Similar projects have been repeated in Camden (Camden healthy cities project), Belfast (Blackstaff community health project) and Liverpool (Croxteth Health Action Area) (WHO 1994). These projects share a concern for the relationship between poverty and health and for encouraging multi-agency responses to address this relationship (Laughlin and Black 1995).

Various initiatives in East London have tried to address some of the particular issues for health and health care in the area. For example, a GP has been employed with a special brief to care for the health of homeless people in the area, and a multidisciplinary homeless help team has been set up for East London. Health advocates for travellers and migrant populations have been appointed. Sixty GPs in sixteen practices in the area have participated in the healthy East Enders project which aims to collect better information about the populations on local doctors' practice lists and facilitate delivery of preventive care to the local population of working age. Healthy eating

Plate 9.2 The 'Health for All' mobile service in East London
Source: East London and City Health Authority Health Promotion Department

awards for school tuck shops are being established by the health promotion service and the community dietician to promote healthier diets for the young population of the area. School nurses and teachers have co-operated in special campaigns to increase levels of immunization against infectious diseases in the population of school age (especially measles and rubella).

There are also many examples in East London of non-statutory agencies taking innovative action. One example is the Tower Hamlets Health Strategy Group which has been undertaking and promoting community development work in areas ranging from housing infestation, men's health, continence promotion and housing resettlement advocacy (THHSG 1995). Coombes (1993) highlights the ways in which this group was able to act as a catalyst for communities to help themselves. Such initiatives are rooted in a 'bottom-up' philosophy that encourages health action at a community level and gives an impetus for interagency planning. They are, however, vulnerable in terms of funding, and may become frustrated by trying to accommodate the different agendas of the organizations involved. The variety of healthy cities initiatives reflects the diversity of conditions pertaining to health in our cities as well as the plurality of urban populations with whom the initiatives try to engage.

CONCLUSION

We began this chapter by contrasting nineteenth-century and twentieth-century concerns about health in the city. Divisions that existed in the nineteenth century are still evident today but in different forms. Inequalities of health persist and are clearly related to levels of poverty so that deprived urban areas often have quite distinctive health profiles and health care needs that differentiate them from wealthy urban areas and the nation generally. These inequalities are marked in inner city areas experiencing demographic and social changes related to decentralization which contributes to the formation of pockets of populations made up of the most vulnerable and marginalized sectors of society. Concomitant to this is the development of national government policies designed to rationalize health care provision in the cities, to develop a market-based NHS, to allocate resources on the basis of capitation funding and to assess the progress being made towards improving health. These policies create pressure for radical change in inner city areas that are also having to deal with acute demand for services from vulnerable populations. Alternative responses to the health problems of inner cities can be found in the work of the healthy cities movement which works towards addressing the structural determinants of ill health in a collaborative manner. Urban health divides are often striking because of the extreme inequalities that are evident over quite small geographical areas. It is increasingly being understood that the processes producing these problems are only partly situated in the localities and populations affected; they also depend on the wider social and economic structures that determine geographical differences in life chances. Thus urban health divides cannot be seen just as a local problem but must be viewed as a national and international issue

GUIDE TO FURTHER READING

Ashton, J. (1992) *Healthy Cities,* Milton Keynes: Open University Press is a collection of twenty-four articles documenting the progress made in 'healthy cities' initiatives in Europe, Australia and North America. It is a useful introduction to the history and practical experiences of the healthy cities movement. Davies, J. and Kelly, M. (1993) *Healthy Cities: Research and Practice*, London: Routledge is a good collection of papers focusing on research as part of healthy city work and documenting the practical and theoretical difficulties involved. Smaje, C. (1995) *Health, Race and Ethnicity: Making Sense of the Evidence*, London: Kings Fund Institute is an excellent review of research and policy in which a critical approach to the evidence is developed in chapters on health status, specific diseases, access to health services and health policy. Harrison, P. (1983) *Inside the Inner City: Life Under the Cutting Edge,* London: Penguin is a readable account of the experiences of the urban poor in the early 1980s that remains relevant to the 1990s. Curtis, S. and Taket, A. (1995) *Health and Societies: Changing Perspectives,* London: Arnold is not focused solely on issues of health but provides a recent review of the ways in which such issues are situated in wider social, economic and political contexts, both nationally and globally.

REFERENCES

Access to Health (1992) *Mental Health and Homelessness: A Critical Review*, London. Access to Health.

——(1993) *Weighting for Homelessness: Developing a Shared Approach Across the Four Thames Regions*, London: Access to Health.

Ashton, J. (1992) 'The origin of healthy cities' in J. Ashton *Healthy Cities*, Milton Keynes: Open University Press.

Balazs, J. (1993) 'Health care for homeless people' in J. Collins and K. Fisher *Homelessness, Health Care and Welfare Provision*, London: Routledge: 51–93.

Beck, U. (1992) *Risk Society, Towards a New Modernity*, London: Sage.

Benzeval, M., Judge, K. and Solomon, M. (1992) *Health Status of Londoners: A Comparative Perspective*, London: King's Fund Institute.

Bhatti, N., Law, M., Morris, R., Halliday, R. and Moore-Gillon, J. (1995) 'Increasing incidence of tuberculosis in England and Wales: a study of the likely causes', *British Medical Journal*, 310: 967–9.

Blaxter, M. (1990) *Health and Lifestyles*, London: Routledge.

Breheny, M. (1995) 'The compact city and transport energy consumption', *Transactions of the Institute of British Geographers*, 20(1): 81–101.

Britton, M. (1990) 'Geographic variation in mortality, 1973–1983' in M. Britton *Mortality and Geography: A Review in the Mid-1980s, England and Wales*, OPCS Series DS No. 9, London: HMSO.

Brownill, S. (1990) *Developing London's Docklands: Another Great Planning Disaster?*, London: Paul Chapman.

Bryce, C., Curtis, S. and Mohan, J. (1994) 'Coronary heart disease: trends in spatial inequalities and implications for health care planning in England', *Social Science and Medicine*, 38(5): 677–90.

Byrne, D. (1995) 'Deindustrialisation and dispossession: an examination of social division in the industrial city', *Sociology*, 29(1): 95–115.

Calnan, M., Cant. S. and Gabe, J. (1993) *Going Private: Why People Pay for Their Health Care*, Milton Keynes: Open University Press.

Collins, K.J. (1993) 'Cold and heat related illnesses in the indoor environment' in R. Burridge and D. Ormandy *Unhealthy Housing: Research, Remedies and Reform*, London: E. & F. N. Spon.

Coombes, Y.J. (1993) *A Geography of the New Public Health*, Ph.D. thesis, Queen Mary and Westfield College, University of London.

Craft, N. and Dillner, L. (1994) 'GPs urgently need help in inner cities says college', *British Medical Journal*, 309: 1186.

Davey Smith, G. and Morris, J. (1994) 'Increasing inequalities in the health of the nation', *British Medical Journal*, 309: 1453–4.

Dear, M. and Wolch, J. (1987) *Landscapes of Despair: From De-institutionalization to Homelessness*, Oxford: Polity Press.

Department of Environment (1994) 'Households found accommodation under the Homelessness Provisions of the 1985 Housing Act: England', *Information Bulletin, No. 517,* London: DoE.

Department of Health (1992) *The Health of the Nation: A Strategy for Health in England*, London: HMSO.

Donovan, J. (1986) 'Black people's health: a different approach' in T. Rathwell and D. Phillips *Health and Race and Ethnicity*, London: Croom Helm: 117–36.

Edwards, N. and Raftery, J. (1995) 'Bedtime stories', *Health Service Journal*, 2nd March: 26–8.

ELCHA (1994) *Health in the East End, Annual Public Health Report 1994/95*, London: The Directorate of Public Health, East London and The City Health Authority.

——(1995) *Health in the East End, Annual Public Health Report 1995/96*, London: The Directorate of Public Health, East London and The City Health Authority.

Engels, F. (1844[1969]) *The Condition of the Working Class in England*, London: Panther Books.

Eyles, J. (1987) *The Geography of the National Health*, London: Croom Helm.

Foreman, C. (1989) *Spitalfields: A Battle for Land*, London: Hilary Shipman.

Godlee, F. (1992) 'Air pollution I – from pea souper to photochemical smog' in A. Goodlee and A. Walker *Health and the Environment*, London: British Medical Journal Publications.

Harrison, P. (1983) *Inside the Inner City: Life under the Cutting Edge*, Harmondsworth: Penguin.

Harvey, D. (1989) *The Condition of Postmodernity*, Oxford: Basil Blackwell.

Herzberg, J.L. (1987) 'No fixed abode: a comparison of men and women admitted to an East London psychiatric hospital', *British Journal of Psychiatry*, 150: 621–7.

Hinton, T. (1994) *Battling Through the Barriers: A Study of Single Homelessness in Newham and Access to Primary Health Care*, London: Health Action for Homeless People.

Hobsbawm, E.J. (1962) *The Age of Revolution: Europe, 1789–1848*, London: Weidenfeld & Nicolson.

Hyndman, S. (1990) 'Housing, dampness and health among British Bengalis in East London', *Social Science and Medicine*, 30: 131–41.

Jacobson, B. (1993) 'Public health in inner London' in J. Smith *London after Tomlinson: Reorganising Big City Medicine*, London: British Medical Journal Publications.

Jarman, B. (1993) 'Is London overbedded?', *British Medical Journal*, 306: 979–82.

Jarman, B. and Bosanquet, N. (1992) 'Primary health care in London – changes since the Acheson report', *British Medical Journal*, 305: 1130–3.

Jowell, T. (1992) 'Community care in London: the prospect', *British Medical Journal*, 305: 1418–20.

Kelly, M.K., Davies, J.K. and Charlton, B.G. (1993) 'Healthy cities: a modern problem or a post-modern solution?' in J.K. Davies and M. Kelly *Healthy Cities, Research and Practice*, London: Routledge: 159–67.

Laughlin, S. and Black, D. (1995) *Poverty and Health: Ideas Analysis, Information, Action*, Birmingham: Public Health Trust.

Leavy, R. (1985) *Access to GPs*, Department of General Practice, University of Manchester.

Lowry, S. (1989) 'Housing and health: an introduction to housing and health', *British Medical Journal*, 299: 1261–2.

McCarron, P.G., Davey Smith, D. and Womersley, J. (1994) 'Deprivation and mortality in Glasgow: increasing differentials from 1980 to 1992', *British Medical Journal*, 309: 1481–2.

McCloone, P. and Boddy, P. (1994) 'Deprivation and mortality in Scotland, 1981 and 1991', *British Medical Journal*, 309: 1465–70.

Macintyre, S., Maciver, S. and Sooman, A. (1993) 'Area, class and health: should we be focusing on places or people?', *Journal of Social Policy*, 22(2): 213–34.

Mangtani, P., Jolley, D., Watson, J. and Rodrigues, L. (1995) 'Socio-economic deprivation and notification rates for tuberculosis in London during 1982–1991', *British Medical Journal*, 310: 963–6.

Marshall, M. (1989) 'Collected and neglected: are Oxford hostels for the homeless filling up with disabled psychiatric patients?', *British Medical Journal*, 29: 706–9.

Mathew, G.K. (1992) 'Health and the environment: the significance of chemicals and radiation' in P. Elliot, J. Cuzick, D. English and R. Stern (eds) *Geographical and Environmental Epidemiology: Methods for Small-area Studies*, Oxford: Oxford University Press: 22–23.

Mohan, J. (1995) *A National Health Service? The Restructuring of Health Care in Britain since 1979*, Basingstoke: Macmillan.

Moon, G. (1990) 'Conceptions of space and community in British health policy', *Social Science and Medicine*, 30(1): 165–71.

Morris, R. and Carstairs, V. (1991) 'Which deprivation? A comparison of selected deprivation indexes', *Journal of Public Health Medicine* 13(4): 318–26.

NHS Executive (1994) *HCHS Revenue Resource Allocation, Weighted Capitation Formula*, Quarry House, Leeds: NHS Executive.

Office of Population Censuses and Surveys (OPCS) (1989) *General Household Survey, 1987*, London: HMSO.

—— (1994) *Undercoverage in Great Britain*, London: HMSO.

Ogden, P. (1992) *London Docklands: The Challenge of Development*, Cambridge: Cambridge University Press.

Patterson, K. and Roderick, P.J. (1990) 'Obstetric outcome in homeless women', *British Medical Journal*, 301: 263–4.

Phillimore, P. and Moffat, S. (1994) 'Discounted knowledge: local experience, environmental pollution and health' in J. Popay and G. Williams (eds) *Researching the People's Health*, London: Routledge: 134–53.

Phillimore, P. and Reading, R. (1992) 'A rural advantage? Urban–rural health differences in Northern England', *Journal of Public Health Medicine*, 14(3): 290–9.

POST (1994) *Breathing in our Cities – Urban Air Pollution and Respiratory Health*, Parliamentary Office of Science and Technology.

Read, C. (1991) *Air Pollution and Child Health*, London: Greenpeace.

Rivett, G. (1986) *The Development of the London Hospital System, 1823–1982*, London: Kings Fund.

Roderick, P., Jones, I., Raleigh, V.S., McGeown, M. and Mallick, N. (1994) 'Population need for renal replacement therapy in Thames regions: ethnic dimension', *British Medical Journal*, 309: 1111–14.

Salter, B. (1995) 'The private sector and the NHS: redefining the welfare state', *Policy and Politics*, 23(1): 17–30.

Sampson, A. and Phillips, C. (1992) 'Multiple victimisation: racial attacks on an East London estate', *Crime Prevention Unit Series: Paper No. 36*, London: Police Research Group, Home Office Police Department.

Scheuer, M., Black, M., Victor, C., Benzeval, M., Gill, M. and Judge, K. (1991) *Homelessness and the Utilisation of Acute Hospital Services in London*, London: Kings Fund.

Schwartz, J. (1994) 'PM10, ozone and hospital admissions of the elderly in Minneapolis-St Paul, Minnesota', *Archives of Environmental Health*, 49(5): 366–74.

SCOPH (The Standing Conference on Public Health) (1994) *Housing, Homelessness and Health*, London: The Nuffield Provincial Hospitals Trust.

SEIPH, London Boroughs Association and Association of Local Authorities (1994) *Air Quality in London (2nd report)*, Tunbridge Wells: South East Institute of Public Health.

Smaje, C. (1995) *Health, 'Race' and Ethnicity: Making Sense of the Evidence*, London: King's Fund.

THHSG (1995) *Annual Report*, London: Tower Hamlets Health Strategy Group.

Tomlinson, Sir B. (1992) 'Inquiry into London's health service, medical education and research', *Report*, London: HMSO.

Townsend, P., Davidson, N. and Whitehead, M. (1988a) *Inequalities in Health*, Harmondsworth: Pelican.

Townsend, P., Phillimore, P. and Beattie, A. (1988b) *Health and Deprivation, Inequality and the North*, London: Croom Helm.

Tudor-Hart, J. (1971) 'The inverse care law', *The Lancet*, (i): 405–12.

Wallace, D. (1990) 'Roots of increased health inequality in New York', *Social Science and Medicine*, 31(11): 1219–27.

Wallace, R. (1990) 'Urban desertification, public health, and public order: planned shrinkage, violent death, substance abuse and AIDS in the Bronx', *Social Science and Medicine*, 31: 801–13.

Warren, A. (1993) *Can Mobile Inner City Lists be Accurate and at What Cost?*, London: Gill Street Health Centre.

Weller, M., Hollander, D., Tobiansky, R.I. and Ibrahimi, S. (1989) 'Psychosis and destitution at Christmas 1985–88', *The Lancet*, 23rd–30th December: 1509, 1511.

WHO (1992) *Twenty Steps for Developing a Healthy Cities Project*, Geneva: World Health Organisation.

——(1994) *Action for Health in Cities*, Geneva: World Health Organisation.

Widgery, D. (1991) *Some Lives! A GP's East End*, London: Sinclair-Stevenson.

Wilson, W.J. (1994) 'Citizenship and the inner-city ghetto poor' in B. van Steenbergen (ed.) *The Condition of Citizenship*, London: Sage: 49–65.

Wood, J. (1983) 'Are the problems of primary care in inner cities fact or fiction?' *British Medical Journal*, 286: 1109–12.

10

CRIME

Nicholas Fyfe

INTRODUCTION

In 1989, in an essay on 'The Challenge of Urban Crime', Smith observed that 'the streets of Britain are getting meaner' (Smith 1989: 271). Recent research suggests that this view needs critical re-examination. To be sure, the 1988 British Crime Survey revealed that the streets of England and Wales were certainly getting meaner, with a 22 per cent increase in household offences and 3 per cent increase in personal offences between 1981 and 1987. But in Scotland over the same period, household offences showed no change and personal offences fell by 12 per cent (Kinsey and Anderson 1992). Furthermore, the results of the most recent crime surveys in Britain indicate that the gap between Scotland and England and Wales is widening, with Scottish victimization rates now lower than those south of the border for all categories of crime. Rates for assault and robbery in Scotland in 1992, for example, were 380 and 31 incidents per 10,000 adults respectively, compared with figures of 586 and 45 in England and Wales (Anderson and Leitch 1994: 5). In addition to these differences, Scottish cities have not witnessed the outbreaks of violent public disorder experienced in other areas of urban Britain. The summer of 1991, for example, saw rioting in the suburbs of Oxford, Newcastle and Cardiff resulting in 500 arrests and damage estimated at £12 million (Campbell 1993: ix). Against this background, the importance of a geographical sensitivity to the uneven distribution of crime should be clear. However, the spatial differentiation in crime evident at a national level is even more marked at the intra-urban scale, which is the focus of this chapter. Within the city some streets are quite simply far meaner than others, as the first section of this chapter, on the risks of crime, will illustrate. This, in turn, affects the quality of urban life, with crime and the fear of crime having an important impact on people's lifestyles within the city. The second section of the chapter addresses these issues, revealing how social groups have different perceptions and experiences of cities' mean streets. Finally, attempts to make the streets less mean via a variety of crime prevention initiatives are considered.

Table 10.1 Crime reported to the police

	Merseyside crime survey (1984/5)(%)	*Islington crime survey (1985/6)(%)*	*Edinburgh crime survey (1988/9)(%)*
Vandalism	27	42	30
Theft from MV	44	ND	55
Theft of MV	100	54[a]	93
Housebreaking	73	78	81
Bicycle theft	53	47	69
Theft in dwelling	16	ND	40

ND = No data
Note (a) This relatively low figure may be due to residents of Islington having their vehicles stolen outside the Islington police area and therefore reporting the theft elsewhere
Sources: Kinsey 1985; Jones *et al.* 1986; Anderson *et al.* 1990

MAPPING THE CONTOURS OF URBAN CRIME

The problems of measuring crime are well known. Police statistics of recorded crime largely reflect police practice and public propensity to report offences, yielding a partial and distorted impression of the level and distribution of crime. In an attempt to overcome these problems there have been a proliferation of national and local crime surveys in Britain since the early 1980s measuring the incidence and impact of victimization among a random sample of the population. While not without their drawbacks, crime surveys do provide a more accurate alternative to police statistics in assessing the dimensions of the problem of urban crime, not least by indicating the volume of crime that goes unreported to, and therefore unrecorded by, the police. Table 10.1 indicates the extent of underreporting and how this varies between urban areas and, more significantly, with the type of offence. Given the greater degree of accuracy provided by crime surveys, this chapter uses both national and local surveys to map the contours of urban crime. In particular, extensive use is made of findings from the most recent British Crime Survey (Mayhew *et al.* 1993) and, to provide case study material at the city level, the Edinburgh Crime Survey (Anderson *et al.* 1990).

At a national level, the four sweeps of the British Crime Survey (in 1982, 1984, 1988 and 1992) in England and Wales have identified several important temporal and spatial trends. The British Crime Survey (BCS) shows that although police recorded crime nearly doubled between 1981 and 1991, crime measured by the BCS increased by only 50 per cent over the same period (Mayhew *et al.* 1993). This discrepancy most probably reflects increased reporting of offences to the police as a result of wider insurance cover and declining public tolerance of crime. Geographically, a valuable product of the successive sweeps of the BCS is confirmation that the probability of experiencing crime is neither evenly nor randomly distributed over space

Table 10.2 Crime risk and ACORN neighbourhood group

High risk	*Medium risk*	*Low risk*
Mixed inner metropolitan areas	Better off council estates	Agricultural areas
High status non-family areas	Older terraced housing	Better off retirement areas
Poorest council estates	Less well-off council estates	Modern family housing, higher incomes
		Affluent suburban housing
		Older housing of intermediate status

Source: Mayhew *et al.* 1993

(Smith 1989). Based on victimization patterns for all crimes, and using the ACORN classification scheme (which assigns respondents' homes to neighbourhood groups according to demographic, employment and housing characteristics of the immediate area), the BCS identifies high, medium and low crime risk areas, as shown in Table 10.2

This table clearly illustrates the significant intra-urban variations in crime risk that exist within Britain's cities. High risk neighbourhoods include the multi-occupancy older housing areas found in inner metropolitan areas and the poorest council estates situated in both inner city and peripheral locations, while low risk areas include the affluent suburbs of inter-war semi-detached housing and the more modern, private, family oriented housing estates. This locational pattern is significant not least because it illustrates unequivocally that 'crime is experienced more by people who are already suffering from other social problems' (Kinsey *et al.* 1986: 70).

The differential risks of crime within and between these neighbourhoods will, of course, vary with the type of offence, as Table 10.3 illustrates. In terms of burglary, the chances of being a victim in high risk neighbourhoods are more than double the national average and more than three times that of low risk neighbourhoods. While area of residence is not the only factor that influences the vulnerability of households to burglary (dwelling type, the presence of security devices and household composition are also important), the strongest correlates of burglary after all other factors have been taken into account were measures of the type of area in which people lived (Mayhew *et al.* 1993: 47). This conclusion is supported by evidence from local crime surveys. The Edinburgh Crime Survey (ECS) (Anderson *et al.* 1990) covering central Edinburgh and two contrasting outlying areas, Corstorphine (a mainly middle-class, owner-occupied suburb) and Craigmillar (a poor, peripheral council estate), found that the most important factor in determining the level of risk of housebreaking (the Scottish equivalent of burglary but in Scots law the term burglary has no meaning) was the area

Plate 10.1 A typical high-risk crime area, Blackhill in Glasgow
Source: Robert Perry

Plate 10.2 A typical low-risk crime area of affluent suburban housing, Corstorphine in Edinburgh
Source: Robert Perry

Table 10.3 Relative rates of crime and attempted crime by neighbourhood type

	Burglary	*Auto crime around home*	*Theft from person*
Low risk			
Agricultural areas	20	20	50
Modern family housing	60	70	70
Older housing – intermediate status	70	100	60
Affluent suburban housing	70	70	70
Better off retirement areas	70	80	70
Medium risk			
Older terraced housing	120	160	100
Better off council estates	90	110	120
Less well-off council estates	150	160	100
High risk			
Poorest council estates	280	240	200
Mixed inner metropolitan areas	180	190	340
High status non-family areas	220	150	250
Indexed national average	100	100	100

Source: data from British Crime Survey 1992; table from Loveday 1994

in which people live. Housebreaking and attempted housebreaking were most common in the peripheral council estate with nearly 1 in 5 of respondents reporting either a burglary or an attempted burglary in the previous twelve months, compared to roughly 1 in 11 in central Edinburgh and 1 in 33 in Corstorphine (Figure 10.1a). As Anderson *et al.* (1990: 22) conclude, people are more or less likely to be victims of burglary not because of *who* they are, but because of *where* they are. Nevertheless, when the spatial scale of analysis is narrowed even further, evidence of important intra-neighbourhood variations in crime become apparent. The 1988 BCS for Scotland included a Small Area Survey of two areas within a poor, peripheral council estate. Although both areas were in the same ACORN category, the reported housebreakings in one area were not only considerably above the national average for poor council estates but more than twice as high as in the other area of the same estate (Kinsey and Anderson 1992: 26). What this and similar evidence (see Jones and Short 1993) suggests is the importance of geographical scale in detecting the clustering of crime within neighbourhoods.

Like burglary, vehicle crime also displays strong intra-urban variations. High risk areas have theft rates for cars from around the home of twice the

Figure 10.1 Crime in Edinburgh
Source: Anderson *et al.* 1990

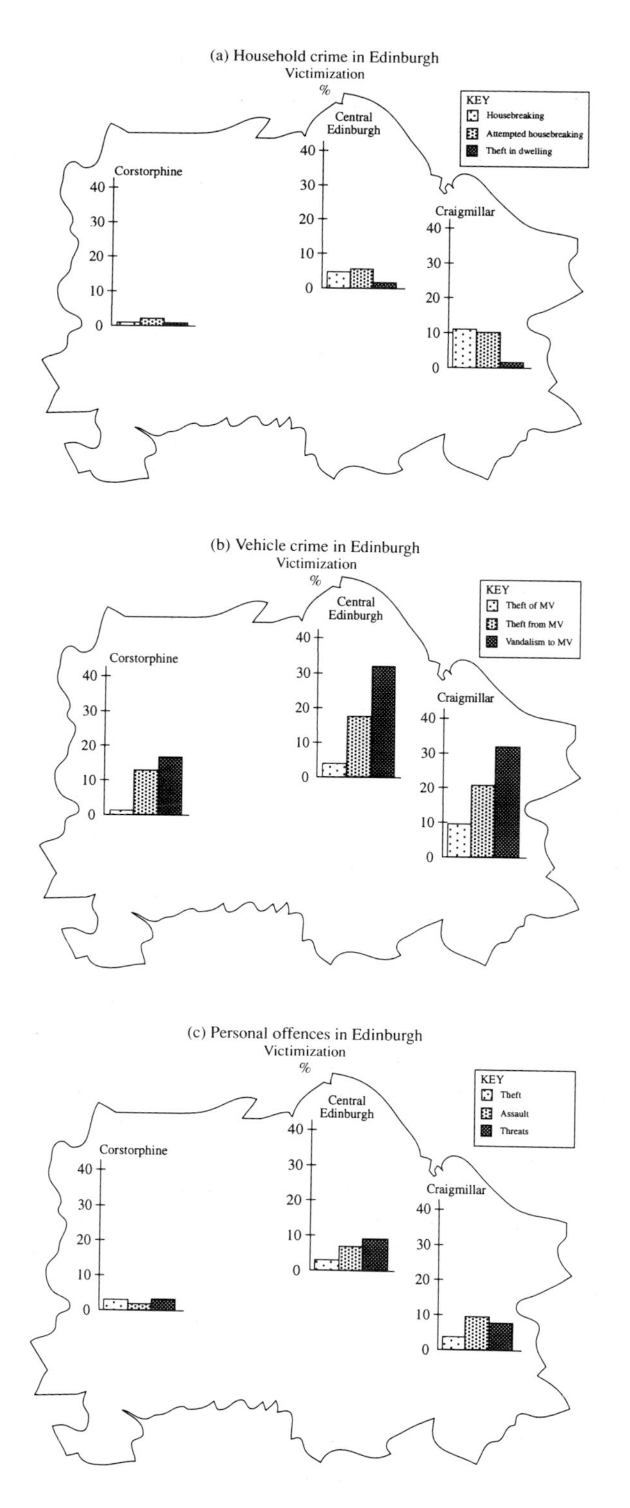
(a) Household crime in Edinburgh
Victimization
%
KEY
Housebreaking
Attempted housebreaking
Theft in dwelling
Corstorphine
Central Edinburgh
Craigmillar
40
30
20
10
0
(b) Vehicle crime in Edinburgh
Victimization
%
KEY
Theft of MV
Theft from MV
Vandalism to MV
Corstorphine
Central Edinburgh
Craigmillar
(c) Personal offences in Edinburgh
Victimization
%
KEY
Theft
Assault
Threats
Corstorphine
Central Edinburgh
Craigmillar

national average, and regression analysis reveals that the risk of having a car broken into or stolen is higher in inner city areas even when other factors, such as the availability of off-street parking and the value of the car, are taken into account (Mayhew *et al.* 1993: 70). Evidence from the ECS broadly reinforces this conclusion but also highlights the importance of retaining a sensitivity to local urban conditions. In Edinburgh there are relatively high levels of vehicle crime both in the inner city, which unlike other British cities is a relatively affluent area, and in Craigmillar, the peripheral council estate (Figure 10.1b). The nature and probable causes of vehicle crime in these two areas is, however, rather different. In central Edinburgh, there are high levels of car vandalism committed most probably by people passing through the area on their way to and from places of entertainment, while in Craigmillar high levels of car vandalism and car theft are likely to have been committed by local people (Anderson *et al.* 1990: 26).

The risks of violent crime, which includes wounding, common assault and mugging, are also unevenly distributed over space, but as Table 10.3 shows, the differences at the neighbourhood level are not as marked as for burglary or vehicle crime. Indeed, it appears that demographic and lifestyle characteristics are perhaps more strongly associated with the risks of violent crime than geography. According to the BCS, the older people are the less they are generally at risk, with 16 to 29 year olds having a risk of being the victim of violent crime almost three times the national average. Gender, too, is important, with men aged 16 to 29 more than one and half times more likely to be the victim of a violent offence as women in the same age group. In terms of lifestyle, those who go out four or more times a week or who regularly visit the pub were found to be twice as likely to be the victim of a violent offence compared with the national average (Mayhew *et al.* 1993: 81–6). Again the results of the ECS broadly confirm this pattern at a local level with personal crime being spread relatively evenly across the city (Figure 10.1c) but strongly differentiated by age and gender. Taking the example of the risk of assault, overall 7.4 per cent of the central Edinburgh sample had been victims in the previous twelve months but for men aged 16 to 30, the figure was 20.8 per cent compared to 8.1 per cent of those aged 31 to 45 and less than 1 per cent of those over 46.

Despite the undoubted importance of age and gender however, questions of location are still vitally important in understanding the distribution of violent crime. The high figure for men in Edinburgh city centre, with offences concentrated in the period after 9 pm, clearly suggests that the risks of victimization are closely linked to the lifestyle and culture of young people who use the city centre for entertainment. This prompts the ECS researchers to argue that 'these young people *are* at greater risk, but they are at risk because of where they are and when they are there, not simply because of their age and gender' (Anderson *et al.* 1990: 46). What this suggests, in turn, is that the ACORN classification scheme is too coarse to differentiate the micro-

Table 10.4 Risks of violence by age, sex and location – individuals victimized once or more, 1991

	Domestic (%)	*Home-based (%)*	*Street (%)*	*Pub/Club (%)*	*Work-based (%)*
Men					
16–29	0.4	1.2	3.8	2.1	1.5
30–59	0.2	0.6	1.1	0.3	0.6
60+	0.1	0.1	0.2	0.1	–
Women					
16–29	2.2	0.8	1.0	1.1	0.6
30–59	0.7	0.4	0.2	0.1	0.3
60+	0.1	0.1	0.1	–	–

Source: Mayhew *et al.* 1993

environments in which violent crime occurs (Davidson 1988). As Table 10.4 reveals, the risks of violence do vary over space but this is in terms of distinctions between the public space of the street and the private space of the home, and between spaces of recreation and spaces of work. Men aged 16 to 29, for example, are nearly three times more likely to be the victim of violence on the street as women in that age group but the latter are far more likely to be the victims of violence within the home than men.

This more detailed view of the uneven distribution of crime is particularly important because the street, the home, and places of recreation and work are key sites of social and economic activity, and therefore the impact of crime at this level has significant implications for the quality of urban life.

CRIME AND THE QUALITY OF URBAN LIFE

Just as the risks of victimization vary over space, so too does the degree to which people see crime as a problem. In central Edinburgh, for example, less than 10 per cent of those asked thought crime was a major problem (Anderson *et al.* 1990: 11) whereas in Islington (an inner city area of London) over 80 per cent thought it was (Crawford *et al.* 1990: 22–3). Superimposed on these inter-urban differences are equally marked intra-urban variations. The ECS contrasted affluent, suburban Corstorphine, where about 5 per cent of respondents see crime as a big problem, with Craigmillar, the relatively poor council estate where the figure is over 70 per cent. As these figures suggest, anxiety about crime is closely associated with the real risks of victimization in an area, something confirmed by the results of the most recent British Crime Survey. Table 10.5 shows clearly that those living in high crime risk areas, the inner city, and places where there are signs of 'disorder', such as vandalism and graffiti, are generally far more anxious about crime than those living elsewhere. Local crime surveys reinforce the findings from the BCS, as Figure 10.2, showing fear of crime in Edinburgh, illustrates. Those living in Craigmillar are not

Table 10.5 Area of residence and anxiety about crime

	Feeling unsafe out at night (%)	*Feeling unsafe at home (%)*	*Very worried about burglary (%)*	*Very worried about mugging (%)*
Risk of crime in ACORN area:				
High	41	15	26	23
Medium	39	13	24	23
Low	28	9	15	15
Area:				
Inner city	43	15	27	25
Non-inner city	30	10	17	16
Signs of disorder in area:				
More of a problem	41	15	25	23
Less of a problem	27	8	15	15
All BCS	32	11	19	18

Source: Mirrlees-Black and Maung 1994

only more often victims of crime than those living in some other parts of the city but they also tend to be more worried about crime. Thus 'it is not only crime but fear of crime that is a problem for poor communities, where it both compounds and is compounded by other problems associated with deprivation' (Kinsey 1993: 32).

In making sense of these anxieties about crime, however, it is important to recognize that spatial variability in the fear of crime cannot be accounted for only by variations in the risks of crime. There are a variety of non-crime, place-based factors that also generate fear (Smith 1987). Indeed, several researchers have highlighted the relationship between fear and the urban landscape, focusing on how features of the 'situational' environment, comprising both physical and social elements, appear to prompt feelings of unsafety (Bannister 1991, 1993). These features include the run-down and vandalized appearance of a neighbourhood, poor street lighting, youths loitering and the perceived existence of places where potential assailants can hide. The presence and absence of these physical and social cues creates a dynamic mental map which defines those places that are safe and unsafe during the day and after dark. Such mental maps will, of course, be socially differentiated, with gender, age and household income all significant variables in understanding reactions to crime. Indeed, as Table 10.6 shows, there are clear variations between social groups in terms of the impact of crime on the spatial practices of everyday life.

Figure 10.2 Fear of crime in Edinburgh
Source: Anderson *et al.* 1990

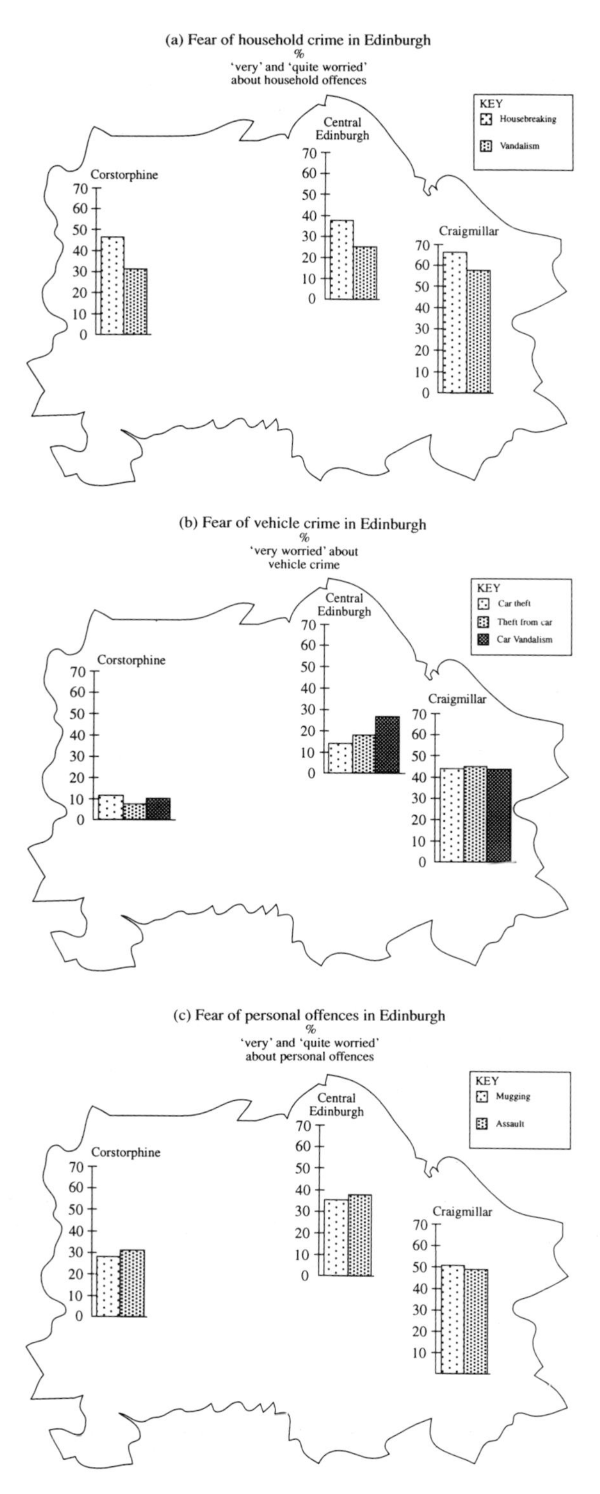
(a) Fear of household crime in Edinburgh
%
'very' and 'quite worried'
about household offences
KEY
Housebreaking
Vandalism
Corstorphine
Central Edinburgh
Craigmillar
70
60
50
40
30
20
10
0
(b) Fear of vehicle crime in Edinburgh
%
'very worried' about
vehicle crime
KEY
Car theft
Theft from car
Car Vandalism
Corstorphine
Central Edinburgh
Craigmillar
(c) Fear of personal offences in Edinburgh
%
'very' and 'quite worried'
about personal offences
KEY
Mugging
Assault
Corstorphine
Central Edinburgh
Craigmillar

Table 10.6 The impact of crime on daily life – precautions taken when out at night

	Ever avoid going out alone because of risks of crime (%)	*Take car rather than walk (%)*	*Go out with someone else (%)*	*Avoid certain streets (%)*
Total sample	26	23	18	26
Men	10	15	61	
5	21			
Women	37	30	29	30
16–24 years	19	18	25	35
25–44 years	23	28	19	29
45–64 years	25	22	16	22
65 and over	44	16	14	17
Under £10K	31	15	18	20
£10K to £19,999	19	30	17	29
£20K and over	16	43	15	42

Source: Kinsey and Anderson 1992

What stands out from this table is the extensive precautionary behaviour adopted by women in order to reduce their risks of victimization from crime and harassment. Nearly four times as many women as men avoid going out alone because of the risks of crime and twice as many take the car rather than walk as a precaution against crime at night. The restrictions on women's spatial behaviour resulting from the fear of crime are therefore severe. Research in inner city areas of London points to women living under a self-imposed curfew after dark as they avoid being alone in public space, avoid public transport and subways and, if they have to go out, often rely on men as escorts (Painter 1992). Moreover, it is women's fear of male sexual violence that particularly curtails their use of public space. As Valentine's (1989, 1992) study in Reading reveals, many of women's 'taken-for-granted' choices of routes and destinations are in fact the product of coping strategies that women adopt to stay safe, given their fears of sexual assault. The irony of these actions, however, is that statistically women are more likely to be assaulted at home by men they know than they are by strangers in public spaces. Explaining this 'spatial paradox', Pain's (1993, see too Pain 1991) interviews with women in Edinburgh indicate that it reflects partly the way in which women are brought up to be fearful of public space and partly women's experience of public places where they are routinely subject to sexual harassment but not assault. Furthermore, it is important to recognize that the large-scale withdrawal of women and, indeed, many men from public spaces after dark not only involves costs to these people themselves, in terms of foregoing social and leisure opportunities, but also has costs for the community more generally: the fewer people there are on the streets at night, the lower the possibility of intervention should a crime occur and the fewer the possible witnesses to incidents, thus increasing levels of anxiety among those who continue to use public spaces after dark.

Anxiety about crime, like crime itself, is thus not randomly distributed across society or over space but affects particular people living in particular places more than others. Moreover, the adaptation and reordering of the routines of everyday life prompted by the fear of crime indicate its negative impact on the quality of life of those living in the city. What, then, is being done to reduce crime and the fear of crime in Britain's cities?

COUNTERING CRIME IN THE CITY

The extensive impact of crime and the fear of crime on social life in British cities emphasizes the importance not only of developing effective measures to reduce crime but also of targeting such measures at those people and places that are most vulnerable. Yet despite more than fifteen years of a Conservative government keen to portray itself as the party of law and order, and which has almost doubled expenditure on the criminal justice system since it came to power in 1979 (Fyfe 1995a), crime and the fear of crime remain at stubbornly high levels in many parts of Britain's cities. Traditionally, of course, the police have been seen as the principal agents in the fight against crime. However, a combination of criminological research indicating the limited ability of police to control crime (Reiner 1992: 146–56, Fyfe 1995b: 762) and a political environment that encourages individual citizens to take responsibility in policy areas previously controlled by the state (Fyfe 1995a, Kearns 1992) means that a much wider range of agencies are now engaged in the fight against crime. Indeed, crime prevention initiatives in Britain's cities currently range from physical measures (such as property marking and improved street lighting) to surveillance schemes (like neighbourhood watch and closed circuit television), and from individualistic approaches (encouraging people to fit locks and bolts, for example) to community-based programmes (involving co-ordinating agencies in the public and private sectors to tackle crime). Against this background, this section focuses on three different initiatives: the role of neighbourhood watch (NW) and improved street lighting in residential neighbourhoods and the increasing use of closed circuit television (CCTV) in non-residential environments.

Countering crime in residential environments: neighbourhood watch and street lighting

The primary aim of NW is to deter would-be offenders by increasing the level of informal surveillance in residential areas by encouraging residents to watch out for and report suspicious incidents to the police (Bennett 1990). In addition, schemes have a secondary aim of increasing social contact and interaction, thereby enhancing social cohesion and reducing the fear of crime

(Rosenbaum 1988). Although introduced into Britain only as recently as 1982, the growth of NW schemes has been spectacular, with 81,000 schemes by 1990 and an estimated 120,000 by 1995 (Johnston 1992) Significantly, however, these schemes are highly unevenly distributed over space. There is substantial overrepresentation of schemes in areas of the city at relatively low risk of crime (such as areas of modern family housing and affluent suburban housing) and underrepresentation of NW in high risk areas of older housing, multiracial areas and on council estates (Husain 1988). Nevertheless, of perhaps greater concern is the fact that despite several evaluations of NW schemes, there is little conclusive evidence that such schemes are effective in reducing either crime or the fear of crime. Indeed, a Home Office study in two areas of London reached the unequivocal conclusion that there was 'no evidence that the [NW] programme resulted in a reduction in levels of victimisation' (Bennett 1992: 282). Moreover, in high crime areas the social interaction that did occur at NW meetings could, according to Rosenbaum (1988), lead to increases in the fear of crime, racial prejudice and other crime-related perceptions or feelings. Against this background, some argue that NW has been 'oversold as a stand-alone strategy in the war against crime' (Rosenbaum 1988: 363). In part this is explained by the way NW dovetails neatly with the government's neo-liberal political thinking about increasing the role of the 'active citizen' in civil society in general and crime prevention in particular (Fyfe 1995a, King 1989). From this perspective, NW is seen as a way of extending active citizenship and a sense of individual responsibility. But even this strategy may be flawed, given research showing that such characteristics are often preconditions for, rather than the consequences of, the establishment of NW schemes (Hope 1988).

A second and increasingly important initiative to reduce crime and the fear of crime in residential environments is improved street lighting. Research in the United States and, more recently, in Britain has pointed to the positive impacts of better street lighting. In her research in London, Painter (1988, 1989) concluded that improved lighting can dramatically, immediately and at low cost reduce specific crimes and fear of crime within a locality. While generally supporting these findings, other studies have been slightly more cautious. In Glasgow, Ditton *et al.* (1993) found that after relighting the area feelings of safety among those outside at night did improve. Fear of crime, however, did not diminish consistently, with respondents claiming that they thought other people were less fearful after relighting but that their own attitudes hadn't changed. Nevertheless, actual crime victimization did fall quite dramatically, particularly car crime which fell to a mere 4 per cent of the total recorded prior to relighting. This is an important finding because, although commonsense might suggest that better lighting must deter crime, it is also quite possible that it might have the opposite impact by illuminating targets and thus attracting offenders (Ditton *et al.* 1993: 101).

Countering crime in non-residential environments: CCTV

Crime is, of course, not confined to residential areas but impacts on the businesses, public facilities (such as schools, hospitals and car parks) and public spaces that make up the urban environment. Although little accurate information on the level of such crime is currently available, estimates of the costs of crime against business have been calculated. The British Retail Consortium's 'Retail Crime Costs' survey revealed that in 1993/4 crimes committed against retailers cost the industry £2.15 billion, a rise of 13 per cent from 1992/3 (Scottish Office 1995). Such crime not only impacts on employees, through physical assault and psychological stress, it also has a cost to the community by deterring inward investment. Against this background, the prevention of crime in non-residential environments of the city is of strategic as well as personal importance.

While some preventive initiatives have targeted crime in specific environments, such as sub-post offices, shops, schools and public transport (Nicholson 1994), it is the rapid expansion of CCTV in city centres that is currently the focus of much attention. A survey of London boroughs, metropolitan authorities and district councils in England found that thirty-nine had CCTV cameras in public spaces in 1993 compared with just two in 1987 (Bulos and Sarno 1994). By 1994, seventy-nine towns and cities had CCTV while by March 1995 the figure was over ninety. CCTV surveillance cameras are, of course, not new. They have been operating in privately owned but publicly accessible places such as shopping malls, banks and football stadia for several years. What is different about the current expansion of CCTV is its extension to the publicly owned spaces of the streets and squares of town and city centres. The reasons for this expansion are partly technological, in that cameras are now available that can operate effectively in the much larger-scale environment of a city centre. But the growth of city centre CCTV also has political and economic rationale. The government has increasingly urged local communities to take responsibility for crime prevention and made it clear that resources for conventional policing are limited, while crime and the fear of crime are increasingly recognized as important reasons for the declining attractiveness of city centres as a location for retailers, service providers and consumers. Against this background, local authorities and the local business community have sought ways of combating crime in order to revive the city centre. The dramatic reductions in crime claimed by pioneering city centre CCTV schemes in places such as Airdrie and Birmingham (Table 10.7) have encouraged the rapid diffusion of schemes to other towns and cities across Britain.

However, while public support for city centre CCTV is undoubtedly very high, doubts remain about its likely effectiveness. There is the possibility, for example, that CCTV will simply displace crime to surrounding areas not in view of the cameras rather than deter crime. There is also concern that it will promote 'bystander indifference', with people no longer caring what

Table 10.7 'Before' and 'after' crime statistics for areas covered by CCTV in Birmingham and Airdrie

Birmingham

	Before CCTV (3 months to 3/91)	*After CCTV (3 months to 9/91)*
Woundings	46	27
Robberies	79	55
Thefts from a person	89	63
Indecency	8	3
Damage	62	80

Source: Birmingham City Centre Development Group 1992

Airdrie

	Before CCTV (12 months to 8/92)	*After CCTV (12 months to 8/93)*
Car break-ins	480	20
Theft of cars	185	13
Serious assaults	39	22
Vandalism	207	36
Break-ins to commercial premises	263	15

Source: Wills 1993

happens around them because they presume that the cameras will 'look after' those on the street (Groombridge and Murji 1994). More generally, the hope that CCTV will revive city centres may also be misplaced. Crime and the fear of crime are not the only causes of city centre decline. The proliferation of out-of-town shopping centres and leisure facilities and the trends towards the domesticization and privatization of social life have all conspired to reduce the significance of urban centres. Against this background it is unlikely that CCTV on its own can reverse the decline of urban public social life (Fyfe and Bannister 1995).

CONCLUSION

In her richly detailed ethnographic exploration of Britain's 'dangerous places', Campbell (1993) examines the events bound up with the violent disorders on the peripheral estates in Oxford, Cardiff and Tyneside during the summer of 1991. One of Campbell's main conclusions, that 'crime was *spatialised* in the nineties' (Campbell 1993: 317, emphasis in the original), clearly resonates with the material presented in this chapter. Cities are unequivocally divided in terms of crime. In part this can be understood in terms of conventional geographical narratives about the spatial differentiation of the city with respect to the risks of crime and the delimitation of high, medium and low risk neighbourhoods. But the divisions also need to be understood in terms of the social constructions of space. The distinctions between the different risks associated with the public sphere and the private domain or the various ways in which social groups read the urban landscape for signs that generate anxiety about crime, are indicative of how the spatial divisions of crime in the city are socially as well as physically based. A sensitivity to these different planes of division is important not least because effective crime prevention and fear-reduction measures depend for their success on a local knowledge both of the spaces in which the risks of crime are greatest and of the places that generate the most anxiety. And, as is clear from the evidence presented in this chapter, these areas are predominantly those where the poorest people live. While the *causes* of crime in such areas continue to be the subject of an increasingly polarized and politicized debate between the theses of social deprivation and moral breakdown, the *consequences* of crime for those living in these areas in terms of anxiety and the reordering of the routines of everyday life are, as this chapter has illustrated, well documented. But such problems must not be seen in isolation. 'The poor suffer from poverty above all', Kinsey *et al.* observe, 'but also from bad housing, unemployment, pollution, nuisance *and* crime' (Kinsey *et al.* 1986: 70). This spatial coalescence of crime with other social and economic ills means that people's experiences of the problems of crime compound and are compounded by other problems that may have nothing to do with crime itself, 'problems which may range from an inadequate bus service or poor street lighting to the social isolation of the elderly, or the simple feeling of yet another problem in an already difficult world' (Kinsey and Anderson 1992: 62).

GUIDE TO FURTHER READING

Mayhew, P., Maung, N.A. and Mirrless-Black, C. (1993) *The 1992 British Crime Survey*, London: Home Office Research and Planning Unit. This is the most recent of the British Crime Surveys and provides detailed information on the temporal and spatial patterns of victimization in England and Wales. A separate Scottish Crime Survey was carried out in 1993 and the first results from this are contained in Anderson, S. and Leitch, S. (1994) *The Scottish Crime Survey 1993: First Results*, Edinburgh: Scottish Office Central Research Unit. At a local level, several crime surveys have now been carried out in British cities; see in particular Anderson *et al.* (1990) *The Edinburgh Crime Survey: First Report*, Edinburgh: Scottish Office Central Research Unit, and Crawford *et al.* (1990) *The Second Islington Crime Survey*, Centre for Criminology and Police Studies, Middlesex University. While such crime surveys can provide a detailed statistical picture of the patterns of crime, they are less effective in terms of exploring the meaning and first-hand experiences of crime for those living in Britain's cities. However, Beatrix Campbell's (1993) *Goliath: Britain's Dangerous Places*, London: Methuen is an exceptionally vivid ethnographic account of the impact of crime and disorder on the lives of people living and working on local authority estates in Oxford, Cardiff and Tyneside.

REFERENCES

Anderson, S. and Leitch, S. (1994) *The Scottish Crime Survey 1993: First Results*, Edinburgh: Scottish Office Central Research Unit Crime and Criminal Justice Research Findings No. 1.

Anderson, S., Smith, C.G., Kinsey, R. and Wood, J. (1990) *The Edinburgh Crime Survey: First Report*, Edinburgh: Scottish Office Central Research Unit.

Bannister, J. (1991) *The Impact of Environmental Design upon the Incidence and Type of Crime: A Literature Review*, Edinburgh: Scottish Office Central Research Unit.

—— (1993) 'Locating fear: environment and ontological security' in H. Jones (ed.) *Crime and the Urban Environment: The Scottish Experience*, Aldershot: Avebury: 85–98.

Bennett, T. (1990) *Evaluating Neighbourhood Watch*, Aldershot: Gower.

—— (1992) 'Themes and variations in neighbourhood watch' in D. Evans, N.R. Fyfe and D.T. Herbert (eds) *Crime, Policing and Place: Essays in Environmental Criminology*, London: Routledge: 272–85.

Birmingham City Centre Development Group (1992) *City Watch*, Birmingham: Birmingham City Centre Development Group.

Bulos, M. and Sarno, C. (1994) *Closed Circuit Television and Local Authority Initiatives: The First National Survey*, School of Land Management and Urban Policy, South Bank University.

Campbell, B. (1993) *Goliath: Britain's Dangerous Places*, London: Methuen.

Crawford, A., Jones, T., Woodhouse, T. and Young, J. (1990) *The Second Islington Crime Survey*, Middlesex University Centre for Criminology.

Davidson, N. (1988) 'Micro-environments of violence' in D. Evans and D.T. Herbert (eds) *The Geography of Crime*, London: Routledge: 59–85.

Ditton, J., Nair, G. and Phillips, S. (1993) 'Crime in the dark: a case study of the relationship between street lighting and crime' in H. Jones (ed.) *Crime and the Urban Environment: The Scottish Experience*, Aldershot: Avebury: 99–112.

Fyfe, N.R. (1995a) 'Law and order policy and the spaces of citizenship in contemporary Britain', *Political Geography*, 14(2): 177–89.

——(1995b) 'Policing the city', *Urban Studies*, 32(4–5): 759–78.

Fyfe, N.R. and Bannister, J. (1995) 'City watching: closed circuit television in public spaces', *Area* 27: 37–46.

Groombridge, N. and Murji, K. (1994) 'As easy as AB and CCTV', *Policing*, 10(4): 283–90.

Hope, T. (1988) 'Support for neighbourhood watch' in T. Hope and M. Shaw *Communities and Crime Reduction*, London: HMSO: 146–61.

Husain, S. (1988) *Neighbourhood Watch in England and Wales: A Locational Analysis*, London: Home Office Crime Prevention Unit Paper No. 12.

Johnston, L. (1992) *The Rebirth of Private Policing*, London: Routledge.

Jones, H. and Short, J. (1993) 'The "pocketing" of crime within the city: evidence from Dundee public housing estates' in H. Jones (ed.) *Crime and the Urban Environment: The Scottish Experience*, Aldershot: Avebury: 85–98.

Jones, T., Maclean, B.D. and Young, J. (1986) *The Islington Crime Survey*, Centre for Criminology and Police Studies, Middlesex Polytechnic.

Kearns, A.J. (1992) 'Active citizenship and urban governance', *Transactions of the Institute of British Geographers*, NS, 17(1): 20–34.

King, M. (1989) 'Social crime prevention à la Thatcher', *Howard Journal*, 28: 291–312.

Kinsey, R. (1985) *The Merseyside Crime Survey: Final Report*, Liverpool: Merseyside Metropolitan Council.

——(1993) 'Crime, deprivation and the Scottish urban environment' in H. Jones (ed.) *Crime and the Urban Environment: The Scottish Experience*, Aldershot: Avebury: 15–44.

Kinsey, R. and Anderson, A. (1992) *Crime and the Quality of Life: Public Perceptions and Experiences of Crime in Scotland – Findings from the 1988 British Crime Survey*, Edinburgh: Scottish Office Central Research Unit.

Kinsey, R., Lea, J. and Young, J. (1986) *Losing the Fight Against Crime*, Oxford: Basil Blackwell.

Loveday, B. (1994) 'Government strategies for community crime prevention programmes in England and Wales: a study in failure', *International Journal of the Sociology of Law*, 22: 181–202.

Mayhew, P., Maung, N.A. and Mirrlees-Black, C. (1993) *The 1992 British Crime Survey*, London: Home Office Research and Planning Unit, Research Study No. 132.

Mirrlees-Black, C. and Maung, N.A. (1994) *Fear of Crime: Findings from the 1992 British Crime Survey*, London: Home Office Research and Statistics Department, Research Findings No. 9.

Nicholson, L. (1994) *What Works in Situational Crime Prevention? A Literature Review*, Edinburgh: Scottish Office Central Research Unit.

Pain, R. (1991) 'Space, sexual violence and social control: integrating geographical and feminist analyses of women's fear of crime', *Progress in Human Geography*, 15(4): 415–31.

——(1993) 'Women's fear of sexual violence: explaining the spatial paradox' in H. Jones (ed.) *Crime and the Urban Environment: The Scottish Experience*, Aldershot: Avebury: 55–68.

Painter, K. (1988) *Lighting and Crime Prevention: The Edmonton Project*, Centre for Criminology and Police Studies, Middlesex Polytechnic.

——(1989) *Lighting and Crime Prevention for Community Safety: The Tower Hamlets Study First Report*, Centre for Criminology and Police Studies, Middlesex Polytechnic.

——(1992) 'Different worlds: the spatial, temporal and social dimensions of female victimization' in D. Evans, N.R. Fyfe and D.T. Herbert (eds) *Crime, Policing and Place: Essays in Environmental Criminology*, London: Routledge: 164–95.

Reiner, R. (1992) 'Police research in the United Kingdom: a critical review' in M. Tonry and N. Morris (eds) *Modern Policing*, Chicago: Chicago University Press: 435–508.

Rosenbaum, D. (1988) 'The theory and research behind neighbourhood watch: is it a sound fear and crime prevention strategy?', *Crime and Delinquency*, 33(1): 103–34.

Scottish Office (1995) 'Mr Lang says crack crime to improve profits', *Information Directorate*, Press Release 23rd June 1995.

Smith, S.J. (1987) 'Fear of crime: beyond a geography of deviance', *Progress in Human Geography*, 11(1): 1–23.

——(1989) 'The challenge of urban crime' in D.T. Herbert and D.M. Smith (eds) *Social Problems and the City: New Perspectives*, Oxford: Oxford University Press: 271–88.

Valentine, G. (1989) 'The geography of women's fear', *Area*, 21(4): 385–90.

——(1992) 'Images of danger: women's sources of information about the spatial distribution of male violence', *Area*, 24(1): 22–9.

Wills, J. (1993) 'Candid cameras', *Local Government Chronicle*, 17th September 1993: 13.

11

ETHNICITY

Graham Moon and Rob Atkinson

INTRODUCTION

In early July 1995, the Commissioner of the (London) Metropolitan Police announced a 'crackdown' on violent street crimes such as mugging. Black youths were singled out as targets for increased police attention as records suggested that the majority of muggers were black. This pronouncement drew criticism from groups concerned that black youth were being stereotyped as a lawless, disaffected stratum of urban society. The months preceding the Commissioner's statement saw other events that indicated the continuing but complex nature of the ethnic divide in Britain's cities. Communities of predominantly Asian ethnic groups which had, again stereotypically, been seen as relatively quiescent in previous periods of urban unrest, experienced riots. Popular and quality newspapers published lists of places where postal workers feared to deliver letters; the areas were uniformly characterized by large black communities. From a more liberal perspective, there was continued acknowledgement that the black and Asian communities in British cities experienced worse unemployment and poorer living conditions than their white fellow citizens.

In this chapter we explore the nature of the divides that partition British cities on the basis of ethnic status. The chapter will be divided into four sections with attention focusing first on a description of the geography of contemporary urban ethnic divides. Second, we will consider the ways in which these divides are perpetuated and the problems that consequently arise. A third section will review the impact of urban policy and the actions of local government on minority ethnic groups; both national and local initiatives will be considered. Finally, consideration will be given to the likely futures in Britain's minority ethnic urban populations. Throughout the chapter we will refer wherever possible to specific population groups. On occasion, however, sources will constrain us to terms such as Afro-Caribbean, Asian and black. We acknowledge that these portmanteau terms may capture a commonality of disadvantage but hide very different ethnic positions.

THE GEOGRAPHY OF URBAN ETHNIC DIVIDES

Though the minority ethnic population of British cities has been present for many hundreds of years, it grew considerably in the years following the Second World War. Its growth and development was such that a question on ethnic group membership was included in the UK census for the first time in 1991 (Sillitoe and White 1992, Robinson 1994). The purpose of this question was to garner information about the housing, employment, economic and social conditions of minority ethnic groups. It sought to counter the mounting deficiencies of an existing question on country of birth that failed to distinguish members of minority ethnic groups who had been born in Britain. The inclusion of the 'ethnic' question in the 1991 census was not unopposed; its eventual inclusion was the result of a period of extensive field-testing and consultation (Teague 1993). Its formulation, though not without problems, allowed a rather more comprehensive view of ethnic divides in British cities than had hitherto been possible.

The results of the ethnic question indicated that 5.5 per cent of the British population were from minority ethnic groups, with the largest groups being Indian (840,000), black Caribbean (500,000) and Pakistani (477,000) in origin. Overall just over 3 million people described themselves as non-white. Overwhelmingly the minority ethnic population lived in urban areas. For most London boroughs and metropolitan areas, the proportion of the population drawn from minority ethnic groups was above 5 per cent, and in the boroughs of Newham and Brent it was over 40 per cent. London boroughs provided the top ten local authorities for populations of both black Caribbean and Chinese origin. Populations of Indian, Pakistani and Bangladeshi origin were more dispersed. High levels of people of Indian origin featured in Leicester, Slough, Wolverhampton and Sandwell, while the percentage of people of Bangladeshi origin was high in Luton, Oldham and Birmingham. The Pakistani group returned high levels in parts of West Yorkshire, Lancashire, Birmingham and Greater Manchester.

Owen (1994) characterizes the distribution of the minority ethnic population as one that is concentrated in two regions. First, he distinguishes a London–Midlands axis bounded by London, Leicester and Birmingham. Second, he cites a trans-Pennine region which includes West Yorkshire, Lancashire and Greater Manchester. Figure 11.1 illustrates the intra-regional distribution of people of Asian origin in the West Yorkshire segment of the second region. As noted by Rees *et al.* (1993), there is a clear inner city concentration, notably in Leeds and markedly in Bradford.

Owen (1994) notes that, while few white people live in a local government ward with no people from a minority ethnic group, most whites live in local government wards where the proportion of people from minority ethnic groups is below the national average. In contrast, the British minority ethnic population tends to live in wards where the minority ethnic population

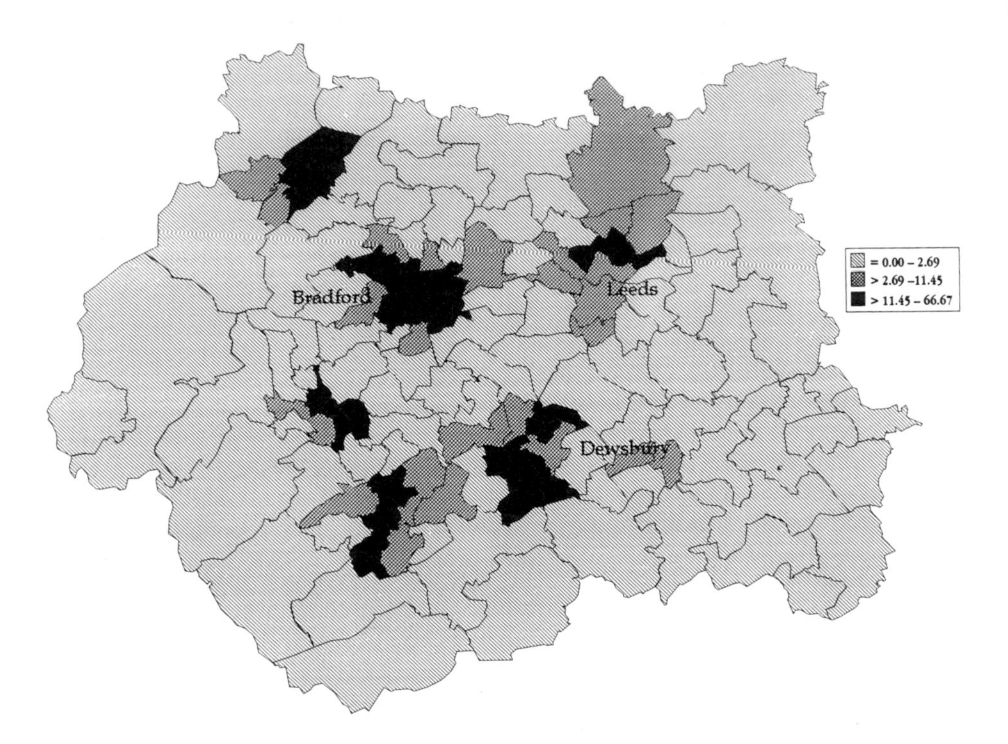

Figure 11.1 West Yorkshire – people from Asian ethnic groups (Indian, Pakistani, Bangladeshi) as a percentage of total population
Source: 1991 Census of Population

is high. This spatial segregation undoubtedly reflects choice and constraint processes in the housing and employment markets which we will return to below. Its consequence is that Britain's minority ethnic population tends to live in local government wards with very specific characteristics. Typically these are more urbanized environments with higher levels of deprivation. There are, however, some marked differences between minority ethnic groups. Indian and, particularly, Bangladeshi groups tend to live in areas of low owner-occupation, high unemployment and overcrowding. Black-Caribbean people are found in similar areas but are more likely to inhabit deprived inner city areas distinguished by older age structures and higher population densities. The group of Chinese origin provides a contrast; it tends to inhabit urban areas but is distributed across a wider range of area types, including relatively high status wards.

It would seem, therefore, that, with the marginal exception of the Chinese population, the spatial distribution of the minority ethnic population reflects both geographical and social segregation. Owen (1994) uses Lieberson's P*

Table 11.1 Isolation of minority ethnic groups

Black	*Indian*	*Pakistani*	*Bangladeshi*
Leeds	Kirklees	Oldham	Oldham
Manchester	Blackburn	Calderdale	Burnley
Trafford	Leicester	Bradford	Rochdale
Liverpool	Bolton	Peterborough	Tameside
Bristol	Ealing	Rochdale	Bradford

Source: adapted from Owen 1994

index (Lieberson 1980) to quantify the extent of segregation at the sub-district level. Table 11.1 summarizes his results and indicates the five local government districts in which the particular minority ethnic group is most isolated from other groups. What is clear is that, despite the general concentration of the minority ethnic population in London, isolation is greater in the more northern of the two areas of minority ethnic residence identified by Owen (1994). The diversity of the minority ethnic population in London generally prevents any one group being particularly segregated. More detailed figures reveal that segregation is greatest for the groups of Asian origin. In Oldham almost 50 per cent of the population of Bangladeshi origin are likely to have a Bangladeshi for a neighbour, while just over 40 per cent of Oldham people of Pakistani origin are likely to have a Pakistani for a neighbour. The highest figure for the black population is 15.5 per cent.

DIVIDING THE CITY: PROCESSES AND PROBLEMS

A simple two-factor model would see the continued ethnic partitioning of British cities in terms of housing and employment opportunities – or, rather the lack, possibly even denial, of such opportunities. The segregated residential patterns described in the previous section reflect, as we have suggested, the operation of twin processes of choice and constraint. In some cases the minority ethnic population may choose to live segregated from other groups; in other cases processes of prejudice and discrimination may be at work.

Two indicator themes may be used to suggest the discrepancies between white and non-white groups: unemployment and overcrowding. Figures 11.2 and 11.3 again use the example of West Yorkshire and reveal an anticipated pattern of spatial concentration. In University ward, Bradford, 65 per cent of the unemployed population are of Asian origin. Similarly, in Thornhill ward, Dewsbury, over 90 per cent of households experiencing overcrowding are of Asian origin. Further analysis of unemployment would reveal that, in West Yorkshire, the local government ward with the highest male unemployment rate among the white population (Little Horton, Bradford) returned a figure of just over 18 per cent in the 1991 census; the figure for the same

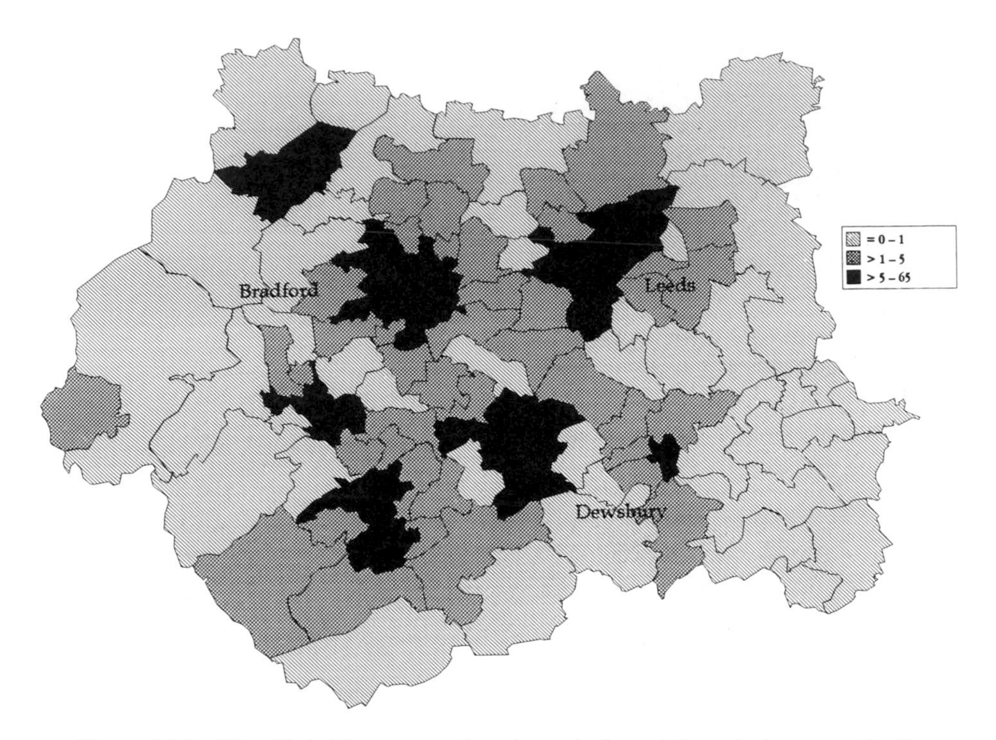

Figure 11.2 West Yorkshire – unemployed people from Asian ethnic groups (Indian, Pakistani, Bangladeshi) as a percentage of total unemployed population
Source: 1991 Census of Population

ward for men of Asian origin was just over 20 per cent. These figures, furthermore, are almost certainly undercounts as they will have been affected by the general census undercount of inner city young men, particularly those from minority ethnic groups. Some evidence for this can be deduced from more recent evidence: in London in January 1995, 62 per cent of young black men between the ages of 16 and 24 were unemployed (*Hansard* Written Answer 17.1.95. col. 461).

We will consider process elements of the housing and the employment aspects of the ethnically divided city in turn. To take housing first, there has long been an acknowledgement that housing markets have not worked in favour of black and minority ethnic groups (Daniel 1968, Rex and Moore 1969). The support that successive governments have given to owner-occupation has assisted this process of discrimination (Karn *et al.* 1985). Black and minority ethnic groups have a history of problems with the mainstream 'high-street' forms of housing finance. They have tended to be viewed as unsafe risks in terms of mortgage loans, and 'redlining' (the demarcation in red of

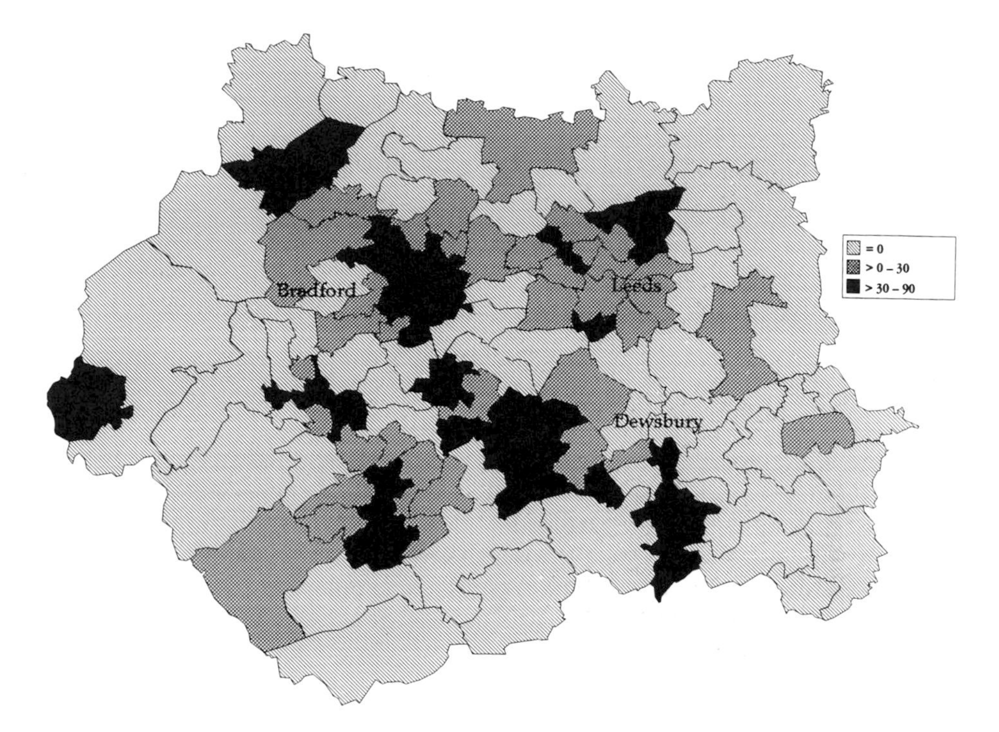

Figure 11.3 West Yorkshire – overcrowded households (>1.5 persons per room) from Asian ethnic groups (Indian, Pakistani, Bangladeshi) as a percentage of total overcrowded households
Source: 1991 Census of Population

areas on a map within which mortgage lenders are not prepared to advance loans) has forced them to utilize less conventional and higher-cost forms of financing. Second, the housing market itself has operated in a racialized manner (Ginsburg 1992); black and minority ethnic potential purchasers have often been routed into particular areas by housing 'gatekeepers' (e.g. estate agents) thus encouraging the creation of specific housing sub-markets as well as segregation. The dwellings thus purchased were often substandard and had higher costs of upkeep and repair. This, in combination with the higher costs of purchase, has meant that owner-occupation has frequently been more financially onerous for members of minority ethnic groups than for comparable white owners. Finally, and partially as a consequence of the previous point, those black and minority ethnic people who managed to purchase a house, frequently could do so only in areas where the benefits of the house price boom of the 1980s were felt less keenly. They were then less able to 'trade-up'. Only recently has any attention been given to empowering black and minority ethnic people in the housing arena through the development of self-help and voluntary housing initiatives (Harrison 1994).

Local authority housing has provided black and minority groups with, at best, a flawed alternative; at worst it has been denied and allocated in an overtly discriminatory fashion that has seen ethnic minorities living in the worst dwellings in the least desirable areas (Solomos and Singh 1990). Access to public housing has been difficult, though it has varied between different ethnic groups, in part due to intra-group attitudes. As a consequence of these pressures in the owner-occupier and public renting sectors, the black and minority ethnic population has tended to utilize disproportionately the private rented sector. In so doing it has tended to rely on its own internal networks with landlords from a particular group servicing the housing needs of that particular group. Taken together these processes have produced what Smith (1993) has termed a 'racialisation of residential space'.

Black and minority ethnic levels of unemployment have become and remain notoriously problematic, a barometer of a society's failure to ensure equal opportunities in the workplace for all its citizens and a probable prime cause of urban unrest (Ohri and Faruqui 1988, Solomos 1988). At root, the history of black and minority ethnic levels of unemployment is a history of discrimination (Brown 1992). The operation of racialized dual or segmented labour markets, via forms of direct and indirect discrimination (Wrench and Solomos 1993), has seen black and Asian workers largely restricted to secondary labour markets where jobs are low paid, unskilled and subject to severe unemployment during periods of recession (Victor Hausner and Associates ([VHA] 1993: 27–43). This process is again amply illustrated by the case of West Yorkshire. Jackson (1992) clearly demonstrates how the operation of the Bradford textile industry became, in the post-1945 period, predicated on the availability of 'immigrant' labour first from central Europe and then from Pakistan.

Such racialized labour markets are clearly vulnerable not only to global shifts in production technology but also to local processes of discrimination and prejudice. The continuing concentration of ethnic minority workers in declining manufacturing industries located in inner urban areas means they have not had access to new manufacturing industries located outside cities; the urban–rural shift has largely by-passed ethnic minorities. As a consequence, black and minority ethnic people are among the first to feel the impact of the recession and among the last to benefit from the occasional mini-booms. According to Newnham (1986) this effect, together with overt discrimination, means that black and minority ethnic unemployment is even higher than would be expected given the industrial sectors and localities involved. Minority ethnic entrepreneurship seldom provides much of a counter to this situation as it too depends on low cost labour and intensification of the labour process (Leman 1992). In those instances where government assistance has created jobs, they have tended to be transient and dependent upon government aid – an attempt to control and manage a problem rather than provide a solution.

THE URBAN POLICY RESPONSE

Given the close relationship between urban areas and the location of minority ethnic populations, race has played a key role in the evolution of urban policy. However, this process has been entwined with what Solomos (1993) has termed the racialization of British politics in which, since the 1950s, the presence of ethnic minorities, particularly people of Afro-Caribbean background, has been defined as a 'problem'. The initial response was to control and limit the numbers of non-white immigrants entering the UK via the *Commonwealth Immigrants Act, 1962* and later the *Commonwealth Immigrants Act, 1968*. These 'control' measures were intended to be counterbalanced by initiatives such as the *Race Relations Act, 1965* designed to outlaw certain forms of discrimination (it excluded housing and employment discrimination) and improve the lot of immigrants already in the UK. At the same time Section 11 Grants were launched by the *Local Government Act, 1966*. These grants were largely intended to aid people who did not have English as a first language to gain the language skills necessary to 'assimilate' into white British society.

Labour governments of the 1960s were unwilling to be seen overtly to favour minority ethnic populations in urban areas for fear of alienating key sections of its white electorate. As a result, when the Urban Programme (UP) was launched in 1968, the then Home Secretary, James Callaghan, announced it in the following terms:

> The purpose of this programme is to supplement the Government's other social and legislative measures to ensure as far as we can that all our citizens have an equal opportunity in life.
>
> (*Hansard*, 22.7.1968, no. 768, col. 41)

By focusing on the most deprived urban areas, minority ethnic populations should inevitably have benefited from any additional resources. In practice, however, the UP lacked a coherent analysis of urban problems (Atkinson and Moon 1994: 44–6). As a result, there was a tendency for ethnic minorities to occupy a marginal position, an outcome that was to be equally evident in relation to later urban initiatives, as we will argue below.

Nevertheless, the plight of ethnic minorities in urban areas did continue to occupy the minds of government ministers, if only because of fears that urban disorders similar to those experienced by US cities in the 1960s might be repeated in the UK. In 1972 Peter Walker, the then Conservative Secretary of State for the Environment, launched the Inner Area Studies, one objective of which was to gather information on the plight of Britain's ethnic minorities in urban areas. These studies influenced a major review of urban policy carried out under the 1974–9 Labour government which was to produce the first coherent government analysis of urban problems – the *Policy for the Inner Cities* (Cmnd 6845) – and lay the foundation for a new urban initiative.

The 1977 White Paper identified racial discrimination as a crucial problem with regard to the decline of Britain's inner cities. Having done so, however, it proceeded to marginalize the issue with regard to new initiatives. It gave the main responsibility for dealing with the problem to the recently created Commission for Racial Equality (CRE); but, unfortunately, the CRE was relatively powerless and had little or no influence over the new initiatives and even less over mainstream programmes such as employment policy. As a consequence, the initiatives that were of most benefit to ethnic minorities remained the UP and Section 11 Grants. This position persisted into the mid-1990s when both programmes were absorbed into the Single Regeneration Budget (SRB), to which we shall return below.

With the advent of the Thatcher government in 1979 the process of racialization took a step forward when there was an explicit attempt to identify black people, and particularly black youth, in the inner cities as a source of disorder and criminality (Solomos 1993: 187–8). Despite electoral rhetoric that had included statements from Thatcher about 'swamping' and the 'enemy within', the post-1979 Conservative government did not put into practice the more radical implications of their statements. Programmes such as Section 11 Grants and the UP continued to operate.

Without the urban unrest of 1981 it is quite likely that the UP would have experienced considerable cuts relatively early in the Thatcher era; in the event it experienced a modest increase. This did not recur after the 1985 unrest and, during the 1980s, the UP experienced considerable internal restructuring as its emphasis was increasingly switched from social and housing projects to economic and environmental ones. As a result, ethnic minority projects tended to suffer, as Munt noted:

> Black projects appear to have been disproportionately affected by both the preference for economic schemes and a marked reduction in available revenue and, unlike the non-statutory sector in total, have not retained the percentage of approved UP expenditure.
>
> (Munt 1991: 190)

In line with the Conservative stress on the need to improve the supply side of the economy, those UP projects affecting black and minority ethnic groups aimed increasingly to enhance employability and business skills. In this the UP was supported by a variety of other projects such as the Ethnic Minorities Business Initiative, launched in 1985, and the Ethnic Minority Grant, launched in 1992.

The Victor Hausner Associates review of the UP, commissioned by the Department of the Environment, confirmed the fragmented and limited nature of the UP. It pointed to a general lack of clarity and strategic guidance

Plate 11.1 Contrasting faces of capital, Palmerston Road, Southsea

Plate 11.2 Micro-scale ethnic division in the restaurant trade, Albert Road, Southsea

in the UP and specifically noted that 'the UP does not deal as explicitly with the problems of ethnic minorities as it might' (VHA 1993: 25). The report went on to review a number of local authority UP schemes, finding a lack of clarity in their design, an undifferentiated approach to the needs of different ethnic minorities and a failure to integrate the needs of ethnic minorities into mainstream council programmes. As a result, it suggested that the needs of ethnic minorities had tended to be confined to the periphery of both central and local government political and policy agendas.

In November 1993 the government announced the launch of a new urban initiative – the SRB – which brought together twenty separate urban initiatives, including the UP and 55 per cent of Section 11 Grant funding, in an attempt to develop a more strategic approach to urban problems and enhance organizational coordination. The intention was that gradually more and more of the SRB's funds would be allocated on a competitive basis with local authorities and/or Training and Enterprise Councils (TECs), in conjunction with the private and voluntary sectors, playing a lead role in organizing bids to a competition for funding.

Among the SRB's objectives are the need to:

> enhance the employment prospects, education and skills of local people, particularly the young and those at a disadvantage, and promote equality of opportunity; promote initiatives of benefit to ethnic minorities.
>
> (DoE 1993: 5)

Whilst this explicit mention for minority ethnic groups is welcome, the relatively marginal position of ethnic minorities in local political processes indicates a need for concern about how minority ethnic groups will fare in the bidding process. This fear was confirmed in a review of the first SRB bidding cycle of 1994/5 which found that 'the needs of ethnic minorities have not figured as a prominent issue across the 201 successful SRB bids' (Mawson *et al.* 1995: 116).

Overall it appears that urban policy has failed to tackle directly and openly issues of racial discrimination and disadvantage. The conclusion reached by Stewart and Whitting (1983) in an earlier review of the UP, with regard to the neglect of minority ethnic needs, still remains salient for urban policy as a whole today. They argued that:

> The major concerns felt by ethnic groups are about racism, discrimination, and relations with the police. Thus for many groups the Urban Programme, if understood at all, is at best a source of funds for useful community based activity and at worst a diversionary smokescreen to divert attention from the absence of a serious commitment to a multiracial society.
>
> (Stewart and Whitting 1983: 18)

THE LOCAL POLICY RESPONSE

At the city level several local authorities have attempted to develop policies to counter racial discrimination both in the services they provide and within the urban area they govern (Gibbon 1990). Perhaps the most obvious examples of such initiatives have been those concerned with local authority housing. As noted above, researchers have clearly documented numerous examples of institutional and individual discrimination that either effectively excluded ethnic minorities or allocated them the least desirable properties in the worst locations. Henderson and Karn (1984) clearly documented these developments in Birmingham as did Philips (1987) in London. As a result, some local authorities attempted to change their housing allocation procedures to ensure that staff were aware of their own racial prejudices. A minority of authorities, notably Lambeth, went further and sought to develop anti-racist strategies to regulate the activities of their tenants; they did this through the introduction of racial harassment clauses into council tenancies that would enable them to evict tenants who engaged in acts of racial harassment. These developments were increasingly the target of bitter attacks from both central government and the media and became entangled in wider conflicts between central and local government. As a result, many of the most proactive authorities in this field were daubed with a 'loony-left' label and forced either to play down or reverse their initiatives.

Even in those authorities that pursued anti-racist strategies there was a tendency for the issue to remain on the periphery of the council in political, policy and organizational terms, Ouseley (1990) noted that local authorities were not prepared to engage in the radical restructuring of their organizational structures and working culture necessary to root out institutionalized forms of discrimination. Even when local authorities experimented with new forms of organization and control such as decentralization, they often encountered considerable problems that allowed racist attitudes to achieve a public platform and an influence over policy, most notably with regard to the allocation of council housing (Burns *et al.* 1994: 229–35).

CONCLUSION

It has become evident during the 1990s that the spurious association between black inner city populations and urban unrest can no longer be maintained. Not only have 'quiescent' Asian populations 'rebelled'; most of the major instances of unrest have involved white youths in the inner cities or on run-down peripheral council estates where people from ethnic minorities, particularly people of Asian background, have frequently been the subject of vicious racial attacks (Campbell 1993). Moreover, although the notion has attracted some attention in the UK, the idea of a racialized underclass of unemployed people living on the margins of society does not as yet carry with it the racist connotations so typical of the USA. While at times in the UK the 'pathology' of the black family and communities has been called upon to explain away examples of urban unrest, overall the underclass has been used to stigmatize the white poor (especially teenage single parents and 'welfare dependants') as much as people from ethnic minority communities.

What cannot be ignored, however, is the growth of social polarization and inequality within British society (Hudson and Williams 1989). Whilst this affects significant sections of the white population, its articulation with race, racism and the residential location of minority ethnic groups threatens to create a racially fragmented and polarized urban structure. Even though there is evidence that some individuals and sections of the minority ethnic population have progressed up the socio-economic ladder, the overwhelming weight of evidence still suggests that, for the foreseeable future, the majority of the residents of these communities will continue to experience the highest rates of unemployment, work in the worst jobs and live in the least desirable dwellings and locations.

Can these problems be tackled? At root this may well be a question of political will. As we have suggested on several occasions in this chapter, there has been a deep-seated unwillingness by the major political parties to be seen to discriminate positively in favour of ethnic minorities. In our view there is little chance of this being reversed in the near future. As a result, race will remain on the periphery of the political and policy agenda. Policy in general, and urban policy in particular, will not prioritize the needs of ethnic minorities. Mainstream programmes such as welfare, employment and education policy will do little to tackle issues of racial discrimination head on. Nor will the CRE be given new powers to pursue individual, let alone institutionalized, forms of discrimination. Its role will continue to be dominated and constrained by a government philosophy that stresses the need to ensure that markets work fairly and eschews tackling deep-rooted individual and organizational forms of discrimination.

With regard to urban initiatives it is likely that race will remain marginal to the main thrust of policy development. Its overt reappearance, however briefly, on the urban policy agenda is most likely to coincide with renewed outbreaks of urban unrest when traditional explanations of criminality, family breakdown and 'unBritishness' are wheeled out to explain away the material conditions that create the context for riotous behaviour. At the local level the continued financial squeeze on local authorities, and the potential for hostile media coverage, will severely limit their actions in what is too often viewed as a marginal, and politically risky, policy area.

If discrimination against minority ethnic groups in the divided British city is to be tackled, two conditions are required. First, a sustained economic recovery must be engineered. On past evidence this will significantly reduce ethnic minority unemployment. Economic expansion alone, however, does not mean that minority workers will move up the employment ladder. This will require a second development: much more positive discriminatory action on the part of government and employers. Government must initiate a series of proactive and properly financed equal opportunity and anti-discrimination policies that will tackle discrimination and disadvantage in an integrated fashion across all areas of social and public policy. As we have noted, both major political parties have been historically unwilling to take up this challenge and, unfortunately, there is little evidence to suggest that they will do so in the future. Thus we remain pessimistic about the prospects for the majority of Britain's urban minority ethnic population.

GUIDE TO FURTHER READING

Mason, D. (1995) *Race and Ethnicity in Modern Britain*, Oxford: Oxford University Press provides a good introduction to the position of ethnic minorities, while Solomos, J. (1993) *Race and Racism in Britain* (2nd edn), London: Macmillan is an excellent overview of the literature. Jones, T. (1993) *Britain's Ethnic Minorities*, London: PSI is an excellent sourcebook for basic information, and Bhat, A., Carr-Hill, R. and Ohri, S. (eds) (1988) *Britain's Black Population. A New Perspective* (2nd edn), Aldershot: Gower, whilst somewhat dated, contains a number of valuable chapters on individual policy areas. Victor Hausner and Associates (1993) *Economic Revitalisation of Inner Cities: The Urban Programme and Ethnic Minorities*, London: HMSO is one of the few overt attempts commissioned by government to assess the implications of an element of urban policy for ethnic minorities, while Ball, W. and Solomos, J. (eds) (1990) *Race and Local Politics*, London: Macmillan offers a series of interesting local policy studies.

REFERENCES

Atkinson, R. and Moon, G. (1994) *Urban Policy in Britain: The City, the State and the Market*, London: Macmillan.

Brown, C. (1992) 'Same difference: the persistence of racial disadvantage in the British employment market' in P. Braham *et al. Racism and Antiracism*, London: Sage: 46–63

Burns, D., Hambleton, R. and Hoggett, P. (1994) *The Politics of Decentralisation: Revitalising Local Democracy*, London: Macmillan.

Campbell, B. (1993) *Goliath: Britain's Dangerous Places*, London: Methuen.

Cmnd 6845 (1977) *Policy for the Inner Cities*, London: HMSO.

Daniel, W. (1968) *Racial Discrimination in England*, Harmondsworth: Penguin.

Department of the Environment (DoE) (1993) *Bidding Guidance: A Guide to Funding from the Single Regeneration Budget*, London: DoE.

Gibbon, P. (1990) 'Equal opportunities policy and race equality', *Critical Social Policy*, 10: 5–24.

Ginsburg, N. (1992) 'Racism and housing: concepts and reality' in P. Braham *et al. Racism and Antiracism*, London: Sage: 109–32.

Harrison, M. (1994) 'Housing empowerment, minority ethnic organisations, and public policy in the UK', *Canadian Journal of Urban Research*, 3: 29–39.

Henderson, J. and Karn, V. (1984) 'Race, class and the allocation of public housing in Britain', *Urban Studies*, 21: 115–28.

Hudson, R. and Williams, A. (1989) *Divided Britain*, London: Belhaven.

Jackson, P. (1992) 'The racialization of labour in post-war Bradford', *Journal of Historical Geography*, 18: 190–209.

Karn, V., Kemeny, J. and Williams, P. (1985) *Home Ownership and the Inner City*, Aldershot: Gower.

Leman, S. (1992) 'Ethnicity, technology and local labour markets in the clothing industry of Northern England', *Urban Anthropology*, 21: 115–36.

Lieberson, S. (1980) *A Piece of the Pie: Black and White Immigrants since 1880*, Berkeley: University of California Press.

Mawson, J. *et al.* (1995) *The Single Regeneration Budget: The Stocktake*, Birmingham: CURS.

Munt, I. (1991) 'Race, urban policy and urban problems', *Urban Studies*, 28: 183–203.

Newnham, A. (1986) *Employment, Unemployment and Black People*, London: Runnymede Trust.

Ohri, S. and Faruqi, S. (1988) 'Racism, employment and unemployment' in A. Bhat *et al. Britain's Black Population*, Aldershot: Gower: 61–100.

Ouseley, H. (1990) 'Resisting institution change' in W. Ball and J. Solomos (eds) *Race and Local Politics*, London: Macmillan, 132–52.

Owen, D. (1994) 'Spatial variations in ethnic minority group populations in Great Britain', *Population Trends*, 78: 23–33.

Philips, D. (1987) *What Price Equality?* GLC Housing Research and Policy Report no. 9, London: GLC.

Rees, P., Phillips, D. and Medway, D. (1993) 'The socioeconomic position of ethnic minorities in two northern cities', *Working Paper*, 93(20), University of Leeds, School of Geography.

Rex, J. and Moore, R. (1969) *Race, Community and Conflict*, Oxford: Oxford University Press.

Robinson, V. (1994) 'The geography of ethnic minorities', *Geography Review*, 7: 10–15.

Sillitoe, K. and White, P. (1992) 'Ethnic group and the British Census: the search for a question', *Journal of the Royal Statistical Society A*, 155: 141–64.

Smith, S. (1993) 'Residential segregation and the politics of racialization' in M. Cross and M. Keith *Racism, the City and the State*, London: Routledge: 128–93.

Solomos, J. (1988) *Black Youth, Racism and the State*, Cambridge: Cambridge University Press.

Solomos, J. and Singh, G. (1990) 'Racial equality, housing and the local state' in W. Ball and J. Solomos *Race and Local Politics*, London: Macmillan: 95–115.

Stewart, M. and Whitting, G. (1983) *Ethnic Minorities and the Urban Programme*, Bristol: SAUS.

Teague, A. (1993) 'Ethnic group: first results from the 1991 Census', *Population Trends*, 72: 12–17.

Victor Hausner and Associates (VHA) (1993) *Economic Revitalisation of Inner Cities: The Urban Programme and Ethnic Minorities*, London: HMSO.

Wrench, J. and Solomos, J. (1993) 'The politics and processes of racial discrimination in Britain', in J. Wrench and J. Solomos *Racism and Migration in Western Europe*, London: Berg: 157–76.

12

AGE

David Phillips and Helen Bartlett

INTRODUCTION

Elderly people form a major part of British society and, in 1991, almost 16 per cent of the population was aged 65 years and over. Like the rest of Britain's population, the majority live in urban environments, although within this context there are some significant distribution differences with other age groups. Increasing life expectancy means that there will be much higher numbers of very elderly people (aged 80 and over) in the coming decades. These include people most likely to be frail and in need of special care and accommodation. The health, welfare, housing and income support of elderly people form major national policy contexts in Britain. Britain's elderly population are not, however, a uniform, quiescent group; neither are they in aggregate any more of a problem than many other demographic groups (Warnes 1989). They represent the range of social, educational, economic, health and ethnicity characteristics of the population. They are, too, a growing force politically and economically and their needs have increasingly to be recognized by policy-makers and service providers. Much interest has focused on the retirement migration of elderly people but the vast majority of them remain and age where they have lived. Their local environments are therefore crucial and policies need to be fostered to make these as appropriate as possible to the needs of older people. It is also likely that Britain's elderly urban residents are better served than their rural counterparts with facilities and care; therefore, urban living (which many seem to prefer, especially in their later old age) may be advantageous, at least in relative terms.

AN AGEING NATION

Fertility reductions and decreased mortality are the two main influences on Britain's population size and structure. Fertility rates in the UK have been below what is considered 'replacement level' fertility of 2.1 since the late 1940s, with the exception of Northern Ireland where this was reached in the early 1970s. The downward trend in total fertility rates has been accompanied by deferred childbearing and, in 1992 for the first time, women in their early

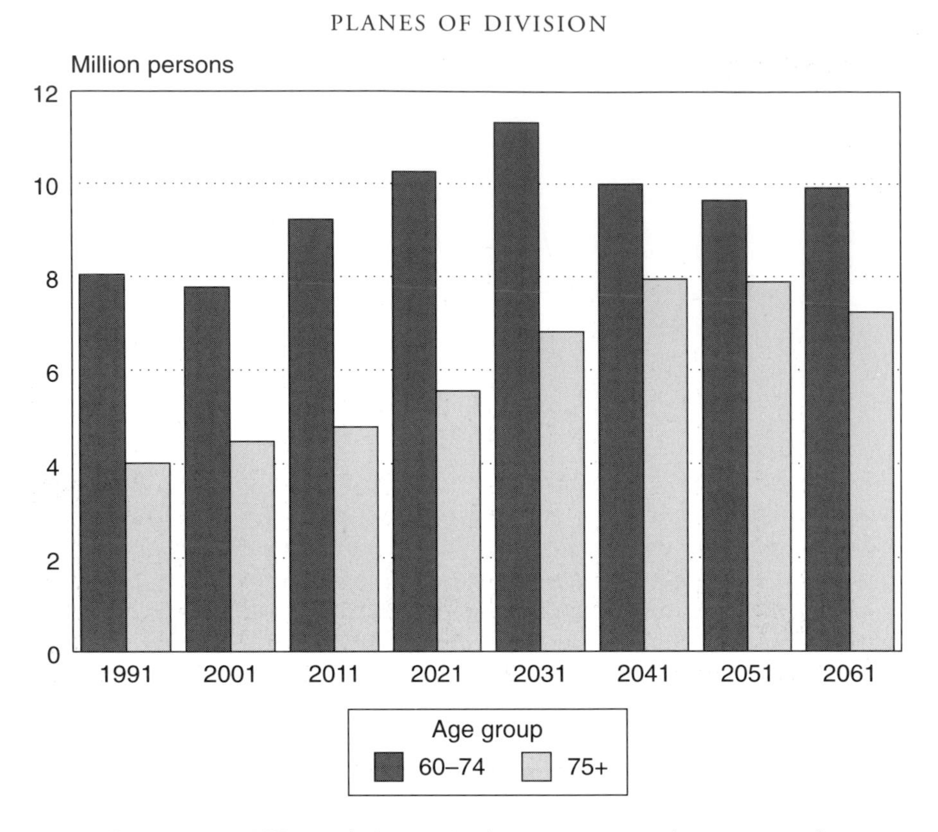

Figure 12.1 UK population projections, age groups 60+, 1991–2061
Source: OPCS 1993

30s were more likely to have a first child than those aged 20–24. This has very important implications for future family patterns and the potential of children to be involved with the care of elderly members. Fewer total children are being born to women and at later ages; women are living longer and there are likely to be fewer economically active persons in the workforce as a whole and fewer in any given family able to be involved with elder care.

In 1911, only about 5 per cent of the population was aged 65 and over. By 1971, this had reached about 13 per cent and, in 1991, 15.7 per cent. Looking far ahead, the proportion of people of pensionable age (currently 60+ for women; 65+ for men) is likely to have reached around 24 per cent by 2051. However, from the early 1990s until about 2005, this proportion will remain fairly static and will increase by only about 2 per cent per annum. The principal demographic features of the elderly population over the next two decades are that the percentage of persons aged 65–79 will stay relatively static at just under 12 per cent of the total population, whilst those

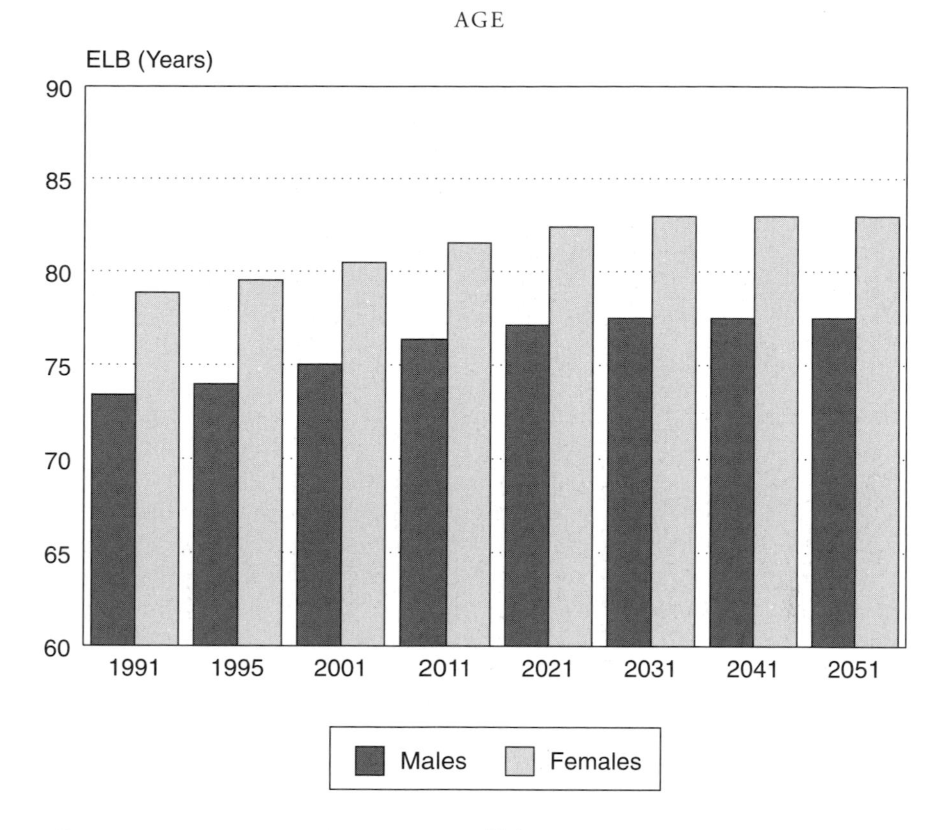

Figure 12.2 UK projected expectation of life at birth (in years), 1991/2–2051/2
Source: OPCS 1993

aged 80 and over will increase slightly from just under 4 per cent to over 5 per cent by 2021. After about 2011, there will be a gradual increase in all elderly groups until 2041, when those aged 65–79 will comprise about 17 per cent of the population and those aged 80+ will comprise about 8 per cent and increase to over 9 per cent in the following decade (Figure 12.1). Longer term, the number of persons aged over 75 is projected to more than double by 2051. However, the number of people aged 90 years and over will increase more than fivefold! This ageing of the elderly population itself, and the preponderance of women in the older age groups, are major current and future policy issues (Grundy 1995). Expectation of Life at Birth (ELB) is gradually increasing, although the gap between that of females and males hardly narrows even into the middle of next century (Figure 12.2). As well as the increase in ELB, differences between male and female longevity have various implications for social policy, housing and pension support for elderly females who might have substantial periods alone in later life.

The relatively small growth in elderly populations until the early years of next century may give the UK a slight respite in which to acquire knowledge and formulate policies (Thane 1989). This is particularly important with respect to devising appropriate policy in housing, income, health and welfare provision and care delivery methods (discussed below). Other trends, such as in types of retail and general service provision and public transport, will also impinge on elderly people, whether in urban or rural areas. For example, it is widely felt that retailing trends such as the growth of large, mainly out-of-town or peri-urban, superstore complexes favour car-borne shopping. Elderly people, particularly single elderly, are the least likely adult group in society to be car owners, and the renewal of driving licences of people aged over 70 years is subject to health criteria. Retail trends such as the growth of superstore chains might reduce the cost of goods for the general public but they can force out of business local convenience stores that may be patronized by elderly people who have reduced transport mobility and who may not wish or be able to afford to purchase goods in bulk at major stores. Therefore, planning policies and retail trends that permit these developments may, albeit inadvertently, disadvantage some elderly consumers (among others).

The 'dependent population' is shifting towards the older age groups, and concerns, some realistic, others ill founded, focus on the increased costs of pensions, social and health care that will fall on national and personal budgets. It is estimated that the numbers over the current pensionable age will peak at around 17 million in 2036. Combined with the projected numbers of children, this gives a dependency ratio of 82 persons per 100 working-age people by 2036. In 1992, this ratio was only 63 per 100. In the UK, the total of 'dependent elderly people' (currently, men aged 65+ and women aged 60+) will outnumber those aged 16 or under at around the year 2011 and the total of elderly dependent population will increase for the subsequent twenty or more years. The principal worries about this shift in the dependency ratio revolve around there being fewer working-age people to provide economic support and also physically to care for any dependent elderly people.

WHAT ARE THE MAIN DISTRIBUTIONAL FEATURES OF BRITAIN'S ELDERLY POPULATION?

As with any group, elderly people have macro-scale, regional distributions and local, intra-urban distributions. Both are likely to be affected by a number of factors including population ageing trends, in-migration and out-migration rates of elderly and younger age cohorts, and by economic opportunities, housing availability and perceptions of the environment on the part of various age groups. Macro-level county distributions of elderly persons have for some time been studied by social geographers and gerontologists. The pre-eminence

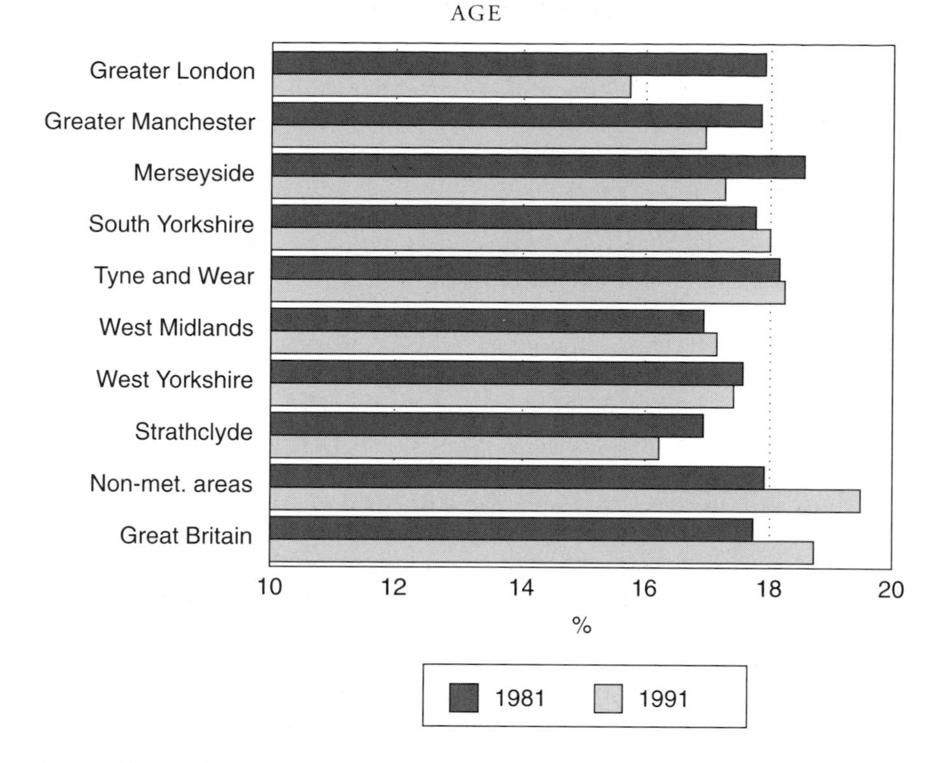

Figure 12.3 Relative presence of pensioners, metropolitan counties of Britain, 1981 and 1991. Pensionable ages = 60 (female) and 65 (male)
Source: Based on data in Warnes 1994

of elderly populations in parts of some southern and south-eastern 'retirement' counties of England, such as south Hampshire and parts of Sussex, and the relative ageing of areas such as Devon and coastal Lincolnshire has been a feature. This has been influenced both by earlier in-migration of young retirees, past out-migration of younger age groups and by subsequent ageing in place of locals and in-comers. Within cities, there is a slightly lower concentration of elderly people in inner city areas which tend to feature more young and ethnic groups (Boddy *et al.* 1995). However, there are important and potentially needy elderly groups in some deprived city areas, as noted below.

DEMOGRAPHIC AGEING AND BRITAIN'S CITIES

At the city level, the pattern of distribution of elderly people is fairly complex and has been changing in recent years. Using census data for 1971, 1981 and 1991, Warnes (1994) has shown that, during the 1990s, the metropolitan counties and other large cities have aged on average somewhat less than other parts of the country, although many cities had age structures quite close to the national average (Figure 12.3). London (particularly Inner London) had the most distinctive age structure in 1991 with an approximately 17 per cent underrepresentation of pensioners in its profile. The only urban areas with substantial overrepresentations of elderly people were in the mixed group of resort and retirement towns; they were also to be seen in the remoter, largely rural areas of the country. In London, this decline in elderly population has resulted from retirement out-migration and 1980s economic development in the inner areas, a trend discernible but less so in earlier periods. Rapid rates of increase were also apparent due to cohort ageing of the earlier residents of a number of New Towns who by 1990 were approaching retirement age. Warnes did not find much evidence of a substantial return to the London metropolitan area or to other large cities. Nor was there much support for the generalization that elderly populations are to be found in the urban cores of Britain's cities.

During the two decades since 1971, London has changed from having a quite elderly age structure to a very young one. In 1971, the 'expected' growth pattern was discernible, with a relatively elderly population in the established core and middle ring and a younger periphery. By 1991, Warnes reports that whilst there was still quite a high density of elderly people in the Inner London boroughs, this had been muted and there had been rapid growth of elderly populations in some peripheral districts and particularly in the surrounding New Towns. This is attributed to both cohort and migration effects. Economic and social rejuvenation of parts of Inner London during the 1980s provided opportunities for out-migration of older people and in-migration of young adults. Elsewhere, ageing in place was to be seen in the public housing estates of the 1960s, in which elderly residents often remain.

However, London is not wholly typical of the rate of change of elderly populations in all large British cities. Its core dominated by economic development and working-age population and its trend for peripheral growth of older population is not replicated as clearly in, for example, the West Midlands. There, Warnes points to more rapid ageing in Birmingham and in the other parts of the West Midlands. Ageing has also been somewhat more even amongst West Midlands local authorities than in London, although some features such as the attraction to elderly people of some pleasant surrounding semi-rural areas (as may be found in Hereford, Worcestershire and Warwickshire) are to be seen. Two factors are likely to underpin the West Midlands distribution of elderly people. First, the rapid industrial growth of its metropolitan area in the 1950s and 1960s means that the then young workforce is now reaching retirement age. Second, there is a slower out-migration at retirement from the West Midlands cities than there was from London, which could have counterbalanced the ageing of their populations.

Retirement migration has been a phenomenon long influencing the distribution of elderly people within cities and between areas of the country. The lack of a need to be located near to work has given many older people less reason to be tied to inner urban locations. The spur to migrate might come from the desire to have a smaller house, to realize a capital sum from the sale of a family dwelling and to be in a more congenial location. The ability to sell a house at an acceptable price and the availability of suitable retirement housing are influenced by the housing market and by housing policy but, on average, retirement or approaching retirement can be a time of mobility for a number of elderly people. However, this should not obscure the fact that most ageing does occur *in situ*; the image of footloose pensioners is not altogether accurate.

DEPRIVATION AND ELDERLY URBAN RESIDENTS

Whilst there is no clear-cut and agreed definition of deprivation, it is generally accepted that it is a state that is characterized by a lack of access to material resources and opportunities. A deprived area is one in which a substantial proportion of the population exhibit such characteristics, and there have in Britain been a number of indices constructed to attempt to measure relative deprivation. Unemployment, crowding, lone pensioners and low social class tend to be major variables included in such measures. The NHS uses a Jarman Underprivileged Area Index to indicate areas in which GPs are paid additional remuneration in wards that meet certain levels of deprivation on that index. The Townsend index, comprising four variables, is also popular for identifying areas of underprivilege and deprivation. However, these indices tend to identify deprivation only in urban areas, where it is more concentrated, and they are less effective in identifying rural deprivation (Brigham *et al.* 1995, Moore 1995).

Deprivation and elderly people: the Nottingham Health District

Analyses of deprivation in British cities by these and other indices do not generally highlight elderly people (as a population group) as being generally overrepresented in particularly deprived areas. Deprived areas (in part because of the variables included in the deprivation indices) tend to be those with higher proportions of children aged 0–4, single parents, persons from ethnic minorities and unemployed people. Many local authorities and health authorities have analysed recent census and survey data by deprived areas. A useful example is the Nottingham Health District, where the 1995 Report of the Director of Public Health (DPH) has a focus on the spatial and social expressions of deprivation (Figure 12.4). This mainly urban district (which is wider than the city itself) has a total population of some 635,000 persons. The DPH Report focuses on the relationships between deprivation, health care need and policies to tackle resulting problems, a number of which have implications for elderly residents (Nottingham Health 1995).

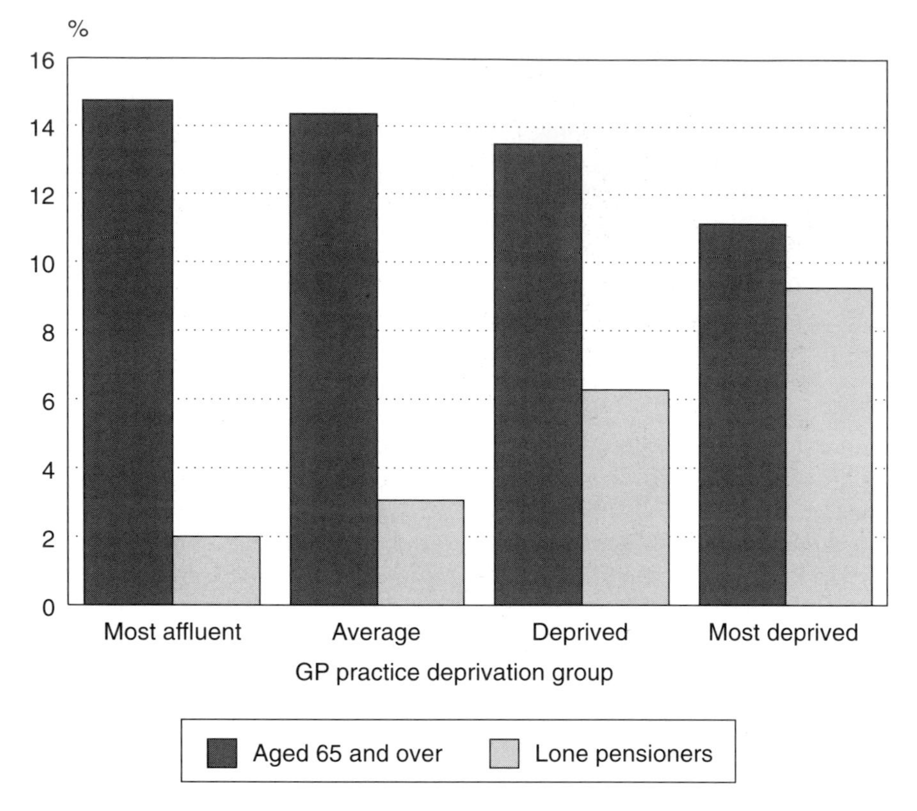

Figure 12.4 Nottingham Health District – elderly population and deprivation. GP practices grouped by Townsend index scores
Source: based on data in Nottingham Health 1995

In national terms, Nottingham does not suffer unduly from the effects of deprivation, although its spatial incidence is fairly concentrated, in terms of Townsend scores, as shown in Figure 12.5a. It is fairly typical of towns of similar size such as Newcastle, Leicester, Hull and Bolton, and is regarded as being in the same typical cluster as towns such as Southampton, Leeds and Coventry. The two categories of highest deprivation (Figure 12.5a) are mainly confined to the inner city areas such as Lenton and Radford, spreading south to St Anns and east to Forest Fields. Disadvantage does not tend to be concentrated only in inner city or peripheral council estates but is also seen in various pockets in the central parts of Nottingham. The implications of having a high Townsend score are important; for example, the most deprived wards of Nottingham have an unemployment rate that is five times that of the most affluent wards and even double that of average score areas,

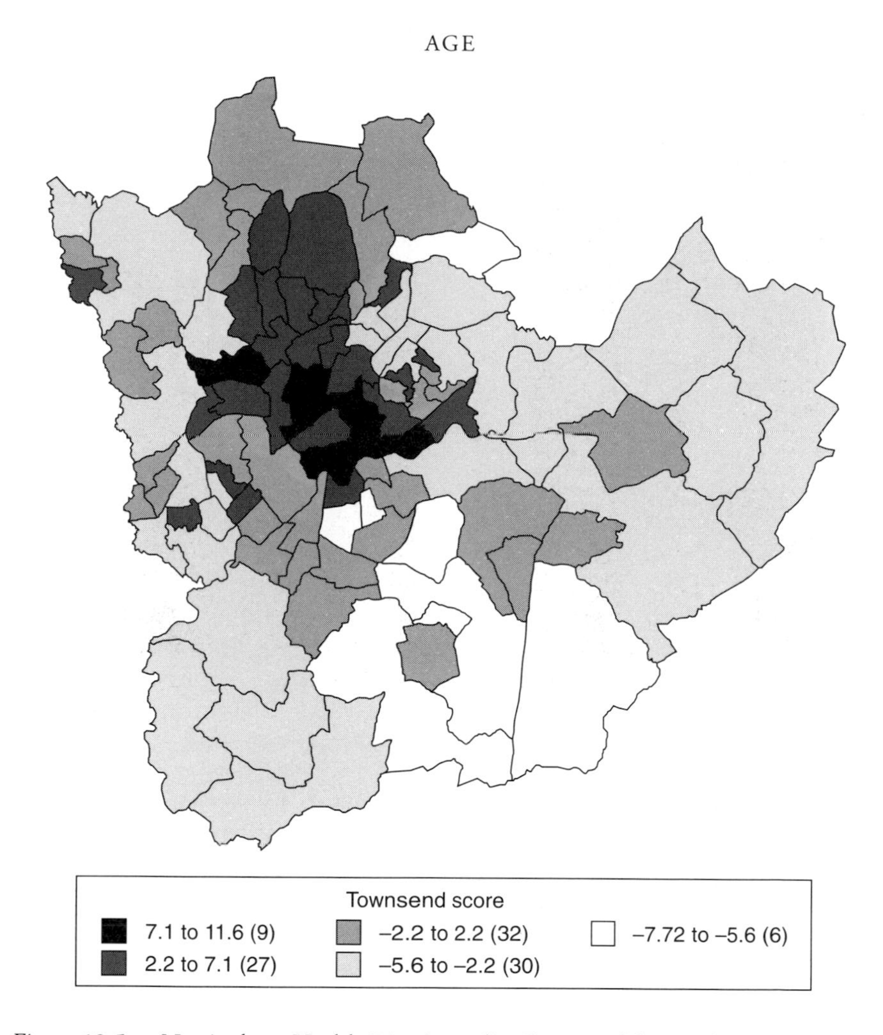

Figure 12.5a Nottingham Health District – distribution of deprived areas. A higher positive score on the Townsend index indicates areas of greater deprivation
Source: Nottingham Health 1995

and as many as one in four of their workforce is out of a job. Over half of the population in the most deprived areas live in rented accommodation and over half do not have a car. Whilst the areas with the highest deprivation scores are most likely to affect black and ethnic minority residents, young children and lone parents, the conditions in such areas also impact on elderly people, particularly lone pensioners and ethnic elders (Figure 12.5b).

Elderly people in Nottingham, as for the country as a whole, are not overly concentrated in the areas of greatest material deprivation. Indeed, the age structure of the most deprived GP practice areas in Nottingham Health

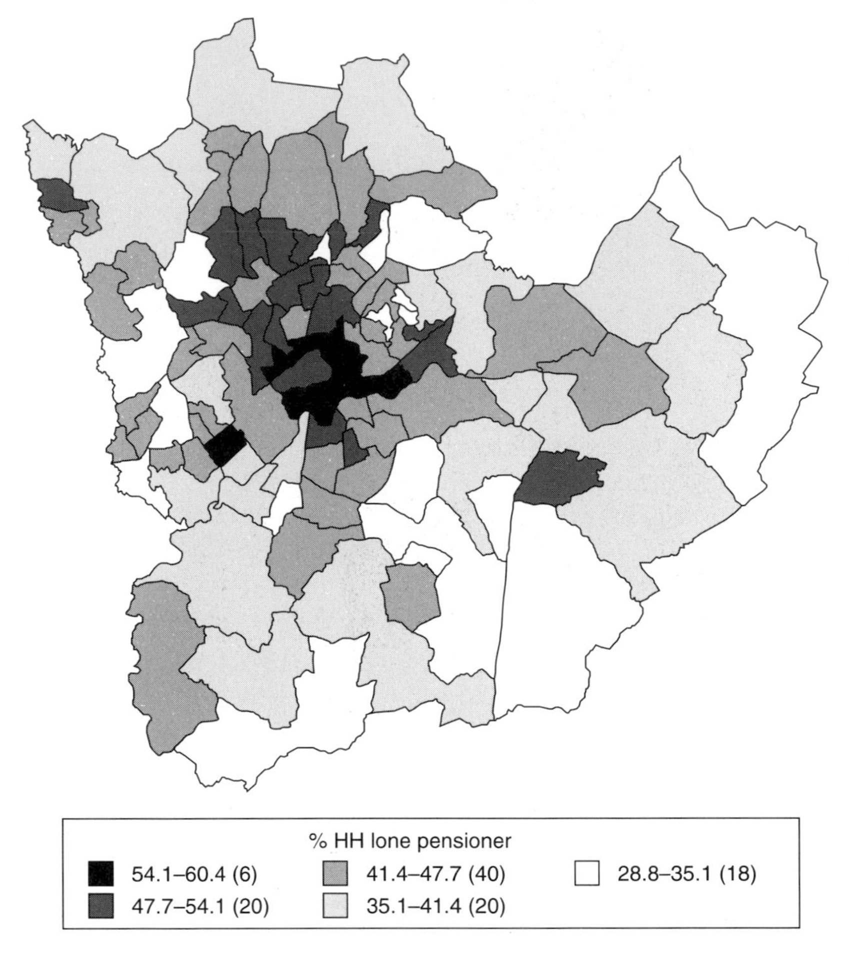

Figure 12.5b Nottingham Health District – households with lone pensioners as a percentage of all households with pensioners
Source: Nottingham Health 1995

District tends towards a higher proportion of young persons, and the proportion of elderly people tends to be lower than in more affluent GP practice areas. However, *in terms of need for health and social care provision*, the most deprived areas of the city tend to have higher percentages of elderly people living alone (Figures 12.4 and 12.5). This is important for services and policies, as the 'old and alone' in deprived environments are arguably more vulnerable than elderly people in other areas. For example, if they fall ill, they are, as a result of living alone, more likely to require in-patient care; their discharge home may be more problematic and requires greater co-

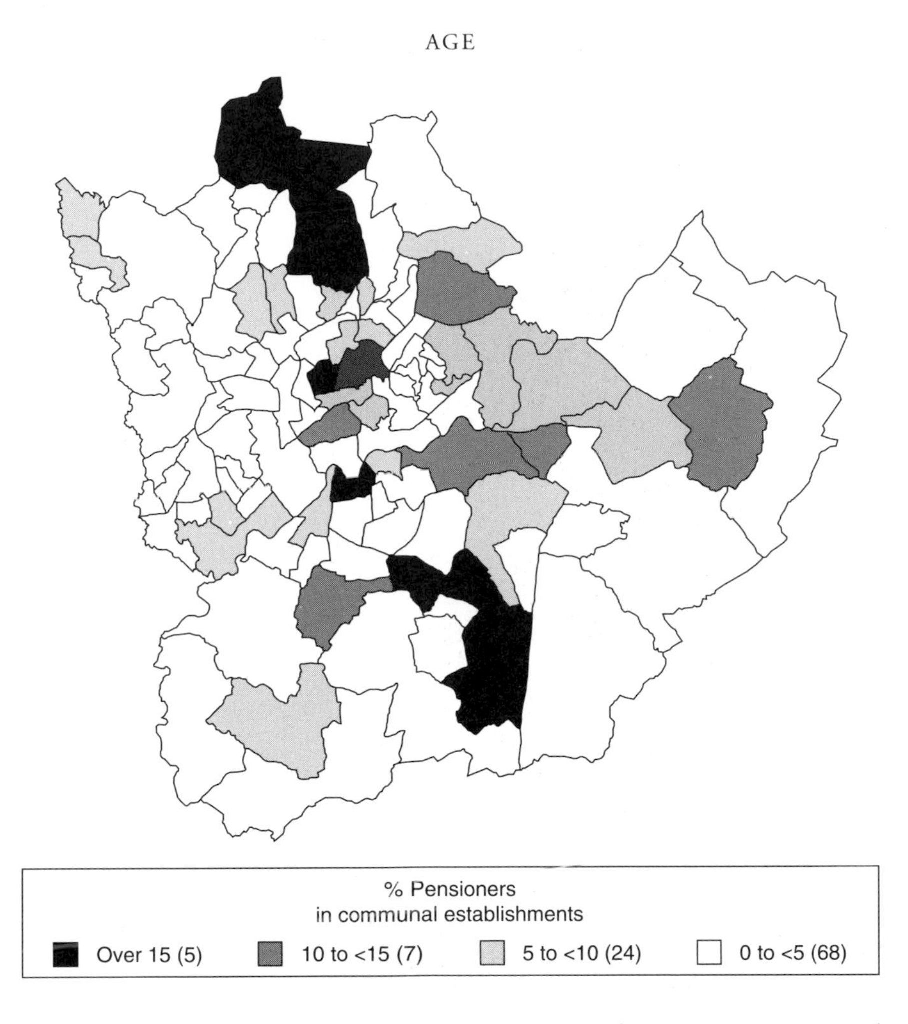

Figure 12.6 Nottingham Health District – percentage of pensioners in communal establishments
Source: Nottingham Health 1995

ordination between medical and community care services; and they may face nutritional problems as well as problems related to accessing shops and services. It is interesting to note that it tends not to be the inner areas of the city that have the highest proportions of elderly people living in communal establishments (such as residential and nursing homes). Figure 12.6 shows that, for Nottingham, it tends to be the more suburban wards that have higher proportions of elderly people in such homes; there are, however, a few inner city areas in which such concentrations are also evident.

Plate 12.1 Conflicting priorities – traffic and the elderly

Health and social policies in Nottingham have targeted certain inner city areas with serious problems of decline, in particular the St Anns and Sneinton districts, which have a resident population of some 20,000. Healthy living and City Challenge projects have aimed to help people in these areas identify and respond to their health needs. The City Challenge project is very much a pilot which, if successful, could be extended to other areas of Nottingham Health District that have similar problems and disadvantage. A major task has been to align a range of services so that they are known and accessible to local people, and to co-ordinate a range of statutory and voluntary services in 'healthy alliances' to provide overall health gain to those in greatest need

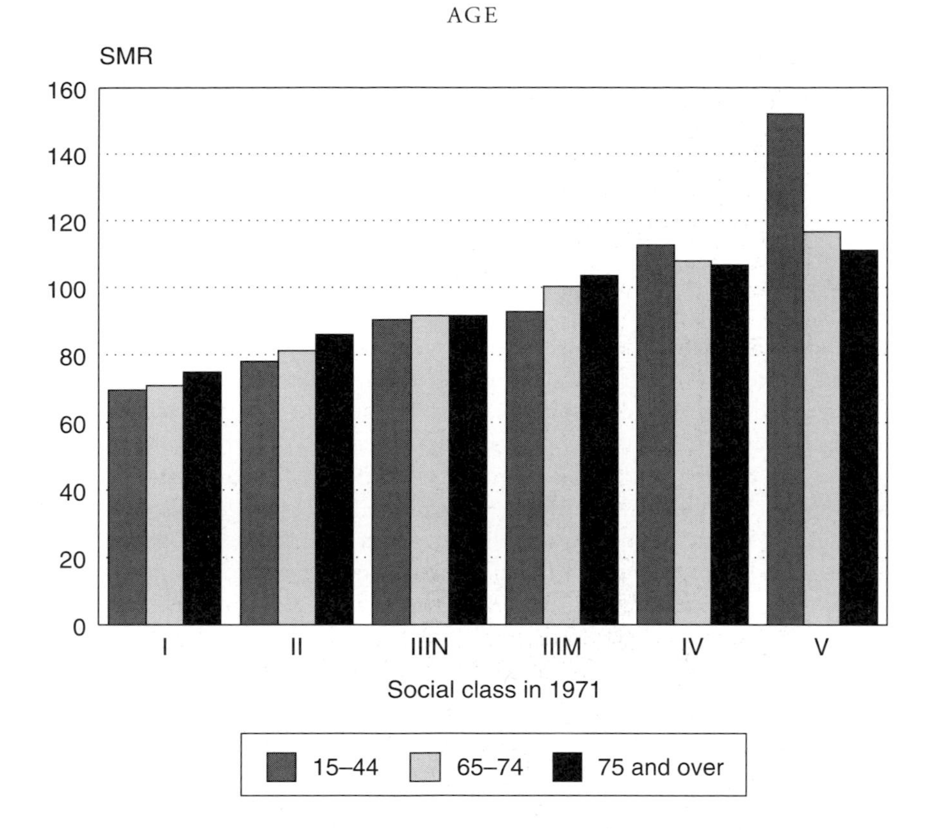

Figure 12.7 Mortality of males by social class, standardized mortality rates (SMRs), England and Wales, 1967–89
Source: based on data in Harding 1995

(Nottingham Health 1995). Within these particularly deprived communities, the Nottingham City Challenge goes beyond the provision of health care and attempts to change social and environmental circumstances that impact on health. These deprived areas have increased morbidity across the range of health indices, and the areal approach targets local groups whose current take-up of services is low, including members of the black community, those with mental health problems, substance misusers and elderly people. In conjunction with the environmental initiatives of the City Challenge, the aims are to reduce crime and to make neighbourhoods generally more pleasant and safer to live in. Overall, this should help the physical and psychological well-being of elderly people, particularly single elderly persons. Elsewhere in Nottingham, various attempts have been made to enhance access to health care and information for black and ethnic minority elder groups. These include a district nurse and interpreter attending an Indian Community Centre lunch club, to provide targeted and culturally appropriate health care advice and to liaise as necessary with elderly people's primary care providers. A similar scheme involves a district nurse and nursing auxiliary, who give advice and direct treatments, provided for older people at an African Caribbean Centre (day centre and lunch club) in the city.

Social class and health status

A further important feature relates to social class variations in health, in which the professional groups have almost always been shown to have better general health status and lower mortality than manual workers, particularly the semi-skilled and unskilled groups (social classes IV and V). Figure 12.7 indicates that a social class gradient, in which standardized mortality rates (SMRS) increase from social classes I to V, persists amongst elderly groups although the differentials are somewhat narrower than in earlier life (Harding 1995). As social class divisions remain spatially quite marked in many British cities, there are likely to be areas in which higher rates of morbidity and mortality prevail among elderly persons in these social class groupings, highlighting additional need and demands for care in some urban areas.

AGE AND RACE: DOUBLE DISCRIMINATION

> Older people from ethnic minorities are often described as facing a double, and even triple, jeopardy growing older in Britain; they are at risk because of age, because of racial and cultural discrimination, and because they cannot obtain health, housing and social services.
>
> (Age Concern 1995a: 1)

In 1995, Age Concern England published a discussion and review of life in Britain for the elderly members of ethnic minorities. The 1991 Census enumerated approximately 3 million people from ethnic minorities (itself an imprecise and potentially misleading collective term), which is about 5.5 per cent of the population. Only some 3.22 per cent were identified as being 65 and over (compared to almost 16 per cent for the whole population, noted earlier). However, these groups are ageing and some quite rapidly; black Caribbeans, Chinese, Indians and other Asians tend to be oldest; Bangladeshis, Pakistanis and black-other and other-other census groups tend to be youngest. Black Caribbeans, for example, have 5.65 per cent aged 65+ and 10.9 per cent aged over 60. Whilst a relatively small proportion, many among these ethnic minority elders are first generation immigrants; some have the least good knowledge of English and maintain cultural traditions that are diverse. Indeed, this diversity amongst and between groups is of great importance and should not be overlooked in policy, practice and

Plate 12.2 A purpose-built nursing home for the elderly in Nottingham

research. In addition, whilst current numbers of ethnic minority older people may be relatively small, there is a middle-aged bulge which means that there will be rapid increases in the future and as much as a tenfold increase in elderly percentages over the next thirty to forty years. This has important implications for the social and spatial provision of appropriate services, which was evident in the Nottingham case study noted above.

A key feature is that although people from ethnic minorities are to be found in all parts of Britain, they do tend to be highly concentrated in the more urbanized areas of the country. Spatial concentrations of ethnic minorities have been identified as follows: 44.6 per cent live in Greater London, 14.1 per cent in the West Midlands and 4.9 per cent in Greater Manchester (Age Concern 1995a, Owen 1994). By contrast, some ethnic minority residents do live in more isolated areas and their needs may, ironically, not be met in the same ways that they may be in areas of greater concentration.

Ethnic elderly people in cities and elsewhere can face a number of problems. In particular, in the current generation, elderly females from some communities may have a poorer command of English and may face some isolation from health and social services. However, in some areas, successfully targeted campaigns have brought health and social care in appropriate languages and culturally acceptable presentations to many ethnic groups. Age

Concern (1995a) identifies a number of potential problem areas for ethnic elderly people in housing, health, income, social welfare and education and leisure. They also recognize the importance of overcoming myths surrounding ethnic elderly groups, such as: that because of their small numbers, there are no problems; or, that ethnic minority elders return to their country of origin in old age; or, that all ethnic minority families support their older relatives. These can be injurious both to formulation of practical policies and to relationships with other groups of British citizens. Income may be a problem and access to and take-up of benefits can be low in some ethnic minority communities. Studies indicate that this does vary between groups, however. Some might feel that it is shameful to take up state benefits; others strongly believe that claiming is a right for those who have paid National Insurance contributions. Nevertheless, for many of the current elderly members of some ethnic minority groups, who have had shorter than average working records in Britain, both state and occupational pensions (if any) will be modest and financial need therefore greater (Age Concern 1995b). This may be closely connected with residence in more deprived urban areas as contrasted with the non-ethnic elderly groups who, it was noted, are not generally over-represented in deprived areas.

CARE PROVISION AND ACCOMMODATION

Accommodation

The vast majority of elderly people in Britain do not live in institutions and, indeed, only about 6 per cent of the 65+ age group do so. This means that there is a very big range of types of accommodation that elderly people occupy. Housing is particularly important for older people; if it is appropriate to their needs in terms of size, facilities, condition, location and costs, this can help to keep them in the community and reduce the need for entry to some type of residential care (Means 1993). Community care policies in Britain imply the need for greater support for people to 'stay put' in their existing accommodation. This can depend on a number of things and McCafferty (1995) identifies these as ranging from staying put with no additional support; staying put with necessary repairs only; staying put with repairs and adaptations or with domiciliary care support; a move to a smaller or mainstream house or a move to specialized housing such as sheltered accommodation. Thereafter, there is the possibility of the need for residential accommodation (living in a private old people's home, or a local authority or private nursing home; hospital long-stay beds are a further but diminishing option). Specialized housing comprises various degrees of sheltered housing with warden support as well as very sheltered housing for frailer elderly people, which can provide extra care services and, for example, meals and additional communal facilities.

Specialized housing for elderly people

The key providers of subsidized specialized housing for older people are local authorities, Abbeyfield, housing associations and almshouses. With the exceptions of London and the West Midlands, local authorities are the major providers across all regions. Four main types of specialized housing are available, depending on the applicant's level of activity. Category 1 accommodation caters for the more active elderly and is specially designed, with or without communal facilities. Sheltered housing is provided in the form of Category 1.5, 2 and 2.5 accommodation. Category 1.5 is similar to Category 1, but an alarm system and warden support is also provided. Category 2 accommodation caters for less active elderly people and has a resident or non-resident warden. 'Very sheltered' schemes are known as Category 2.5 accommodation and are designed for frail elderly people, so they may provide extra services such as meals, additional wardens, care assistants and extra communal facilities. Recent evidence suggests an oversupply of Category 2 accommodation in each Department of the Environment Region in England (McCafferty 1995). By contrast, there is a shortage of Category 2.5 accommodation nationally and regionally. The regional planned provision of subsidized specialized housing ranges from 1.1 per cent (East Midlands) to 3.2 per cent (Yorkshire and Humberside), but variations in the type of accommodation are apparent within these projections.

Details are not readily available for intra-urban provision of such accommodation. However, the Department of the Environment survey of accommodation for older people reveals clear differences between DoE northern regions and southern regions. For example, the provision of Category 1.5 accommodation is greater in the northern regions, whereas the proportion of sheltered housing stock in Categories 2 and 2.5 is greater in southern regions. These differences can in part be explained by the southern geographical bias of housing associations, Abbeyfield societies and almshouses, which have traditionally offered sheltered housing with both warden support and one or more communal facilities. Differences in the distribution of specialized housing stock are also evident according to types of local authority. Metropolitan and district councils outside London are twice as likely to have Category 1 and 1.5 accommodation than are London boroughs.

The profile of housing types (bungalows, houses, self-contained flats) is fairly consistent across the DoE regions, with the exception of London, where a higher proportion of specialized housing units consists of flats rather than bungalows. Bedsits and one-bedroom units account for 95 per cent of all provision in the capital compared to 80 per cent nationally. London is particularly underprovided with bungalows, which comprise only 5 per cent of specialized housing stock compared to 33 per cent nationally.

Other regional differences that will impact on the suitability of accommodation in urban areas relate to age and tenure of properties. Compared to the national average of 55 per cent of owner-occupancy among elderly households, only 47 per cent in London and 41 per cent in the northern region own their property. This age group is most likely to rent council property in these regions. Of all regions, London has the highest proportion of elderly people living in pre-1945 (56 per cent) and pre-1919 (25 per cent) properties. The age of dwelling is of considerable importance as the 1991 English House Condition Survey indicates that the majority of unfit dwellings are in the pre-1919 stock (Department of the Environment 1993).

At the intra-urban level in Britain, the housing geography of individual cities means that certain types of accommodation tend to be concentrated in specific areas. This can limit choice of appropriate housing that elderly people might like to occupy and enhance their chances of 'staying put' in their own home. However, as Tinker (1995) illustrates, the actual type of housing occupied by many elderly people did not match their preferences. For example, 64 per cent of 60–74 year olds and 56 per cent of the 75+ groups in General Household Survey data stated a wish to live in a bungalow. Yet, the majority were living in terraced, detached or semi-detached houses, which would in many cases have been less suitable for them. In addition, elderly householders, particularly owner-occupiers, are likely to face problems and costs of maintenance and, as a result, their houses have often been in poorer repair than those of younger age groups. Schemes such as home improvement agencies run by local authorities, housing associations and the private sector have, in recent years, been established to try to support elderly people to stay in their own homes. In 1992, elderly people in one in three local authority areas in England had access to home improvement agencies which are co-ordinated by Care and Repair Ltd (an officially sponsored umbrella organization for co-ordination of home improvement agencies in England). A crucial point is that the ability of elderly people to 'stay put' is only partly determined by the housing they occupy. Solutions can require a combination of housing adaptation and the availability of social, health and family support (McCafferty 1995). It is apparent, although detailed intra-urban research is lacking, that variations are considerable both in the range and quantity of such support that is available and the effectiveness of such support in different parts of cities.

Residential, continuing care and community care

Major changes have occurred in the balance of long-term care provision between the NHS, local authority and the independent residential and nursing home sector. Government policy prior to the 1990 community care reforms had clearly encouraged the growth of private nursing and residential care home provision by allowing easy access to social security funds to support elderly people in private sector homes, with a resultant burgeoning of the sector in the 1980s (Bartlett 1987, Phillips *et al.* 1988b). This guaranteed support was replaced by more stringent assessment of need for residential care, with responsibility shifted to local authorities to assess financial and need aspects of residential care. There is now an emphasis on keeping elderly people in the community, and a reduction in long-stay NHS beds commenced in anticipation of the reforms. Of the total 556,000 places available across all sectors in 1994, the largest share (39 per cent) was in the private and voluntary residential care sector, followed by 35 per cent in the private and voluntary nursing home sector. Local authority residential home places accounted for 16 per cent of the total and NHS long-stay hospitals just 10 per cent. The reduction in public residential home and NHS long-stay places commenced in the mid-1980s. By contrast, places in private and voluntary nursing homes have been multiplying rapidly since 1970, when the figure was 20,300, to the 1994 level of 194,800 – an increase of over 800 per cent. While not as dramatic as the private nursing home sector, places in the private and voluntary residential care homes have tripled from the 1970 level of 63,800 to 218,200 in 1994 (Laing 1994).

The decline in the number of NHS continuing care beds has various consequences but of particular concern is the inequity of access to long-term care that is likely to result from geographical variations in this provision. It is believed that the establishment of locally based eligibility criteria by health authorities is likely to increase this inequity (Royal College of Nursing 1995); local decisions may mean very different opportunities being available in different cities and parts of cities. Not only is there a concern about geographical variation in provision but there are also worries about uneven standards of care, particularly in the long-stay residential sector (Bartlett 1993). The efficiency of the regulatory system has certainly been questioned and inconsistencies in inspection procedures and requirements across the country noted (RCN 1994).

Intra-urban variations in support and provision for continuing care and for elderly people to remain in their own homes exist. Research has also identified wide geographical variations both within cities and between different types of area in Britain in long-stay care, particularly in private and public residential and nursing home provision (Larder *et al.* 1986, Phillips and Vincent 1986, Challis and Bartlett 1987, Phillips *et al.* 1988a, b, Corden 1992, Hamnett and Mullings 1992). There are variations in local authority residential homes and nursing homes by urban areas and settlement types. For example, Warnes (1994) found that, in 1991, whilst 15.9 per cent of the 85+ population nationally were in nursing homes, the range was from 6.3 per cent and 9.3 per cent in Inner and Outer London to 25.1 per cent in retirement resorts. Virtually all principal cities (the core authorities of metropolitan areas, such as Birmingham, Glasgow, Leeds, Liverpool and Manchester), large non-metropolitan districts (which include settlements such as Aberdeen, Bristol, Cardiff, Hull, Leicester, Plymouth, Portsmouth and Southampton), New Towns and industrial towns had fewer than the national average of their 85+ age group in nursing homes and were underprovided with these facilities. Indeed, only the retirement resorts had relatively high percentages of their 85+ age group in nursing homes. By contrast, some metropolitan areas, Inner London and industrial towns and some remoter rural areas tended to be slightly better provided than retirement resorts, large non-metropolitan cities and principal cities with local authority residential homes. These differences do have significance for the choices available to local elderly residents seeking particular types of residential care. They may also place differing demands on care in the community in the various urban settings.

In the private residential care sector, significant local as well as national variations in availability of homes have been noted. In Corden's (1992) North Yorkshire study, the attraction of coastal resorts and established local markets for residential care homes was evident. In some towns and villages, the growth of homes was demonstrated where there was already an overprovision of facilities (for example, in the Harrogate area). The question has been posed here and in other studies whether this concentration of facilities was causing or responding to local concentrations of elderly people. In Devon, a similar coastal town concentration of homes for elderly people has been noted, leaving towns such as Plymouth and Exeter with relatively fewer homes than might have been expected. By contrast, the coastal resort of Torbay attempted policies to restrict the further growth of homes for elderly people in some parts of the city that were deemed to be overprovided or of maximum tourist potential (Phillips *et al.* 1988b). These differential levels of provision impact on the choices elderly people can have in their own urban areas if and when they require long-stay or even shorter, respite care in a residential or a nursing home. There are important questions of equity and equality to be asked that have not yet been thoroughly addressed at the intra-urban level.

One of the main goals of the NHS and Community Care Act of 1990 (implemented in 1993) was to reduce reliance on residential care solutions to increasing dependency. However, the support necessary to enable people to remain in the community is insufficiently developed (Lewis *et al.* 1995). Packages of domiciliary care may be arranged but co-ordination is often poor. Whilst much detailed research remains to be done, it is clear from anecdotal evidence that the pattern of provision is not yet robust. Many urban areas vary in the nature and extent of support that their care packages will provide. This is likely to mean that the quality and quantity of support will increasingly depend on where you live.

CONCLUSION

This chapter has raised a number of issues about the nature of ageing in Britain's cities and the provision of care and accommodation for elderly people. Space does not permit discussion of the findings of research into the lives and behaviour of elderly urban residents, their hopes and needs, attitudes to services, fear of crime and the like. However, it is clear that national policies are likely to influence the levels of public sector provision that will be provided in a range of services and facilities. Policies are increasingly stressing the need for personal provision of, for example, pensions and housing. This will impact on poor elderly residents in cities in particular and also on single, ethnic or disadvantaged elderly people.

It should not be automatically assumed that urban elderly residents are worse off than their rural counterparts; indeed, research suggests that many elderly people, especially the very old, would prefer to be in the town rather than the country. In addition, the vast majority would prefer to stay where they are rather than move from their present home, suggesting that the Third Age is a time for stability (Askham 1992). This highlights the importance of policies and practical help to enable elderly people to live in their homes, and stresses the need for community development to make local environments as usable and friendly as possible for elderly people. As Means (1993) points out, housing is therefore a crucial (even if often neglected) component of community care. Other policies should be oriented to support and enhance the quality of life of elderly people, to help them overcome fear of crime, worries of financial impoverishment and of inadequacies in health and social care. In addition, in the urban environment, we need to remain aware that the vast majority of elderly people are not institutionalized. Yet there is a type of segregation, which can increase with older age, as elderly people can be hived off into sheltered accommodation, residential care or nursing homes. In the order of those three types of provision (ostensibly community based but increasingly effectively institutional), elderly people can become less and less visible in the neighbourhoods in which they live. They may be regarded as set apart from the rest of society. The spatial concentration of residential care facilities that is apparent in many of Britain's cities also has crucial implications for the provision of urban health and social care. Thus, effective policies to co-ordinate a number of services and sectors to meet the care needs of Britain's elderly population are essential.

GUIDE TO FURTHER READING

Allen, I. and Perkins, E. (eds) 1995 *The Future of Family Care for Older People*, London: HMSO.
A useful edited collection of readings that cover a range of issues from demographic change, migration, housing and care.

Means, R. (1993) 'Housing and community care', in J. Johnson and R. Slater (eds) *Ageing and Later Life*, London: Sage: 246–53.
Provides an overview of housing as an essential component of community care, particularly relevant to elderly people.

Warnes, A.M. (1994) 'Cities and elderly people: recent population and distributional trends', *Urban Studies* 31(4/5): 799–816.

A useful summary of major changes in urban elderly population distributions with an international comparison.

REFERENCES

Age Concern (1995a) *Age and Race: Double Discrimination. An Overview of Life for Ethnic Minority Elders in Britain Today*, London: Age Concern England and Commission for Racial Equality.

——(1995b) *Age and Race: Double Discrimination. Income*, London: Age Concern and Commission for Racial Equality.

Askham, J. (1992) 'Attitudes of people in the Third Age' in A.Warnes (ed.) *Homes and Travel: Local Life in the Third Age*, The Carnegie Inquiry into the Third Age, Research Paper 5, Dunfermline: The Carnegie United Kingdom Trust.

Bartlett, H. (1987) 'Social security policy and private sector care for the elderly' in M. Brenton and C. Ungerson (eds) *Yearbook of Social Policy*, London: Longman: 66–83.

——(1993) *Private Nursing Homes for Elderly People: Questions of Quality and Policy*, Melbourne: Harwood Academic Publishers.

Boddy, M. with G. Bridge, P. Burton and D. Gordon (1995) *Socio-demographic Change and the Inner City*, London: HMSO.

Brigham, P., Phillips, D.R. and Priestnall, G. (1995) *A Methodology for Identifying Demographic, Socio-economic and Health Status Characteristics of General Practice and Health Locality Areas: A Study of General Practices and Localities in Devon*, Working Paper 31, Department of Geography, University of Nottingham.

Challis, L. and Bartlett, H. (1987) *Old and Ill: Private Nursing Homes for Elderly People*, London: Age Concern England.

Corden, A. (1992) 'Geographical development of the long-term care market for elderly people', *Transactions of the Institute of British Geographers*, 17: 80–90.

Department of Health (1995) *NHS Responsibilities for Meeting Continuing Health Care Needs*, HSG (95) 8, LAC (95) 5. London: DoH.

Department of the Environment (1993) *English House Condition Survey 1991*, London: HMSO.

Grundy, E. (1995) 'Demographic influences on the future of family care' in I. Allen and E. Perkins (eds) *The Future of Family Care for Older People*, London: HMSO: 1–17.

Hamnett, C. and Mullings, B. (1992) 'The distribution of public and private residential homes for elderly persons in England and Wales', *Area*, 24(2): 130–44.

Harding, S. (1995) 'Social class differences of mortality in men: recent evidence from the OPCS Longitudinal Study', *Population Trends*, 80: 31–7.

Laing, W. (1994) *Care of Elderly People: Market Survey*, London: Laing & Buisson.

Larder, D., Day, P. and Klein, R. (1986) *Institutional Care for the Elderly: The Geographical Distribution of the Public/Private Mix in England*, Bath Social Policy Paper No. 10, Centre for the Analysis of Social Policy, University of Bath.

Lewis, J., Bernstein, P. and Bovell, V. (1995) 'The community care changes: unresolved tensions in policy and issues in implementation', *Journal of Social Policy*, 24(1): 73–94.

McCafferty, P. (1995) *Living Independently: A Study of the Housing Needs of Elderly and Disabled People*, London: HMSO.

Means, R. (1993) 'Housing and community care' in J. Johnson and R. Slater (eds) *Ageing in Later Life*, London: Sage: 246–53.

Medical Research Council (1994) *The Health of the UK's Elderly People*, London: MRC.

Moore, A. (1995) 'Deprivation payments in general practice: some spatial issues in resource allocation in the UK', *Health and Place*, 1(2): 121–5.
Nottingham Health (1995) *Aiming for Health in the Year 2000: Annual Report of the Director of Public Health*, Nottingham: Nottingham Health District.
Office of Population Censuses and Surveys (1993) *National Population Projections 1991-Based,* Series PP2 no. 18, London: HMSO.
Owen, D. (1994) 'Spatial variations in ethnic minority group populations in Great Britain', *Population Trends*, 78: 23–33.
Phillips, D.R. and Vincent, J.A. (1986) 'Private residential accommodation for the elderly: geographical aspects of developments in Devon', *Transactions of the Institute of British Geographers*, 11(2): 155–73.
——(1987) 'Spatial concentrations of residential homes for the elderly: planning responses and dilemmas', *Transactions of the Institute of British Geographers*, 12: 73–83.
Phillips, D.R., Vincent, J.A. and Blacksell, S. (1988a) 'Privatising residential care for elderly people: the geography of developments in Devon, England', *Social Science and Medicine*, 26(1): 37–47.
——(1988b) *Home from Home? Private Residential Care for Elderly People*, Social Services Monographs: Research in Practice, University of Sheffield Joint Unit for Social Services Research and Community Care.
Royal College of Nursing (1994) *An Inspector Calls? The Regulation of Private Nursing Homes and Hospitals*, London: RCN.
——(1995) *Nursing and Older People. Report of the RCN Task Force on Older People and Nursing*, London: RCN.
Thane, P. (1989) 'Old age: burden or benefit?' in H. Joshi (ed.) *The Changing Population of Britain*, Oxford: Blackwell: 56–71.
Tinker, A. (1995) 'Housing and older people' in I. Allen and E. Perkins (eds) *The Future of Family Care for Older People*, London: HMSO.
Vincent, J.A., Tibbenham, A.D. and Phillips, D.R. (1988) 'Choice in residential care: myths and realities', *Journal of Social Policy*, 16(4): 435–60.
Warnes, A. (1989) 'Social problems of elderly people in cities' in D.T. Herbert and D.M. Smith (eds) *Social Problems and the City*, Oxford: Oxford University Press: 197–212.
——(1994) 'Cities and elderly people: recent population and distributional trends', *Urban Studies*, 31(4/5): 799–816.

13

GENDER

Liz Bondi and Hazel Christie

INTRODUCTION

Since the mid-1960s or so, feminist urban geographers have documented the nature of gender divisions in British cities. The evidence gathered demonstrates how gender inequalities are played out in urban life. For example, women are disproportionately represented among the urban poor (Fainstein *et al.* 1992). Moreover, the planning and organization of cities often serve to limit women's opportunities (McDowell 1983, Foord and Lewis 1984, Tivers 1985). Recently, a number of commentators have argued that the disadvantages of cities have been overemphasized. They argue that urban lifestyles offer women the greatest opportunities, and that cities provide scope for equality between women and men (Wilson 1991, Young 1990). They do not deny that gender inequalities exist, but they suggest that cities provide possibilities for eroding rather than exacerbating such inequalities.

This chapter explores the impact of economic restructuring on gender inequalities in urban Britain and we consider whether or not recent changes indicate an amelioration of gender inequalities. We begin with the national picture, examining how the transition to post-Fordism has affected the relative positions of women and men. We then focus on one urban centre in more detail, selecting Swindon, which exemplifies with particular clarity the broader elements of post-Fordist economic change in Britain. Finally, we review the policy dimension of gender inequalities in urban Britain.

NATIONAL PATTERNS

Women have long comprised a significant proportion of the labour force: between 1850 and 1950 this proportion remained at least 30 per cent (Siltanen 1994). However, the nature of women's participation in the labour force has always been linked closely to domestic divisions of labour and to ideologies about gender roles and family forms. Thus, in the inter-war period, any woman employed by the state, for example as a teacher or a nurse, had to leave her job if she married: the state actively enforced the view that a married woman's place was in the home.

In the immediate aftermath of the Second World War, this ideology was also very powerful, but by the early 1950s married women were beginning to re-enter the workforce in substantial numbers. Between 1951 and 1971, the percentage of married women in paid employment increased from 22 per cent to 42 per cent. However, this increase did little to challenge traditional gender roles, which associated women with home-making and the care of their kin. Rather, the nature of the employment women entered dovetailed closely with their domestic responsibilities. It did so in two ways. First, the great majority of married women in the workforce during this period worked part-time (legally defined as any post requiring less than 30 hours per week). This enabled them to combine wage-earning with child-rearing and home-making. It also ensured that married women remained financially dependent on men: whilst their wages from part-time work contributed to household income, their earnings were rarely sufficient to support themselves let alone their families. Second, women in general and married women in particular were concentrated in a limited number of occupations, most of which included attributes rather similar to those associated with home-making. Examples include social work, cleaning, waitressing, hairdressing and so on. These jobs also tended to be relatively poorly paid: in 1970, the average hourly pay rate of employed women was only 63 per cent that of employed men (Beechey 1986).

Against this background, we examine the consequences for gender inequalities of the recent movement towards a post-Fordist economy. One of the key elements of this movement involved a very substantial decline in traditional sectors of the economy, most especially in heavy manufacturing such as iron and steel and ship building, and extractive industries such as coal mining. This was accompanied by a two-stage expansion in the service sector. Public and welfare employment opportunities grew in the 1970s and were followed by private sector expansion in the 1980s and 1990s, most notably in business (or producer) services such as banking, finance and related services and advertising. Financial deregulation has been crucial to these developments, as have innovations in working practices, particularly those described as 'flexible' (including flexible working hours and flexible job demarcations).

Many feminists, including geographers, have demonstrated the intrinsically gendered nature of these shifts in the composition of the labour market (Adkins 1995, McDowell 1991, Walby 1989). Industrial restructuring has led to a redistribution of jobs between women and men. Large numbers of blue collar jobs, traditionally occupied almost exclusively by men, have disappeared as a result of the decline in the manufacturing and extractive industries. Meanwhile, the expansion of the service sector has created new jobs many of which have tended to be viewed as 'women's work'. All of this has led to substantial changes in employment, the most dramatic aspects of which have been the development and persistence of high levels of male unemployment, and the creation of a labour market for women that is more regular in its demand than that for men. Thus, employment restructuring has been a deeply gendered process (Bradley 1989).

The extent of these changes in industrial composition can be seen by looking at the employment patterns of women and men since the early 1970s. Perhaps the most striking change in the labour market in this period has been the steady rise in the number of women, especially married women with children, who undertake waged work. In 1972, women constituted 38 per cent of the total labour force; in 1996, women now make up almost half of the employees in employment and look set to become the majority participants in the next few years. Some 70 per cent of working age women are now engaged in the formal labour market, the highest percentage this century, and by 1990, for the first time, working women were drawn equally from the married and single sections of the potential workforce (*General Household Survey* 1992).

This 'feminization' of the workforce was not achieved solely by an increase in the number of women in employment. During the same period the demand for male employees declined as jobs were shed through the intensification and rationalization of manufacturing (and nationalized) industries. In the period 1974 to 1994 alone the number of men in full-time employment fell by 3.4 million (*Employment Gazette* 1994: 10).

At first sight these changes would appear to have reduced differences between women and men in that employment is now more equally distributed between the sexes. However, in various ways most women's experience of employment remains qualitatively different from that of most men. Important facets of this include the incidence of part-time working, gender segregation at work and income differentials between the sexes. We examine each of these in turn.

Part-time employment

The creation of part-time employment has been a central feature of post-Fordist restructuring in Britain. The number of women in full-time work has remained almost constant since 1950, whereas the number working part-time has risen steadily (Hakim 1993). Thus, of the 2.5 million women entering the labour market from 1972 to 1990, 1.8 million (71 per cent) took up part-time work. By 1995 part-time work accounted for 27 per cent of all officially recorded jobs, with women undertaking 80 per cent of these jobs.

Part-time work remains an important means by which women combine wage-earning with family responsibilities. Of all mothers in employment in 1993/4, 63 per cent worked part-time, compared with 44 per cent of all working women and 6 per cent of men (Sly 1994). The proportion of women with dependent children under the age of 16 who are in paid work has increased steadily, with the most marked increases occurring amongst women

Table 13.1 Employment, gender, family and relationship status

	Population in employment (%)	*Those in employment working part-time (%)*
All men (16–64)	75	6
All women (16–59)	65	44
All mothers (16–59)	58	63
Married or cohabiting women with dependent children under 16	62	64
Female lone parents with dependent children under 16	39	57

Source: Sly 1994

with children under 5 (see Table 13.1). Thus, women are tending to take less time out of the labour market to raise children. This suggests some loosening of the gender role ideology that positions mothers as the full-time carers of pre-school children. It also suggests that at least some households cannot afford to forgo women's part-time earnings for more than very brief periods.

Participation in part-time work is further stratified by the relationship status of the mother. Lone parenthood has increased substantially in recent years, principally because of an increase in the number of marriages ending in separation and divorce (Bradshaw and Millar 1991). This is a strongly gendered phenomenon: 90 per cent of lone parents are women. As Table 13.1 shows, mothers in couples are more likely to be in employment than lone mothers: in 1994, 62 per cent of mothers in couples were in employment compared to 39 per cent of lone mothers. That 61 per cent of lone mothers were entirely dependent on state support reflects the paucity of well-paid jobs available to women and the discriminatory effects of a welfare state designed to bolster the nuclear family (Glendinning and Millar 1992). It is only feasible for lone mothers to support themselves and their children through labour market participation if they can command a breadwinning wage; hence those lone mothers who do work are more likely to be in full-time positions.

Gender segregation

Gender segregation within the labour force is very pronounced, and the great majority of people work in environments strongly dominated by either women or men. There are two dimensions to this. First, horizontal segregation refers to the concentration of particular groups within particular industries and sectors of the economy. Women are concentrated in the service sector, especially in public sector services such as health and education, and in personal services such as hairdressing. Second, vertical segregation refers to the concentration of particular groups in particular grades. Women are concentrated in the lower ranks of occupational hierarchies, whether engaged in manual work (for example as cleaners) or non-manual work (for example as clerical assistants and secretaries). Thus, there is a strong correlation between gender and skill: less skilled occupations (both manual and non-manual) are more likely to be dominated by women than more skilled occupations (see Phillips and Taylor 1980).

Income differentials

The concentration of women into lower status jobs is reflected in income inequalities between women and men. Despite the implementation of the Sex Discrimination Act and Equal Opportunities policies in the mid-1970s, and the Equal Pay for Equal Value Report in 1984, women on average earn only 78 per cent of the gross hourly pay received by their male counterparts, a percentage that has remained largely static for the past twenty years (*New Earnings Survey* 1995). One of the factors underlying the persistence of these income differentials is the association between part-time jobs and lower status. But among those in full-time employment the differentials are even greater. For example, in 1995 the average full-time male employee received £369.50 per week compared with £269.30 for women (73 per cent) (*New Earnings Survey* 1995). This is a function partly of gender differences in job status, but also in the availability and distribution of overtime, which is almost entirely taken up by men.

In recent years there have been some shifts in these patterns. In particular women have increased their share of higher level jobs, most especially in professions such as medicine and law, and to some extent in managerial positions (Crompton and Sanderson 1990). However, even in these contexts gender hierarchies remain, as Lindley observes:

> Even where women are increasing their shares of employment in higher level occupations there remains a danger that such broad occupational categories conceal further patterns of segregation with, for example, women managers tending to be concentrated in the relatively low status service sector areas rather than in higher status posts in major manufacturing corporations.
>
> (Lindley 1994: 27)

Thus, any narrowing of gender inequalities amongst the more affluent members of the workforce is proceeding more slowly than is sometimes acknowledged (see also McDowell and Court 1994).

Perhaps more significant are changes at the lower end of the occupational hierarchy. Most notably, more men are moving into poorly paid and part-time jobs. Hourly wages of men in the bottom tenth of the earnings league were lower in real terms in 1992 than in 1975 (Barclay 1995, Gosling *et al.*

1994). Further, the number of men in part-time jobs rose in the early 1990s, albeit from a low base. Together these trends have contributed to some reconfiguration of the gendered division of low-paid workers. Whereas the percentage of women working full-time and earning below the Council of Europe's decency threshold fell from 57.6 per cent to 49.5 per cent between 1979 and 1995, the proportion of men affected increased from 14.6 per cent to 30.4 per cent. Overall, the incidence of low pay increased dramatically in the period from 1979. In 1995 nearly 10 million people in the UK earned less than the Council of Europe's decency threshold wage of £221.50 a week or £5.88 an hour (Low Pay Unit, *The New Review,* Nov/Dec 1995: 48).

This growth in poverty and low pay has been associated with changes in the characteristics of contracts of employment. The movement towards a post-Fordist economy has been associated with intense competition, narrow profit margins and a generally volatile economic context for many businesses. Employers have attempted to diminish the costs of labour in a number of ways, including the increased use of temporary contracts, contracting out business and personal services, and greater use of seasonal or part-time workers. Through these means, employers reduce their liability for national insurance and pension contributions, and their responsibilities for training. The costs of these new forms of employment are borne by the workers themselves. This trend has sharpened a distinction between 'core' or 'primary' workers, who enjoy permanent employment, structured career progression and relatively high remuneration, and 'secondary' or 'peripheral' workers, who are restricted to insecure and low-paid work. The number of peripheral workers has also grown, with many more people dependent upon insecure forms of employment, whether limited fixed-term contracts or self-employment. People employed in these ways are likely to experience recurrent spells of unemployment and highly unpredictable incomes. In addition to the problem of low total earnings, therefore, many households face the difficulty of budgeting for regular and predictable outgoings from erratic and unstable incomes.

The expansion of insecure forms of employment has been accompanied by high levels of unemployment, both short term and long term. Official unemployment figures are notoriously problematic because of frequent changes in methods of counting (thirty changes between 1979 and 1990) (Oppenheim 1993). Data relating to gender differences are even more difficult to interpret because most women with caring responsibilities are not officially defined as unemployed, whether or not they are seeking paid work.

So far we have described income differentials entirely at the level of individuals. However, for many people, household income is more salient than individual income and it is at the level of households that the full consequences of trends in wage levels and conditions of employment are felt. The relative fall in the value of wages earned by men positioned towards the bottom of the income distribution, together with the increased incidence of

Table 13.2 Number of earners per household

	2+ incomes (%)	*1 income (%)*	*no income (%)*
1973	39	36	25
1982	34	32	34
1991	47	28	25

Source: Williams and Windebank 1995

job insecurity, combine with real increases in the costs of living to make many two-parent family households dependent on the income of both spouses if they are to remain above the poverty line. The overall effect of the movement of women into paid employment has not been to increase the standard of living of all households. Indeed, the Low Pay Unit has estimated that the numbers in poverty would increase by 50 per cent if women's earnings were discounted (Low Pay Unit, *The New Review* Jul/Aug, 1995: 8). This is further compounded by social security regulations that militate against wives of unemployed spouses taking paid work unless they can secure well-paid employment because of the pound for pound loss of the husband's benefit beyond a £5 per week disregard. We have already indicated how lone mothers are also often unable to take any paid work because of the operation of the benefit system (see Table 13.1). Consequently, households are increasingly concentrated into two categories: multiple-earner or 'work-rich' households and no-earner or 'work-poor' households. This is illustrated in Table 13.2, which also shows how the recession of the early 1980s propelled more households into the no-earner category. Clearly, the traditional model of family households consisting of a male breadwinner, a female care-giver and dependent children has become outmoded in a post-Fordist economy.

In summary, national trends suggest little change in the pattern of gender inequalities except among the poorest sectors of society, who are themselves increasingly impoverished. In less well-off households, women and men increasingly share in either insecure and poorly paid employment or unemployment and dependence on welfare benefits. To interpret this as a significant amelioration of gender inequalities is misplaced; rather, substantial numbers of both women and men (and children) are suffering as a result of widening income inequalities associated with a shift towards post-Fordism. In other words, the effect of post-Fordist economic restructuring on gender inequalities varies between social classes, with little change evident among the more affluent and a 'levelling-down' evident among the least affluent.

LOCAL CASE STUDY – SWINDON

Behind the national patterns we have described lie distinctive regional variations. The loss of blue collar, male jobs associated with the decline in

Table 13.3 Regional variations in earnings – full-time employees earning below the Council of Europe's decency threshold, by region, 1995

	Male (%)	*Female (%)*	*All (%)*
Greater London	15.0	22.6	18.0
Rest of the south-east	26.7	45.9	33.8
North-west	33.0	55.2	41.1
South-west	34.0	55.9	42.0
East Anglia	34.7	56.5	42.1
Scotland	34.0	56.2	42.6
West Midlands	34.7	57.3	42.7
Yorkshire and Humberside	35.8	59.6	43.9
East Midlands	35.9	60.3	44.1
Wales	36.9	57.9	44.4
North	35.6	60.1	44.5

Source: Low Pay Unit *The New Review* (Nov/Dec, 1995: 10)

heavy industry has been felt principally in the north, which lost twice as many jobs as the south in the recession of the early 1980s (Martin 1995). At the same time, new employment opportunities in the service sector, especially new part-time jobs for women, were much more evenly distributed across space. Since then, there has been some narrowing of regional differences in unemployment, particularly when the recession of the early 1990s resulted in a shake-out of professional and service sector jobs in Greater London and the south-east. However, the income gap between north and south is still very wide (Martin 1995). Indeed, Table 13.3 indicates that only in Greater London and the south-east do the majority of women in full-time employment earn more than the Council of Europe's decency threshold.

The regional scale is itself too aggregated to capture effectively the ways in which economic change and economic inequalities are experienced. Much more salient is the scale of local labour markets, which, through most of Britain, correspond closely to urban areas. Local labour markets vary in occupational structure and therefore in the opportunities for work available to men and women. Different occupational structures, in turn, are associated with different forms of gender relations. For example, local labour markets dominated by heavy industry provide blue collar jobs for men, fewer job opportunities for women, and a tradition of highly differentiated gender roles within households. By contrast, local labour markets dominated by service sector jobs tend to provide more employment opportunities for women (see Massey and McDowell 1984). It is at this scale that the processes at work in maintaining and reproducing gender and household inequalities are played out. Accordingly, we turn now to a local case study in order to examine in more detail the impact of economic restructuring on gender inequalities. We have selected the town of Swindon for this purpose because it is in several

Plate 13.1 Gender segregation at work in 'old' Swindon: skilled manufacturing jobs for men in the railway industry in 1950

ways emblematic of the processes of economic change and polarization described in the preceding section. We begin by summarizing the impact of restructuring on the local economy, and we then examine the impact of these changes on gender inequalities.

The impact of restructuring on the local economy

Swindon is often described as very 'average' because it epitomizes the rise and fall of traditional manufacturing industries, and the shift to post-Fordism. In the early twentieth century Swindon was a single industry railway town with a gendered division of labour based on skilled male manufacturing employment and low female participation in the labour market. This picture began to change in the post-war period when Swindon was designated a New

Town. Overspill population from London and a sustained effort by the local council to diversify the industrial structure resulted in rapid expansion of the town.

Recently the character of the town has changed more radically. In the 1980s Swindon earned for itself a reputation as a 'boom town' and was credited with having the lowest rate of unemployment in Western Europe (*Financial Times* 16.4.1992). Certainly economic performance was good compared to the national position and suggested that the town had made a highly successful transition into post-Fordism (Martin 1989, Marsh and Vogler 1994). Public sector employment in the railway industry was replaced by private sector employment brought by the relocation of national company headquarters from London and by the concentration of 'sunrise' industries along the M4 corridor. Firms that have relocated their headquarters in Swindon include Burmah Castrol, Allied Dunbar, Intel, Nationwide, Readers Digest, Anchor Butter, Galileo, W.H. Smith and Book Club Associates. These trends have been bolstered by the subsequent relocation of some central government agencies and quangos out of London, including the Natural and Environmental Research Council, the Engineering and Physical Science Research Council, the Economic and Social Research Council, Cranfield Institute and a subsidiary of the Ministry of Agriculture, Forestry and Food. Most recently, English Heritage has moved into one of the old railway warehouses.

The impact of restructuring on gender inequalities

Turning to changes in the gender composition of the workforce, Swindon illustrates broader trends very clearly. During the 1950s and the 1960s, the workforce changed considerably in gender composition as substantial numbers of women were drawn into waged work within the diversifying local economy. Thus, census data show that in 1951 under 30 per cent of women were in employment whereas this had risen to 42.5 per cent in 1966 (Harloe 1975). This was higher than the national average rate and very high given that the town also recorded above-average numbers of married women with children. Expressed in terms of the rate of increase of women in the workforce, the figures are even more dramatic: between 1959 and 1965 the number of women in waged work doubled; during the same period, the number of men in waged work increased by a third.

Since 1970, Swindon has experienced a continuing 'feminization' of the local labour force. Between the two last census counts local employment grew at 23 per cent (compared to a national figure of 7 per cent). Overall growth was accounted for by a 15 per cent increase in the male labour force and a 37 per cent increase in the female labour force. Within the latter, the number of economically active married women increased by 26 per cent, whilst for single, widowed and divorced women the increase was 61 per cent.

Table 13.4 Economic activity rates by sex, Swindon (Thamesdown), 1991

	Total persons	*Male*	*Female*
	135,021	66,094	68,927
Economically active	92,598	52,625	39,973
Employees full-time	62,227	39,592	22,635
Employees part-time	15,009	1,552	13,457
Self-employed with employees	2,222	1,751	471
Self-employed without employees	5,667	4,616	1,051
Government scheme	725	427	298
Unemployed	6,748	4,867	2,061
Economically inactive	42,423	13,469	28,954
Students	3,458	1,720	1,730
Permanently sick	3,836	2,175	1,661
Retired	21,550	9,202	12,348
Other inactive	13,579	372	13,207

Source: Population Census 1991

Table 13.4 summarizes gender differences in economic activity in Swindon in 1991.

As elsewhere in Britain, this major shift towards more equal participation in the workforce has been accompanied by significant gender differentiation in the characteristics of paid work. In the last section we identified the main elements of this in terms of the incidence of part-time working, gender segregation at work and income differentials between the sexes. We now examine each of these in relation to Swindon.

Part-time working has been a very major aspect of the feminization of the Swindon workforce. The proportion of economically active women in part-time employment rose from 32 per cent in 1951 to 53 per cent in 1970 (Harloe 1975). Turning to the more recent past, of all new jobs created for women in Swindon between the last two census counts, more than one in three were part-time. In 1991, 16 per cent of employees worked part-time with women accounting for 91 per cent of all people in this position (see Table 13.4). Although the percentage of jobs that are part-time is low compared to the national figure of 27 per cent, the trend in Swindon is upward and this seems set to continue.

Turning to the question of gender segregation at work, the (early) decline in manufacturing and the very rapid increase in service sector employment in Swindon meant that by the mid-1980s the town had lower proportions of highly gender-segregated jobs than many other urban areas (Scott 1994). To some extent this was a product of the particularly masculine workforce when the town was dominated by the railway industry. As the railways declined, so too did the most sharply gender-segregated workplaces. However, gender segregation remains intense both in its horizontal and vertical forms.

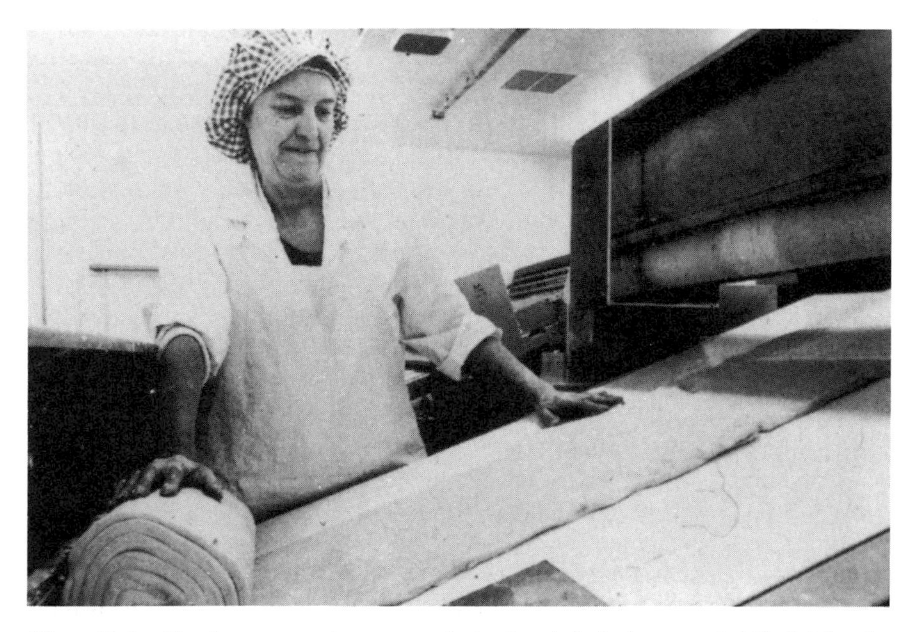

Plate 13.2 Gender segregation at work in 'new' Swindon: semi-skilled jobs for women in the baking industry in 1993

Thus women in Swindon are overrepresented in the more routine, less skilled jobs at the bottom of the occupational hierarchy and often work in occupations dominated by their sex. In case study material drawn from both manufacturing and services, Lovering (1994) found patterns of gender segregation that were remarkably conventional, whether it was in an old-fashioned bakery or engineering firm or the more modern pharmaceuticals or electronics firms. He found that the factors involved in the creation and maintenance of this intense gender segregation included the stereotyping of jobs as 'women's work' or 'men's work', the effects of word-of-mouth recruiting and a preference by employers for single-sex work teams.

Both part-time working and gender segregation are associated with income differentials that position women as a group unfavourably compared to men as a group. On average, women in Swindon earn less then men. In a study of those in full-time jobs, Sloane (1994) found that women's earnings were, on average, only 73.2 per cent of their male counterparts. In a comparison of earnings in the non-manual sector, which has been the major area of growth in Swindon's labour market, women fared even worse, earning on average only 58 per cent of the hourly rates of their male counterparts (Sloane 1994). Sloane's study set Swindon against other local labour markets, some of which are not so strongly post-Fordist in character. The findings suggest that women in non-manual work in Swindon do particularly badly: in

Coventry, women in non-manual jobs were found to earn 75 per cent as much as their male counterparts.

These average figures hide complex patterns in which gender differences are cut across by class inequalities. These patterns are particularly marked in the service sector. On the one hand, Swindon has seen a growth in highly paid, professional employment; on the other hand, the town has also experienced growth in low-wage consumer services and light manufacturing jobs typified by 'flexible', insecure working conditions (Sassen 1991, Lovering 1994, Massey and Allen 1995). Some establishments, such as the research institutes, have a bias towards the professional classes, others to low-wage, highly routinized jobs such as the fast food and retail industries. Yet others, in the hospital and local authority, have a variety of professions, semi-professions and clerical work. New forms of gender and class inequalities are related to the diverse nature of the service sector. One of the hallmarks of economic prosperity in the post-Fordist era has been increased divisions between core or primary and peripheral or secondary workers. Thus, underpinning Swindon's success story are the numerous unskilled workers, many of whom are women, who are restricted to part-time, low-paid and flexible jobs. Economic prosperity has brought rewards for some in Swindon but not for others.

Swindon exemplifies the tension between job creation and employment opportunities for women on the one hand and decline and unemployment for men on the other. Although unemployment rates are low relative to national levels, the numbers officially out of work doubled between 1990 and 1993 to reach a maximum of 8 per cent. Since then, unemployment has fallen to 6 per cent (Thamesdown Borough Council 1995). Men account for three-quarters of the unemployed population. As in Britain more generally, men in skilled manual jobs in Swindon have been particularly vulnerable to unemployment. Although new manufacturing jobs are available in Swindon, they tend to go to men living outside the local area (Thamesdown Borough Council 1993); and, in common with the rest of the UK, downward pressure has been exerted on the wages paid to men in these industries. In a survey of prospects for Swindon in the 1990s a *Guardian* report outlines new forms of stratification in the labour market by taking Honda as an example:

> it [Honda] has been steadily recruiting a youthful management team, will pay less than Rover, and is said to be unwilling to give jobs to redundant Rover workers. . . . And there are no unions.
>
> (*The Guardian* 19.8.1991)

While production of cars has kept alive manufacturing industry, the state of the car industry gives some clues about (poor) socio-economic prospects for men in Swindon in the 1990s.

In summary, this detailed study of one urban area demonstrates that economic boom and prosperity have been felt unevenly by workers of different genders and classes. The gender relations embodied in the 'new' employment contracts associated with the movement towards post-Fordism involve a double dependency for many working-class households: not only are women dependent on having a (male) partner to ensure the reproduction of themselves and their family unit but, increasingly, men's earning capacity is not sufficient for them to support a family. Any amelioration of gender inequalities has been difficult to achieve when women's income is a necessary component of a family budget rather than a passage to economic independence.

POLICY ISSUES

In the preceding sections of this chapter we have argued that significant gender inequalities have persisted in the transition to a post-Fordist economy. Despite the expansion of opportunities for well-educated women, especially in professional and managerial occupations in the service sector, gender hierarchies remain entrenched among the urban middle class. The only notable narrowing of gender inequalities has occurred through the movement of men into unemployment or poorly paid and often insecure forms of employment. The effect of this has been to increase economic inequalities overall and to create two kinds of highly disadvantaged households, namely those without any income other than welfare benefits and those with two earners, both of whom are in poorly paid, insecure employment, often including individuals who combine multiple part-time jobs. In this concluding section we consider some policy issues associated with these trends. In particular, we draw attention to a tension between policies concerned with welfare and policies concerned with equal opportunities.

Traditionally, welfare provision has been intended to provide a 'safety net' to catch those who might otherwise slip into poverty. However, the evidence we have examined indicates that welfare provision sometimes serves to exacerbate inequalities. In particular, the pound for pound loss of benefit (beyond a £5 disregard) when any member of a household takes up employment sharpens divisions between households. As we have indicated, this propels lone parents to choose between full-time work and no work, with the majority of lone mothers forced into the latter category. And it results in many women relinquishing employment if their husbands become unemployed. The fundamental problem is that social security regulations continue to rely on a model of the nuclear family in which a wife and dependent children are 'normally' supported by a male breadwinner. This model bears less and less relation to reality across all sectors of society and the effects are particularly detrimental for the least well-off. This is further aggravated by the erosion in the total value of welfare benefits received by many such households as a result of efforts to reduce public expenditure which have placed enormous pressure on welfare provision.

One of the reasons for the persistence of the model of the asymmetrical nuclear family within welfare provision is that it is regarded as endorsing the institution of marriage. But alongside this model is a body of legislation that enshrines equal rights for women and men and that prohibits sex discrimination in the workplace. This legislation has had some limited success. For example, between 1970 and 1980 women's gross hourly earnings increased from 63 per cent of men's to 74 per cent of men's (Beechey 1986). However, since then the differential has narrowed only very slowly, to reach 78 per cent in 1995 (*New Earnings Survey*, 1995). It seems that little more progress can be made when gender segregation at work remains entrenched and when this social separation of men's work and women's work is associated with such clear differences in status and pay.

The uneven division of responsibility for children also militates against greater equality in earnings. It is almost exclusively women who interrupt their participation in paid work to raise families, which substantially reduces their opportunities and prospects for career progression. It is also mainly women who schedule their working hours to enable them to combine waged work with child care. This all too often limits women's employment opportunities and confines many women to occupational positions markedly lower than those occupied by men with similar educational qualifications. In theory, such gender inequalities could be ameliorated through the enhancement of child care provision and especially through a very substantial increase in the availability of full-day child care. This would have the effect of releasing women to devote themselves more fully to paid employment. In practice, however, it is still women who have primary responsibility for children, and ideologies about motherhood have proved surprisingly entrenched. For example, in dual-income middle-class households where most change might be expected, there has not been a widespread shift in the gendered dynamics of domestic organization. Indeed the practice of wealthy women employing poor women to 'service' them itself contributes to the widening and deepening of gender and class inequalities (Gregson and Lowe 1994).

While policies that enhance child care provision might reduce gender inequalities among those with career prospects, it would be of limited benefit to those who lack the qualifications or opportunities to escape the insecure and low-paid sector of the labour market that has expanded with the shift to post-Fordism. For these people, the majority of whom are women, but who also include substantial numbers of men, welfare policies, including employment protection and the introduction of a statutory minimum wage, are likely to be more significant than equal opportunities policies. Consequently, gender inequalities in urban Britain today can be addressed effectively only in conjunction with class inequalities.

GUIDE TO FURTHER READING

Debates about the gendered nature of industrial restructuring are addressed in many studies. For an overview see McDowell, L. (1991) 'Life without father and Ford: the new gender order of post-Fordism', *Transactions of the Institute of British Geographers*, 16: 400–19.

For detailed empirical investigation of current employment differences between women and men see Lindley, R. (1994) 'Economic and social dimensions' in R. Lindley (ed.) *Social Structures and Prospects for Women*, Manchester: Equal Opportunities Commission: 72–81, and Sly, F. (1994) 'Mothers in the labour market', *Employment Gazette* November, London: HMSO: 403–12.

A useful study of the role of child care in reworking divisions of gender and class is contained in Gregson, N. and Lowe, M. (1994) *Servicing the Middle Classes*, London: Routledge.

Gender differences at the level of local labour markets are examined in a set of texts drawing on the Social Change and Economic Life Initiative. The active creation of gender inequalities in the Swindon labour market is addressed in Lovering, J. (1994) 'Employers, the sex-typing of jobs, and economic restructuring' in A. Scott (ed.) *Gender Segregation and Social Change*, Oxford: Oxford University Press: 329–55.

The argument that urban living has emancipatory possibilities for women is well represented by Wilson, E. (1991) *The Sphinx and the City*, London: Virago.

REFERENCES

Adkins, L. (1995) *Gendered Work: Sexuality, Family and the Labour Market*, Buckingham: Open University Press.

Barclay, P. (1995) *Inquiry into Income and Wealth*, vol. 1, York: Joseph Rowntree Foundation.

Beechey, V. (1986) 'Women's employment in contemporary Britain' in V. Beechey and E. Whitelegg (eds) *Women in Britain Today*, Milton Keynes: Open University Press: 721–31.

Bradley, H. (1989) *Men's Work, Women's Work: A Sociological History of the Sexual Division of Labour*, Cambridge: Polity Press.

Bradshaw, J. and Millar, J. (1991) *Lone Parent Families in the UK*, London: Department of Social Security, HMSO.

Crompton, R. and Sanderson, K. (1990) *Gendered Jobs and Social Change*, London: Unwin Hyman.

Fainstein, S., Gordon, I. and Harloe, M. (1992) *Divided Cities: New York and London in the Contemporary World*, Oxford: Basil Blackwell.

Foord, J. and Lewis, J. (1984) 'New Towns and new gender relations in old industrial regions: women's employment in Peterlee and East Kilbride', *Built Environment*, 10: 42–52.

Glendinning, C. and Millar, J. (1992) (eds) *Women and Poverty in Britain in the 1990s*, Basingstoke: Macmillan.

Gosling, A., Machin, S. and Meghir, C. (1994) *What Has Happened to Wages?*, London: Institute of Fiscal Studies.

Gregson, N. and Lowe, M. (1994) *Servicing the Middle Classes*, London: Routledge.

Hakim, C. (1993) 'The myth of rising female employment', *Work, Employment and Society*, 7(1): 97–120.

Harloe, M. (1975) *Swindon: A Town in Transition*, London: Heinemann.

Lindley, R. (1994) 'Economic and social dimensions' in R. Lindley (ed.) *Social*

Structures and Prospects for Women, Manchester: Equal Opportunities Commission: 72–81

Lovering, J. (1994) 'Employers, the sex-typing of jobs, and economic restructuring' in Scott, A. (ed.) *Gender Segregation and Social Change*, Oxford: Oxford University Press: 329–55.

McDowell, L. (1983) 'Towards an understanding of the gender division of urban space', *Environment and Planning D: Society and Space*, 1(1): 59–72.

——(1991) 'Life without father and Ford: the new gender order of post-Fordism', *Transactions of the Institute of British Geographers*, 16: 400–19.

McDowell, L. and Court, G. (1994) 'Gender divisions of labour in the post-Fordist economy: the maintenance of occupational sex segregation in the financial services sector', *Environment and Planning A*, 26: 1397–1418.

Marsh, C. and Vogler, C. (1994) 'Economic convergence: a tale of six cities' in D. Gallie, C. Marsh and C. Vogler (eds) *Social Change and the Experience of Unemployment*, Oxford: Oxford University Press: 31–65.

Martin, R. (1989) 'Regional imbalance as consequence and constraint' in F. Green (ed.) *The Restructuring of the UK Economy*, London: Harvester Wheatsheaf: 80–97.

——(1995) 'Income and poverty inequalities across regional Britain: the North–South divide lingers on' in C. Philo (ed.) *Off The Map: The Social Geography of Poverty in the UK*, London: Child Poverty Action Group: 128–47.

Massey, D. and Allen, J. (1995) 'High-tech places: poverty in the midst of growth' in C. Philo (ed.) *Off The Map: The Social Geography of Poverty in the UK*, London: Child Poverty Action Group.

Massey, D. and McDowell, L. (1984) 'A woman's place is in the home' in D. Massey and J. Allen (eds) *Geography Matters!*, Cambridge: Cambridge University Press: 128–47.

Oppenheim, C. (1993) (2nd edn) *Poverty: The Facts*, London: Child Poverty Action Group.

Phillips, A. and Taylor, B. (1980) 'Sex and skill: notes towards a feminist economics', *Feminist Review*, 6: 79–88.

Sassen, S. (1991) *The Global City: London, New York and Tokyo*, Princeton NJ: Princeton University Press.

Scott, A. (ed.) (1994) *Gender Segregation and Social Change*, Oxford: Oxford University Press.

Siltanen, J. (1994) *Locating Gender*, London: UCL Press.

Sloane, P. (1994) 'The gender wage differential and discrimination in the six SCELI local labour markets' in A. Scott (ed.) *Gender Segregation and Social Change*, Oxford: Oxford University Press: 157–204.

Sly, F. (1994) 'Mothers in the labour market', *Employment Gazette* November, London: HMSO.

Thamesdown Borough Council (1993) '1991 Census: main findings and implications for economic development committee', *Report to Economic Development Committee*, Swindon: Thamesdown Borough Council.

——(1995) *Unemployment Bulletin*, Swindon: Community Development Department, Thamesdown Borough Council.

Tivers, J. (1985) *Women Attached: the Daily Lives of Women with Children*, London: Croom Helm.

Walby, S. (1989) 'Flexibility and the changing sexual division of labour' in S. Wood (ed.) *The Transformation of Work*, London: Routledge: 127–40.

Williams, C. and Windebank, J. (1995) 'Social polarization of households in contemporary Britain: a "whole economy" perspective', *Regional Studies*, 29(8): 723–8.

Wilson, E. (1991) *The Sphinx and the City*, London: Virago.

Young, I.M. (1990) *Justice and the Politics of Difference*, Princeton NJ: Princeton University Press.

14

POLITICS AND GOVERNANCE

Ronan Paddison

INTRODUCTION

Few would deny that the nature of urban politics in Britain has changed substantively during recent decades. Postmodernists, elevating the politics of difference, are able to point to the declining significance of traditional class-based cleavages and to the emergent new identities, coalescing around gender, the marginalized, environmental activism and other issues leading to the generation of new social movements and local oppositional politics. Those seeking to understand the changing nature of local governance and local democracy are able to point to the radical changes to the institutional structure through which urban politics is conducted, to the shift in policy emphasis within urban arenas towards ensuring the production of proactive local economic development strategies and to the creation of quasi-public agencies alongside the formal machinery of local power. Not unexpectedly, such changes have engendered conflict and resistance reflected in the volatile relationships between the central state and city governments, particularly in the 1980s, and within the attempt by the (central) state to redefine the relationships between itself and civil society. Clearly, such representations of the nature of urban politics emphasize different aspects of it, reflecting the contrasting theoretical understandings of how to conceptualize the key bases of urban politics, power and the purposes to which it is (and should be) directed. As disparate as are these alternative perspectives, they share a common perception of the increasing salience, and often conflictual nature, of urban politics.

To the extent that urban political arenas have become more emphatically the sites of contest and opposition, the contexts for such changes are rooted in the restructuring of the economy and the development of the neo-liberal state. The years since the early 1970s have been witness to radical changes in the British economy, the rapid erosion of the 'old' industrial base accompanied by the burgeoning of the service sector in particular, both of which are linked to the deepening processes of globalization. Nowhere have these changes been more significant than in the city, producing, as Sassen (1993) has argued, the fragmentation of the city into three separate, but interconnected, sectors: the corporate city, the dying industrial city and the immigrant city. Though Sassen's model is a representation intended to describe conditions in the global city, the polarization of urban economies between prospering and decaying sectors, accompanied by the growth of the informal economy, is commonplace amongst British cities.

The state, through the spread of neo-liberalism, has also sought to redefine its relationships with society, championing the role of market forces, the individual as opposed to the collective, and introducing quasi-markets as the means of distribution and allocation. Effectively, such a repositioning sought to redefine the social rights of citizenship, those in particular that had been the mainstay of state welfarism and had been an essential pillar of the post-war long boom. Within a Marshallian sense, such repositioning was not confined to the question of social rights; as Marshall (1950) argued, social rights were closely allied to those of the political and civic spheres, which collectively defined the conditions of citizenship. While the cutting back of the welfare state impinged most directly on the redefinition of social rights, other changes, such as the recasting of the role of local government and that of local democracy, and the implementation of reforms such as the Criminal Justice Act (1994), have been widely interpreted as altering the political and civil rights of the individual.

The impacts of these economic and political shifts have been variable between social groups. In Britain, as in the other advanced economies, the negative impacts of economic and political restructuring have become widely associated with social exclusion. As an identifier of incomplete citizenship, social exclusion – the denial to minorities of those rights enjoyed by the majority – is to some degree a characteristic, if not inevitable, outcome of capitalist society. Yet, there is a wealth of data to show that, following restructuring, the conditions defining social exclusion – the onset of mass unemployment and of the flexible labour market and the creation of 'new' poverty in particular – have become more salient. Allied to, and often contributory to, these conditions of social exclusion have been demographic and social changes, including the growing number of lone parent families in which unemployment or casualized employment is commonplace, the feminization of the labour force and the growth of the elderly population living on fixed incomes, each of which has brought new social groups into poverty. Social exclusion cannot be limited to poverty, although this is of critical importance not only to being able to participate in the consumption society but also in affecting life-chances in so many other ways. The socially excluded include the politically disenfranchised, youth, especially those without work experience, immigrants precluded from access to the formal labour market, women and the disabled. Indeed, the process of defining social exclusion, and the disparate groups that it embraces, underlines the socially fragmenting and polarizing effects of restructuring, as well as the different publics that city politics addresses.

Plate 14.1 Conflictual participation in government: protesting about the M77 motorway in south-west Glasgow

The years since the late 1970s have been witness to a pronounced rupture from the consensualist (though not conflict-free) politics of the post-war boom years, and this contestation has frequently been sited within the cities. One reason for this is obvious – cities have borne the brunt of the restructuring processes. Yet it is precisely because these effects, and the responses of local state agencies to them, have varied that the revival in interest of 'place politics' is justified. Other explanations are perhaps less obvious. Thus, where (urban) local governments were to become a, if not the, primary vehicle for delivering the welfare state, inevitably they were to become deeply implicated in the process of post-welfarism. Yet devolving responsibility for delivering the welfare state to local government harboured the potential for central–local conflict. In the territorially pluralist state it is not only central government that can claim electoral legitimacy for its actions – an argument that was to have particular force with the situation over the 1970s and 1980s in which for the most part the centre has been controlled by the Right and the cities by the Left. In some cities, Liverpool being among the better known examples (Parkinson 1985), Left-wing local councils used the local power base to support the social wage against a reforming central state intent on reducing the welfare burden and the culture of state dependency.

A third explanation is possible – that because of the multiple destabilizing effects of globalization, protecting 'the local' has become all the more important as a means of attaining individual and group security as well as other benefits in an otherwise increasingly insecure, atomizing and competitive world. If the development of local activism was widespread throughout the city, its potential for change has had special significance for those areas in which there were concentrations of the 'new poor' (e.g. the long-term unemployed) and the excluded. The experience of programmes aimed at regenerating peripheral estates and inner cities provides ample evidence of how local activists become engaged in protecting and enhancing the local turf, but that equally their ability to produce change is constrained by the wider structural conditions of power within which local action takes place. How the local should be protected inevitably pits different value systems and representations (as, for example, between actors in the formal and informal political sectors), the interplay between which, as well as possible outcomes, will reflect the differential access to resources and power that characteristically distinguishes between social groups and individuals within the city.

While these explanations are neither exhaustive nor mutually exclusive, they do help to explain the revivification of urban politics. In this chapter the intention is to examine how the changing nature of urban politics has impacted on the conditions of citizenship and social exclusion and the policy responses to these. Inevitably, the discussion cannot consider the city alone – cities, as local systems of governance (just as is true of local economies), lack autonomy. Indeed in both senses, politically and economically, urban governance is an ongoing process that is responsive and proactive to changes in the external environment, whether these conditions were generated by the central state or by shifts in national and global economies. The discussion is organized into three sections, examining in turn the changing configuration of urban governance, the spread of marketization and consumerism, and the role of local state agencies in fostering economic and social regeneration. In each the intention, as elsewhere in this volume, is to draw out the implications of such shifts for the disadvantaged.

THE CHANGING CONFIGURATION OF URBAN GOVERNANCE

Any account of the renewed salience of local politics needs to take into account the changing nature of urban governance. The nature of these shifts is being theorized in different ways: Harvey (1989) describes them in terms of a shift from the managerialism of the (local) welfare state towards the emergence of urban entrepreneurialism, while Jessop (1994) views them in terms of the transition from the welfare to workfare state. In effect, three parallel trends can be identified within the new urban governance: the use of the local arena as the means of co-producing proactive development strategies; the increasing emphasis of local economic development over the 'traditional' concern with social consumption; and the fragmentation of agencies through which change is being organized and conducted. While these identify common trends between cities, they exist in different 'mixes' related to local political conditions and the wider 'structural' position in which cities are located.

In the new urban politics the co-production between local state agencies and the private sector of proactive development strategies has become of critical significance. In an increasingly competitive and globalizing economy, cities have become concerned to ensure their economic vitality. The production of such strategies calls upon the use of new techniques of place promotion and urban marketing, the marshalling of a network of agencies able to meet the infrastructural requirements of potential investors, and the fostering of novel forms of economic activity, such as cultural industries, and of urban tourism, in addition to the attraction of manufacturing enterprises.

Where cities have borne the brunt of the negative impacts of restructuring, the emphasis on economic revitalization is logical. Some theorists take the argument further. Petersen (1981) argued that the overwhelming interest of city politicians was economic growth: quite apart from the benefits the strategy would create, a buoyant economic base was the means by which to reduce the local tax burden for voters. Though the benefits were likely to be distributed unequally, with elites being the chief beneficiaries, others would benefit indirectly through the operation of trickle-down effects. While Petersen's analysis is based on American experience, British cities have increasingly prioritized economic regeneration, none more so than the old, industrial cities in which the task of reconstructing their economic base was the more critical.

It is in these cities – Birmingham and Glasgow are two prime examples – that the debate as to how urban reconstruction should be framed became most politicized. This has highlighted the apparently dual nature of the development of the city in which the plight of the excluded was being compounded by local development policies that were ill-suited to their needs.

Regenerating the city and promoting a new external image for it became linked with capital spending on the construction of major cultural, tourist and convention facilities, or with the development of speciality shopping centres which, critics argued, did little to alleviate the problems of the excluded. Scepticism over the likelihood of trickle-down effects having any real effects for the urban poor were accompanied by criticisms that urban reconstruction, and the imagery underpinning it, was dominated by, and served, particular elite interests. While (as will be seen later) peripheral estates and the inner cities have both been the subject of a variety of policy initiatives, the nature of the debate in cities such as Birmingham raised the wider question of the apparent shift in policy emphasis towards economic regeneration at the expense of social consumption. In fact, such a charge levelled against city governments overlooks the variety of initiatives that they continued to develop to alleviate the problems of multiple deprivation. Rather, the shift

A TALE OF TWO CITIES

DAY THREE OF OUR SERIES ON GLASGOW'S FORGOTTEN COMMUNITIES

The other side of culture city

The other side of Culture City

Beating the drum for Drumchapel

By MARIAN PALLISTER

Shock report on Glasgow's problems

CITY'S GREAT DIVIDE

By ALLAN CALDWELL

A SHOCK picture of Glasgow's future is painted in a major policy document today.

The report is a major acknowledgement by the council that Glasgow's cultural and commercial success is only skin deep.

Cash plea to rescue estates

Smiles better for some

Working-class Glasgow is dying but the wine-bar economy flourishes. Peter Hetherington investigates a Scottish success

Other side of culture city

GLASGOW the DIVIDED CITY

By MAGGIE BARRY

GLASGOW is rapidly becoming two cities.

Dangers lying in deprived schemes

Tenants want to stay put

The cashless society on the edge of culture city

City's rebirth leaves poor on the margins

By Helen Hague
Labour Correspondent

THE GREAT DIVIDE

Culture City's deserts of despair

By AUSLAN CRAMB

Plate 14.2 Media representations of the dual city

in policy emphasis formed an important part of the agenda of the *central state* in the 1980s – the fostering of the enterprise society – to which local government was considered largely an obstacle. This, in turn, led to the bypassing of local government and the formation of a variety of local quangos, which would be more responsive to the needs of business and which were made accountable to the local state. In other words, urban governments were considered too inefficient, bureaucratic and wedded to anachronistic, and for the Right contrary, political objectives of using the local power base to achieve social redistribution.

The employment of marketing campaigns and of the imagineering of the city in the shift towards post-industrialism has exacerbated social divisions in urban areas. A much quoted example is Glasgow in which both the city council and local development agencies, working together with local capital, initiated the promotional campaign through a widely circulated advertising strategy. One of the reasons why Glasgow is a frequently quoted example is because amongst British cities it was an early leader in adopting such strategies; another is because of the debate to which it gave rise between the modernizers and the militants.

The debate came to a head during 1990 when Glasgow acted as the European City of Culture. To the modernizers the opportunity created by Glasgow's designation as a cultural capital was to be used aggressively as the means of reaffirming the new identity of the city, which the marketing strategy begun in the early 1980s had sought to establish. Opposition focused around two main issues. First was the question of whose identity(?); whose Glasgow(?), pointing to the misrepresentation of the 'yuppified' image to the working-class roots of the city. Second was the fact that modernizing the economic base of the city was unduly biased to the inner city at the expense of the large post-war estates on the edges of the city. During 1990 particular events were identified as symbolic of the division, notably the flagship concert given by a leading opera singer, tickets for which cost nearly twice as much as the then weekly welfare benefit paid to the unemployed.

There is little doubt that some city councils were using the local power base in the early 1980s to mount social programmes along the lines suggested by the Right, the so-called 'loony Left councils' in particular. Equally, there is little doubt that after a welter of legislation by the end of the decade the ability of local government to determine how social needs should be addressed had been drastically curtailed, not least because much of its finance was allocated by the central state. Local spending and the standards of service provision are determined centrally through Standard Spending Assessments, which determine the amount each council should spend to produce a standard level of service. Other controls limit the ability to spend on capital projects, even in cases where the local authority has the resources to do so, such as the revenue gained from the sale of the public housing stock under the Right to Buy scheme. (Consideration of the shifts in the mode of service provision and their implications for the disadvantaged are discussed in the next section.)

It is not only through the imposition of stricter financial controls that the ability of urban governments to meet the needs of the socially excluded has been reduced. This has also been effected through the territorial restructuring of local government itself. This is particularly so where, during the 1980s and continuing into the 1990s, the cities have become even more trenchantly Labour-controlled. Large Labour-controlled councils, notably the Greater London Council and in Scotland regions such as Strathclyde and Lothian, posed a particular challenge to the more restricted role designated for local government precisely because of their ability to implement programmes of social redistribution. In part, then, the demise of the Greater London Council can be explained as a reaction to a variety of programmes – including the 'Fares Fair' policy (designed to hold down public transport fares) and those directed at women's and gay groups – which were overtly redistributive. This is not to say that smaller authorities do not mount similar types of programmes, rather that, as in the case of Strathclyde, by virtue of their size and resources, larger region- and metropolitan-wide authorities are in a stronger position to do so. The effect of the abolition of the regional councils in Scotland and the metropolitan-wide authorities in England has been to compound the social fragmentation of the city. But whether such fragmentation will be at the expense of those inner city and other councils in which the disadvantages are concentrated will be as much an outcome of decisions by the central state (for example, as to how the local fisc is supported through the distribution of grant) as it will be of local politics.

A much commented upon aspect of the changing system of urban governance is the rapid development of local quangos, which are specific purpose non-elected bodies covering a range of functions including health, social housing, education and local economic development and training. In many of these functions local government continues to play a role. Yet, as Stewart, Greer and Hoggett (1995) demonstrate, local quangos are rivalling local government in terms of the financial support they are given by the central state, where in 1993/4 support for non-elected bodies amounted to slightly over 70 per cent of that given to elected authorities. (Such a figure would move closer to parity were decisions on school opting out to be imposed by the centre with only a residual number of schools left financed by local councils.)

In spite of the heated debate surrounding quangoization of the local polity, the assessment of its social and political impacts is problematic. Difficulties arise partly because of the wide range of responsibilities over which they have control, not to mention the paucity of information on their operation. Taking the contentious question of their composition, and bearing in mind that for some local quangos, such as the Training and Enterprise Councils, their ethos is strongly rooted in the inculcation of private sector and business-led ideas, several studies have pointed to the relative underrepresentation of disadvantaged groups, notably women, the disabled and ethnic minority members. Yet these biases only replicate long-standing problems of local government membership. As the Widdicombe Report demonstrated for local councils in the mid-1980s, only some 18 per cent of elected councillors were women, a figure that was indeed lower than that for local health authorities and housing associations. It is not the case, however, that local quangos need be any less responsive to minority needs than are urban governments (School for Advanced Urban Studies 1994). Yet advocates of local government remain convinced that the intrusion of quangos into local politics has eroded traditional notions of local democracy and accountability (Council for Local Democracy 1995). In the case of the more doctrinaire-driven quangos – the Urban Development Corporations in particular, as a means of 'bypassing' local government – such arguments are difficult to refute.

CITIZENS AS CONSUMERS

The shift towards the marketization of public services in the 1980s and 1990s and the introduction of the language of consumerism, by the Citizen's Charter in particular, and of consumer choice marks an abrupt departure from the language in which the welfare state had been grounded. Rather than citizens enjoying universally accessible rights, the legislation of the 1980s talked in terms of the consumer *exercising choice*. It is argued that increasing emphasis on the question of quality of service provision, its measurement and the publication of performance statistics provides the basis for the consumer to make rational choices. While not all local government services are capable of being marketized – regulatory activities such as town planning are less amenable to the language of consumerism – the provision of a number of the services critical to life-chances, such as housing, education and (outside local authority control) in particular health, has been altered radically through the infusion of New Right ideas. Such changes have had acute consequences for local government which is no longer seen as the sole, or even in some cases the appropriate, provider of essential services; equally, they have important consequences for the consumer-citizen exercising their choice within the marketplace.

For the socially excluded the shift towards the marketization of services is all the more significant precisely because of their greater dependence on the state. Denied full access to private commodity consumption, the disadvantaged rely more on public services to close the 'consumption gap'. As markets are by definition discriminatory, consumption patterns have been affected by the shift. The effects of the privatization of the public housing stock, and the right to buy at a discount given to sitting tenants, demonstrated the divisive effects of marketization: in 'desirable' estates 40 per cent or more of stock was purchased while in the least desirable estates the right to buy uptake was negligible. Notwithstanding the possibility of any negative equity impacts (whereby the market value of the house could decline to below the price paid) which have affected some purchasers, privatization had socially divisive effects. The drive to owner-occupation, a flagship policy of the first Thatcher administration, bypassed substantial numbers of those within the public rented sector. For many of those remaining in the sector this reflected a preference to remain under the landlordship of the local authority (Dunleavy and Weir 1994). Yet the policy also contributed to the polarization of urban socio-spatial divisions, characterized by the extremes of gentrification and the residualization of the social housing sector within the least desirable areas of the city.

The impacts on the disadvantaged of marketization and consumerism of other services are less transparent than is perhaps the case for right to buy and public housing. Education is a good case in point. Prior to the recent round of reforms, the existence of intra-urban disparities in educational provision and outcomes (as crudely measured by examination performance statistics) was widely recognized and, beginning with the creation of Educational Priority Areas in the 1960s, has been the subject of positive discrimination for a considerable period. Recent statistics of educational performance, published as part of the Citizen's Charter initiative, indicate that significant disparities in examination performance remain (Table 14.1), and although longitudinal studies indicating whether the differences have become greater over the last thirty years are lacking, evidence does exist to show that input measures (e.g. the pupil : teacher ratio) worsened during the 1980s in deprived urban areas. Yet, as critics point out, the government's current methods of measuring the quality of delivery – crude measures of indicators such as the examination performance of schools and their truancy rates – fail to take account of the social differences between catchment areas. In other words, both family background and, more problematically, the social characteristics of the neighbourhood, besides the school itself (its resourcing, the quality of teaching and of its leadership, etc.), influence educational outcomes.

It is because of these complex interrelationships, and in the absence of detailed research on the issue, that caution is needed in drawing conclusions as to any direct, additive effects on social exclusion and on intra-urban patterns of deprivation that recent educational reforms, particularly those introduced under the 1988 Act, may have had. Among the more contentious of the Act's provisions was the ability of schools – once a ballot of parents had approved the move – to opt out of local control and become self-governing. Evidence exists to show that during the first years of opt-out there were clear financial advantages for schools making the move (Bush *et al.* 1993). Yet relatively few schools have made such a move, and by June 1994 ballots had been held in only 6 per cent of the 24,706 schools in England and Wales, though of the 1,547 balloting the majority (1,150) did vote for opt-out. An exhaustive analysis of the socio-spatial patterning of school opt-out remains to be conducted, though it would appear that while the number of schools doing so in deprived areas is negligible, and while those that have are among the 'high performers', their patterning is random and has often been undertaken to avoid the threat of closure.

Table 14.1 Examination performance and household poverty – secondary schools in the Greater Glasgow area, 1989/90

	Pupils attaining 3 or more 'Highers'[1] *(%)*	*Pupils receiving clothing grants*[2] *(%)*
Lenzie	57.4	4
Gryffe	53.1	2
Bearsden	52.5	5
Boclair	50.9	5
Williamwood	50.8	2
Douglas	47.8	7
Kyle	47.6	5
Marr	46.8	4
Eastwood	45.2	5
Greenock Academy	45.1	13
Paisley Grammar	43.3	13
Turnbull	42.3	11
Notre Dame, Glasgow	40.9	29
St Pius	*	51
Wellington	*	58
Glenwood	•	50
Cranhill	*	55
Craigbank	*	60
St Leonard	*	56
John Street	*	50
Springburn	*	53
Merksworth	*	48
Garthamlock	*	49
Kingsridge	*	62
St Augustine's	*	53
Possilpark	*	61
Lochend	*	55

Notes: * Below 1 per cent
1 Highers are the main examinations used to gain access to higher education
2 A means-tested welfare payment providing a surrogate measure of household poverty

Source: The Herald 5 March 1992

Prima facie, the consumer rights given to parents under the 1988 Education Reform Act (already established in Scotland in 1981) to be able to place children in schools of their choice is socially divisive, since the ability to absorb the transport and other costs of placing a child outside the community school will vary between households. Once parents know which schools are the 'better deliverers' (information deliberately provided by the Citizen's Charter to enable parents to be able to make their decision), the claim by the Left is that parental choice will reaffirm the hierarchy of urban schools, particularly those at the secondary level, while unfairly rewarding already relatively advantaged households. Intuitively, the claim for uneven advantage between households appears reasonable, though what evidence there is provides support at a more aggregated level. In Glasgow, for example, a few secondary schools, each of them located in middle-class areas, have emerged as magnets for placement requests outside their catchment area; yet in each case the number of requests by households living in deprived areas is lower than from more advantaged areas.

The examples of housing and education, both critical in different ways to life-chances, provide somewhat conflicting evidence as to whether, and how, marketization has had additive effects for the already disadvantaged. By its nature, the introduction of market forces is atomistic, though it is problematic whether the benefits enjoyed by individual households, resulting from the shift towards consumerism in the provision of specific services *per se*, have in turn produced spatially cumulative effects that reinforce the intra-urban map of social disadvantage and exclusion. The key point here is the *separate* effect of service delivery shifts, and particularly in the case of education, whether there has been sufficient time for the changes to have had the spatial effects the rhetoric has suggested for them.

EMPOWERMENT AND LOCAL REGENERATION IN DIFFICULT ESTATES

It is in Britain's 'difficult estates' (of which Power [1991] has estimated that there are approximately 2,000 comprising 500 or more housing units) that there is sufficient evidence to show the extreme effects of restructuring. In some areas more than 80 per cent of the population is unemployed, with most school-leavers having little prospect of gaining work experience, and crime rates are typically several times the national average. Accounts of these sink estates draw graphic descriptions and telling indictments of the failure of the state at both central and local level. Controversially, Campbell (1993) argues that local state agencies and both major political parties have all but abandoned them; they are Britain's 'dangerous places', not just in the stereotyped vision, but as the graveyard potentially for local politicians. If Campbell's interpretation is challenging but contestable, what is far less in doubt is that during the 1980s changes in income and wealth, combined with shifts in housing provision and the delivery of other services together with reductions in welfare benefit payments, had mutually reinforcing effects in such estates.

Current programmes targeted at difficult estates emphasize the ideas of partnership and empowerment. Typically, programmes involve joint action between central and local state agencies working co-operatively with the private sector to attain social and economic objectives. Empowerment involves co-opting local communities into decision-making processes that enable them to voice their needs. Programmes include City Challenge, begun in 1991, the Single Regeneration Budget, the Priority Estates Project and European Union projects such as Poverty 3.

Particular interest has focused on the effectiveness of these programmes in empowering the local community. In part, this is because previous inclusion of participatory democracy had paid lip-service to it, or in the case of earlier projects, such as the Urban Development Corporations, it had been excluded from what were explicitly business-driven initiatives. Yet it also reflects the conditions of political exclusion among the population in these areas. Most studies of empowerment in practice are critical of its achievements on the grounds that the ability of community activists to participate effectively is limited both by bureaucratic procedures and the activists' inexperience of them and by the problems of 'them' and 'us' that exist between professional officers and local elected politicians, on the one hand, and community representatives, on the other. These problems reveal a failure to realize that empowerment involves a transfer of power downwards.

In spite of these problems, empowerment has enabled some groups to realize the potential it offers to attain improvements in service delivery. This has benefited deprived groups and, in particular, women, the unemployed and the young. Women have used the opportunity of partnership and empowerment to identify gaps in service provision most affecting their interests (such as crèche facilities) and to suggest ways in which delivery could be improved. Participation such as this has the potential to deliver not only material benefits but also less tangible ones in raising the self-esteem of the individual and in creating the conditions in which household survival and coping strategies for the individual are dependent on local collective action. In other words, empowering service users and decentralizing the nature of service downwards to the community has the potential for reversing the cycle of disempowerment (Stewart and Taylor 1995). Yet, as in all participatory exercises, expectations may outstrip what are realistic outcomes, particularly where the latter are constrained by power relations and the availability of resources.

CONCLUSION

In her book *Justice and the Politics of Difference*, Young (1990) sets out the normative ideals of city life, the first of which (that it achieve social differentiation without exclusion) has particular bearing for the disadvantaged. Her stance is experiential: cities foster group differentiation and differences of lifestyle, though in the ideal city they should 'not stand in relations of inclusions of exclusion, but overlap and intermingle without becoming homogeneous' (Young 1990: 239). Besides the city's representation as an ideal, the argument celebrates difference and its acceptance as part of city life.

The contrasts between Young's vision of the ideal and the realities of social exclusion in contemporary Britain drawn by Campbell are striking.

Poverty and exclusion may become masked within cities, perhaps because of the very anonymity membership of a large city offers, but the masking is more imposed than sought. Problem estates become subject to 'othering' by most citizens, places that are rarely visited and largely forgotten. Within the terms of Hutton's '30–30–40 society' (Hutton 1995) it is hardly likely to be otherwise; restructuring is imposing economic hardship and/or threat on all, the 30 per cent disadvantaged most of all, but increasing marginalization on the intermediate group and status threats on the 40 per cent relatively privileged.

Exclusion also matters politically, harbouring as it does the potential for social disorder. In the 1980s, urban riots became the most vocal expression of the injustices of the enterprise society and the claim for a more equitable distribution of citizenship rights. They also represented a potential threat to urban society and the ability of the agencies of social order to maintain law and order.

Yet if for the more privileged within urban society the excluded and their communities are 'other places', governments are mindful of the problems. In part, this is because their policies have helped to create the problems, of which the implementation of neo-liberal policies by the central state had the effect of further marginalizing the already disadvantaged. The extension of marketization to public services that were previously considered universal rights, and the distinctions it creates, inevitably provoke political reaction. In the case of the short-lived poll tax, the question of its legacy on the 'disappearing voters' remains, in the effect that the disadvantaged, the unemployed and the young, drawn into the tax-paying net, perhaps for the first time, absconded through self-disenfranchisement.

For the emergent system of urban governance the lesson of the 1970s and 1980s has been the policy interdependencies that exist between the efforts of the different actors, central and local, capital and community. The intractability of the problems of multiple deprivation, precisely because of its very nature, and the shortcomings of previous area-based policy initiatives were widely appreciated. Yet in the changing world of urban governance in the 1980s and 1990s local programmes had to contend with contradictions resulting from policy shifts in the agenda of the central state, and changes to the system of welfare payments in particular. In turn, these spurred new approaches, partnership and empowerment as the possible means of reducing area-based social exclusion. If the prognosis for such policy shifts is itself pessimistic in the sense of being able to realize any substantial redistribution, there is limited consolation in the recognition that the problems of social exclusion are by no means unique to British cities.

GUIDE TO FURTHER READING

Harvey, D. (1989) 'From managerialism to entrepreneurialism: the transformation of urban governance in late capitalism', *Geografiska Annaler*, 71(1): 13–18 provides a theoretically grounded statement of the entrepreneurial nature of the new urban politics, while Cox, K. R. (1995) 'Globalisation, competition and the politics of local economic development', *Urban Studies*, 32(2): 213–24 draws out the linkages between globalization, competition and the politics of local economic development. A thorough statement of the social effects of Thatcherite enterprise-creation policies in British inner cities is provided by Deakin, N. and Edwards, J. (1993) *The Enterprise Culture and the Inner Cities*, London: Routledge. More specific policy accounts are provided in Jones, J. and Lanley, J. (eds) (1995) *Social Policy and the City*, Aldershot: Gower, including their effects on poverty, race, empowerment and the impacts of the Church of England's initiative, *Faith in the City*, on urban deprivation. Finally, Stewart, M. and Taylor, M. (1995) *Empowerment and Estate Generation*, University of Bristol: The Policy Press gives a useful overview of the problems and potential of empowerment for the regeneration of 'problem' estates.

REFERENCES

Boyle, M. and Hughes, G. (1992) 'The politics of the representation of "the real": discourses from the Left on Glasgow's role as European City of Culture, 1990', *Area*, 23(2) 217–28.

Bush, T., Coleman, M. and Glover, D. (1993) *Managing Autonomous Schools*, London: Paul Chapman.

Campbell, B. (1993) *Goliath: Britain's Dangerous Places*, London: Methuen.

Council for Local Democracy (1995) *Taking Charge: The Rebirth of Local Democracy*, London: Council for Local Democracy.

Cox, K.R. (1995) 'Globalisation, competition and the politics of local economic development', *Urban Studies* 32(2): 213–24.

Damer, S. (1990) *Glasgow: Going for a Song*, London: Lawrence Wishart.

Deakin, N. and Edwards, J. (1993) *The Enterprise Culture and the Inner Cities*, London: Routledge.

Dunleavy, P. and Weir, S. (1994) *Local Government Chronicle*, 29.4.1994: 12.

Harvey, D. (1989) 'From managerialism to entrepreneurialism: the transformation of urban governance in late capitalism', *Geografiska Annaler*, 71(1): 13–18.

Hill, D.M. (1994) *Citizens and Cities: Urban Policies in the 1990s*, Hemel Hempstead: Harvester Wheatsheaf.

Hutton, W. (1995) 'The 30–30–40 Society', *Regional Studies*, 29(8): 719–22.

Jessop, B. (1994) 'The transition to post-Fordism and the Schumpeterian workfare state' in R. Burrows and B. Loader (eds) *Towards a Post-Welfarist State*, London: Routledge: 13–37.

Jones, J. and Langley, J. (1995) *Social Policy and the City*, Aldershot: Gower.

Lansley, S., Goss, S. and Wolmar, C. (1989) *Councils in Conflict: The Rise and Fall of the Municipal Left*, London: Macmillan.

Marshall, T.H. (1950) *Citizenship and Social Class and other Essays*, Cambridge: Cambridge University Press.

Paddison, R. (1993) 'City marketing, image reconstruction and urban restructuring', *Urban Studies,* 30(2): 339–50.

Parkinson, M. (1985) *Liverpool on the Brink: One's City Struggle against Government Cuts*, London: Heritage Journals.

Petersen, P. (1981) *City Limits*, Chicago IL: Chicago University Press.

Power, A. (1991) *Housing – A Guide to Quality and Creativity*, London: Longman.

Sassen, S. (1993) 'Rebuilding the global city: economy, ethnicity and space', *Social Justice*, 20(3/4): 32–50.

School for Advanced Urban Studies (1994) *TECs and Provision for People from Ethnic Minorities*, University of Bristol: SAUS.

Stewart, J., Greer, A. and Hogget, P. (1995) *The Quango State: An Alternative Approach*, CLD Research Report No. 10, London: Council for Local Democracy.

Stewart, M. and Taylor, M. (1995) *Empowerment and Estate Generation*, University of Bristol: The Policy Press.

Young, I.M. (1990) *Justice and the Politics of Difference*, Princeton NJ: Princeton University Press.

Part III
PROSPECTIVE

15

THE URBAN CHALLENGE

How to bridge the great divide

Michael Pacione

> The divisions in society are never starker than at Christmas;- bleak beggars in the London subway while in Harrods £1000 hampers are selling out. This inequality is no accident.
>
> (Hutton 1993: 4)

Mounting evidence of the adverse local implications of global economic restructuring, the de-industrialization of the UK economy and market-dominated urban policies has provoked critical reaction. A large number of alternative policies and programmes have been suggested in an attempt to effect a more just distribution of society's resources. These have included radical proposals such as nationalization of the banking system and pension-fund socialism, equity planning, pragmatic radicalism, linked development and leverage schemes as well as a host of local economic strategies and schemes to encourage socially useful production (Boddy and Fudge 1984, Hasluck 1987, Pacione 1990, Kraushaar 1979, Krumholz 1982, Keating 1986, Squires 1989). Other models look for inspiration to the sustainable development achieved in the Third Italy (Hatch 1985), the collective development effort in Mondragon, northern Spain (Cooke 1990), local provision of venture capital as in Germany (Heidenheimer *et al.* 1983), and even some of the aided self-help initiatives found in Third World cities (Gilbert and Gugler 1992).

Successful implementation of any policy, however, is dependent upon the state's ideological stance in relation to social division, poverty and welfare. This governs the type of urban policy deemed acceptable within the prevailing political economy – and in so doing conditions the incidence and intensity of urban division.

In the UK an early response to the urban social and economic policies of Thatcherism was the adoption of confrontational oppositional strategies by local authorities, the most visible being the local economic development policies of the now disbanded Greater London Council. Despite their innovative content, these plans generally failed to realize their main objective of introducing an enhanced social dimension to government urban economic policy. This lack of success was a direct consequence of the ideological gap between their proponents, normally local authorities on the political Left, and central government. This ensured tight control by the latter over the activities and in particular the financial practices of the former. With the exception of small-scale trial projects, the ideological, power and policy differences between the two levels of government have effectively precluded the successful introduction of alternative strategies at the urban level. Realistically, we cannot avoid the conclusion that confrontational economic strategies have failed to effect a significant improvement in the well-being of the urban poor and must be replaced by new approaches.

Idealism must be tempered by realism in the attempt to produce an effective response to the problems of urban economic decline and social division. This does not exclude oppositional critical analysis from the debate over urban restructuring but acknowledges that, in the short term at least, there are barriers to the implementation of more radical alternative policies. This conclusion may be disappointing for those committed to alleviating the position of those disadvantaged by the prevailing political economy, but in a democratic society (particularly where government is dominated by the ideology of the New Right) it is wholly realistic.

Several strategies have been identified that seek to shift the current social and economic imbalance in favour of the disadvantaged, while working, to a greater or lesser extent, within the constraints of the dominant ideology.

LOCAL LEVEL APPROACHES

One of the more radical suggestions takes as its starting point Jacobs' conception of a city as 'a settlement that consistently generates its economic growth from its own local economy' (Jacobs 1969: 48). The notion of local self-sufficiency as the best means of producing long-term, sustainable work was subsequently taken up by Morris and Hess (1975) and Short (1989). This involves replacing the dominant ideology of national and international markets with the creation of stronger local markets. In practical terms it is suggested that this might be promoted by strategies such as selective purchasing by local authorities, adoption of residency requirements for public employees, and encouragement of local investment and partnership between finance capital, local authorities and community groups. All seek to maintain the circulation of capital within the local urban economy. While such strategies have potential to advance this goal, no city is likely to be able to prevent leakage to other parts of the national and international economy.

In a more recent development of the concept of local self-reliance, Johannison (1990) focuses attention on the territorial dimension to local economic development, and contends that the goal of sustainable development may be achieved through 'economies of scope' based on an appreciation of the joint resources distributed among various sectors within the territory. This paradigm regards the entire socio-economic environment as a potential resource bank. Importantly, this perspective does not imply isolation from the wider environment and restriction to internal resources and local markets. In order to be able to cope with changing external conditions, the sustainable community needs members with both a strong local identity and a global outlook. Among the most successful examples of sustainable development of this kind are those in Mondragon, northern Spain, and in the Emilia-Romagna region of central Italy. The closest approach to this model in the UK was the enterprise boards set up by Labour-controlled local authorities to act as regional investment agencies, but these failed to emulate the integrated territorial development strategy characteristic of regions like the Third Italy. The reasons for this are complex. In addition to geographical and structural differences between the two regions, the Italian success was built on three decades of concerted effort by a partnership of agencies with shared cultural and political aims and values at the regional level. This partnership, in practice, comprised a government-sponsored bank to provide low-interest credit to small enterprises, cooperative management agencies, sharing of technical knowledge, and links between schools, training colleges and local employers, all fostered by a significant degree of regional autonomy. The contrasts with the prevailing British political economy are clear and help explain the absence of a comparable example of local development in the UK.

One of the principal difficulties for local attempts to revitalize decayed urban environments is the understandable reluctance of capital to invest in areas perceived as lacking market potential. The evidence from British cities is that appeals for *ethical investment* via initiatives such as Business in the Community have limited relevance for the most disadvantaged neighbourhoods (Metcalfe *et al.* 1990, Pacione 1993). It is possible that a degree of *enlightened self-interest* may convince firms to maintain or increase investment in an area. In some cases, because of the relative immobility of productive capital, as opposed to finance capital, a company with a sizeable investment in a plant in one area may prefer to subsidize initiatives to promote local development as an alternative to relocation. More generally, some economic pressure is often necessary to convince capital to enter into a partnership with local authorities and communities. Two types of strategy offer possibilities within the UK context. The first refers to *linked development* and the second to the concept of *leverage*.

Linked development presents private sector investors and developers with a constrained choice of investment opportunities in an attempt to induce a more socially and spatially balanced pattern of growth. Policies might include, in the housing field, 'inclusionary zoning', which requires developers to set aside low-cost housing units in market rate projects for allocation by a neighbourhood housing association. Linked employment strategies could include a requirement that commercial developers hire labour from specific geographical areas and social groups. This combination of regulation and partnership between public, private and community sectors can effect a more equitable distribution of the benefits of urban growth, but linkage policies are not a panacea for uneven urban growth. The extent to which the public sector can influence private investment decisions depends on the attractiveness of the city for capital; but rarely is there no potential for linkage. Most urban areas possess locations sufficiently attractive to capital for a premium to be exacted for the benefit of the community at large. While the type and value of any premium is a matter for local political leaders to decide, this strategy offers a direct means of redistributing at least some of the profits of central area development to less advantaged parts of the city.

A related strategy employs the concept of leverage to advance the 'popular restructuring' of urban space (Pacione 1992). As commonly employed in the UK the principle of leverage refers to the use of small amounts of public funds to attract larger sums of private investment. In practice, leverage ratios vary considerably between projects (Martin 1990). Most significantly, this form of 'persuasive leverage' does not result in significant private sector investment in deprived urban neighbourhoods. It is clear that to achieve this a more proactive form of leverage is required. This might take the form of social impact analysis by city planning authorities of any proposed development project in order to examine questions such as whether it will provide more jobs at liveable wages for city residents, or whether it will support neighbourhood vitality or contribute to its deterioration. In effect such a policy resembles a radical interpretation of the established practice of 'planning gain'. For some observers this form of action in pursuit of a 'popular restructuring' of urban space smacks of blackmail; for others it is simply an example of sound business practice.

Economic and social strategies aimed at reducing the problems of urban division represent only one part of the equation. Of equal importance is the related issue of the distribution of *political* influence and power in the city. The political power of citizens in the UK is encapsulated in the concept of democracy (equal voting rights), with less explicit attention given to the secondary principle of isogory (the equal right to influence public opinion). While the former is readily available, the latter is effectively unattainable in practice since the vast bulk of information received by citizens is generated at a higher level by a small proportion of the population, not least professional politicians. One way of redressing such 'disenfranchisement by scale' would be to reduce the size of the unit of local government. However, we must treat with caution the suggestion that local government, because it serves a more restricted population than central government, must be more attuned to popular local needs. Decentralization *per se* is not an absolute benefit and could under certain circumstances provide a mechanism for policies of social, economic or racial exclusion. In such circumstances a degree of centralization may be a preferable strategy if it leads to territorial justice or the redistribution of wealth.

The debate over the optimum size of government unit, which characterized the latest reform of the local government system in the UK, hinges on the twin issues of efficiency and local democracy. The first cites scale economies in favour of larger administrative units, with the latter arguing the merits of local accountability and policy sensitivity in support of smaller-scale government. From the viewpoint of hard-pressed city authorities the concept of an 'associational network of local communities' in a *city-region* holds many attractions (Papworth 1988). Carefully demarcated, the city-region can break down the urban–rural divide to reduce spillover and free-rider effects, incorporate affluent and disadvantaged areas to aid redistribution, provide a threshold population of sufficient size to ensure provision of specialized services, and permit devolution of responsibility within the city-region framework. To date in the UK decentralization has been concerned largely with attempts to improve services through local delivery mechanisms (Willmott 1987). Progress towards a more participatory form of democracy through devolution of power has been thwarted by a variety of factors. Some of these relate to concerns over the ability of local communities to accept and manage responsibility, over whether departures from centrally determined service standards can be accommodated, and the need for a higher authority to adjudicate between conflicts at the local level.

Another major obstacle has been resistance from urban professionals, bureaucrats, planners and politicians and the consequent lack of proper resourcing for decentralization. While the absence of real power at the local level limits the effectiveness of much of the work undertaken by community and voluntary agencies, we must also recognize that there is no single best model for populist grassroots decentralization. Central to the empowerment of local communities is the principle of subsidiarity. This maintains that a higher-order community should not interfere in the internal life of a lower-level community to the extent of depriving the latter of its functions, but rather should support it in case of need and help to co-ordinate its activities with those of the rest of society in keeping with the common good. Local communities in turn must be wary of the dangers of introspection and must not neglect their relations with the wider regional, national and international community. Formation of a national confederation of disadvantaged urban communities united in pursuit of a more equitable form of development could act as a political pressure group for distributive justice. Critical analyses undertaken by local groups with the ability to think globally can also perform the important emancipatory function of raising public consciousness about the impact of particular political and economic investment decisions on the local urban environment.

A fundamental difficulty facing alternative strategies for urban regeneration is the inability or unwillingness of government to underwrite the programme with increased funding. Thake and Staubach (1993) have proposed a strategy designed to work within existing levels of expenditure by maximizing benefits through better prioritizing, focusing and integration of programmes. Their first requirement is that national priority be afforded to community regeneration with this commitment reflected in a specific regeneration budget and possibly the creation of a Community Enterprise Corporation (akin to the Housing Corporation for Housing Associations). A second requirement is that the regeneration effort should be targeted on neighbourhoods in which *economic activity rates* are lowest (thereby including the elderly, single parents and other excluded groups in addition to the unemployed). Third, available resources would be concentrated on neighbourhoods (of between 8,000–12,000 people) 'with sufficient critical mass for positive action to take place'. A national programme for community regeneration would, in the first round, seek to cover sufficient areas to include a quarter of those registered unemployed, with other areas being included as successful communities move away from dependence on external support. Fourth, a policy of triage would be appropriate on the grounds that a 'worst first' strategy is not necessarily an effective use of resources. Some neighbourhoods may be irretrievably damaged and investment would deny resources to other more viable neighbourhoods. Fifth, selection of neighbourhoods for inclusion could be made by central government, local authorities or a local community, but once designated the local authority would be responsible for bringing together all groups and agencies with an interest in the neighbourhood to plan its development. Sixth, a Community Enterprise Agency would be set up to operate within the neighbourhood with financial power to set budgets, commission work, employ staff, and trade in land and property in its area, and the potential to develop into a community-based business that could make an impact on the local employment market. Seventh, once integrated neighbourhood programmes had been established, regeneration strategies agreed and CEA identified, the designated neighbourhoods should be granted the special status of Community Enterprise Zones eligible for enterprise zone benefits, such as tax breaks, and employment and training grants. Finally, Thake and Staubach (1993) believe that granting 200 CEZs over a five-year period would permit the formation of a national network of Community Enterprise Agencies that could initiate joint projects, promote national marketing of local products, link companies to local skills and help dissolve the isolation felt by many multiply deprived communities.

A major problem with this proposal is the lack of a detailed analysis of costs to support the contention that the programme could be resourced from existing public expenditure. A significant advantage of the proposal lies in its co-ordination of policy effort at local and national levels, and acknowledgement of the need to recognize the problem of multiply deprived communities as a national priority. Emphasis is also attached to the fact that this urban-scale initiative represents only one component of a programme of national recovery designed to complement and run in parallel with macro-economic programmes.

Reorganization of public expenditure and reappraisal of the goals of urban policy also underlie the recommendations of the Joseph Rowntree Foundation inquiry into income and wealth in the UK (Barclay 1995, Hills 1995). The report criticizes 'short-termism' in public policy and observes that 'public spending appears to have got into a trap, where short-term savings have had long-term costs, in turn creating pressure for later short-term savings, in a continuing spiral'. The principal conclusion is that public spending should be reoriented towards investment for future success rather than paying the price of past failure – as, for example, when social security payments are made to those who are unemployed because their education and training were inadequate to enable them to meet the requirements of modern industry. The report also acknowledged that since the urban crisis has multiple causes, an effective policy response requires action in a wide range of areas. Particular proposals in relation to the labour market include:

1 direct provision of employment opportunities by government through programmes that bring the resources of the unemployed to bear on community and environmental problems;
2 improved child care provision to assist child carers into work;
3 greater flexibility in working time;
4 stronger legislation against discrimination in employment practices against the disabled, women and ethnic minorities;
5 measures to improve the income of the low paid either by means of a statutory minimum wage or an extension of in-work benefits.

In the field of social security the main recommendations were directed towards:

1 annual upgrading of benefits in relation to average incomes (not prices as at present) for those who are unavoidably dependent on benefits;
2 easing the transition between unemployment and work by providing grants (e.g. to cover transport costs) and by easing the rules on returning to social security should employment fail.

In terms of the *geographical concentration* of poverty in marginalized areas the report recommended that:

1 housing allocation policies by local authorities and housing associations should avoid allocations to a particular estate consisting overwhelmingly of those dependent upon benefits, and new developments should be in the form of smaller estates with, if possible, a mix of tenure types;
2 national and local business leaders are encouraged to consider how 'responsible' companies can ensure that deprived communities are not excluded from mainstream economic activity. However, in view of the body of evidence to date, this somewhat pious view ignores commercial reality and probably represents the presence of representatives of the business community on the committee of inquiry;

3 the problem of motivating young people may be eased by establishing direct links between training schemes and local employers;
4 better use of available resources through local management, decentralized budgets and resident involvement in decision-making;
5 improvement in the quality of local schools in order to break the cycle of deprivation;
6 improvements to the physical environment and to transport facilities in order to integrate marginalized outer estates with the rest of the city;
7 finally, and most significantly, the report acknowledged that the continued concentration of multiple deprivation in particular areas required a regeneration strategy that goes beyond initiatives for particular estates.

A key conclusion that emerges from the foregoing analysis of strategies aimed at alleviating problems of urban division is that local initiatives alone are insufficient to overcome the complex of problems affecting deprived urban areas; to be successful such strategies must be undertaken within the framework of a national plan for the regeneration of Britain's cities. Whether such a major reorientation of policy is possible is open to debate, but this does not reduce the need for academic analysis of possible alternative policy structures. In the next section we consider a number of contrasting future scenarios at the national level.

NATIONAL URBAN POLICY

A primary goal of Thatcherism was not only to retain political power but to engineer a fundamental change in British society, including the defeat of socialism. New Right policies – for example, in relation to the sale of council houses, trade union regulation, education, reform of the welfare state, decentralization and privatization, and curtailment of local government – have destroyed the basis of the post-war settlement and have exerted a major impact on the economic, social and political life of Britain. While the future is a terra incognita, it is possible that the doctrine of Thatcherism may well outlive its creator. Certainly, the longer a Conservative government following Thatcherite principles remains in office the more entrenched its values are likely to become and the more difficult it will be to reverse established policies (Thornley 1991). Like the post-war welfare state, the vitality of Thatcherism is dependent upon economic success; to maintain electoral support enough people must feel that they benefit from the mix of economic liberalism and political authoritarianism that characterizes the ideology. In view of this, one may speculate on potential weaknesses in Thatcherism that may undermine its popular support.

The first refers to Thatcherism's dependence on the market. The uneven distributional outcomes of the unrestrained operation of free market forces has resulted in problems of growth (e.g. transport congestion, pollution and escalating land prices) in some areas and problems of decline (e.g. unemployment, poverty and social exclusion) in others. As we have seen, major socio-spatial divisions are apparent both regionally between north and south in Britain and within Britain's cities. The income gap within British society is reflected increasingly in an electoral split between north and south, and between inner city and suburbs. In Inner London and in the cities of northern Britain opposition parties have established a solid political base. Further, the ideological restructuring of the Labour party under Tony Blair, including the jettisoning of traditional policy commitments such as nationalization, has moved the main opposition party to a position in which it appears much less of a high-spending high-taxation *bête noire* to traditional, middle-class Conservative supporters. A social-democratic alternative ideology, involving a degree of state control over the market and emphasizing community values, could prove acceptable to a majority increasingly experiencing the pressures emanating from the individualistic ethos of market capitalism.

A second potential weakness of Thatcherism lies in its marginalization of local democracy and public participation. Despite the introduction of various forms of Citizen's Charters by John Major's government, the growing popular pressure for greater involvement in planning and local decision-making that existed prior to 1979 could re-emerge as discontent grows with central government, especially in those regions and cities that have suffered most under the existing policy regime.

Whether such trends will develop, only time will tell. What is clear is that, as the GLC discovered to its cost, 'conflictual participation' is unlikely to succeed in restructuring British society in the face of contrary forces emanating from the central state and capital, particularly where these interests enjoy a symbiotic relationship and share common economic goals. Put simply, a fundamental reorientation of urban policy in the UK can only follow a fundamental shift in the ideology of government.

ALTERNATIVE IDEOLOGIES

Even prior to the events that heralded the collapse of the Communist ideology in Eastern Europe, such a radical alternative to market society had little prospect of taking root in the UK, where property rights and individual ownership are woven into the fabric of society. Since 1979 in Britain, even traditional socialism has failed to capture the minds and votes of a majority of the population, which has preferred to enjoy the material rewards generated under capitalism.

The attention of those seeking an alternative form of social organization in the UK has focused on the ideologies of market socialism and social democracy. Market socialism is a pragmatic philosophy that acknowledges the power of human self-interest. It attempts to reduce social inequality by retaining private ownership of the means of production and differential wages in accordance with skill and effort but redistributing corporate profits in the form of a 'social dividend' (Roemer 1992). A central feature of this formulation is that it does not seek to impose equality in society. Market socialism is not predicated on an abstract belief in individuals being motivated by a commitment to the common good but is founded on the basis of a fair wage for a fair day's work. In terms of its popular acceptability, an attraction of market socialism may be that it is not just in an abstract sense but it is more just than unrestrained capitalism.

The social-democratic approach lies mid-way between capitalism and market socialism. A social-democratic alternative to a continuation of current urban policy was framed by the report of the Commission on Social Justice (1994). The proposed strategy acknowledged the necessary relationship between wealth production and wealth distribution, the importance of both public and private sector investment in urban regeneration, and the need to combine 'the ethics of community with the dynamics of a market economy' (Commission on Social Justice 1994: 95). Most significantly, the report recognizes that economic and social policy are two sides of the same coin. The essential links between economic and social policy are evident in a number of ways. First, as Glyn and Miliband (1994) have shown, social inequality can *retard* economic growth. It does so directly through the costs both to government (in terms of higher spending on benefits and lower revenue from taxes) and also to business (through higher expenditure on security and on training workers in basic English and arithmetic). It does so indirectly by deterring investors from large parts of Britain's cities and by depressing the demand for goods and services. Most strikingly (and contrary to much thinking on the political Right), evidence from the NICs of the Far East suggests that economic growth has been accompanied by a *decrease* in income inequality and that income inequality is not a precondition for growth. The link between social and economic policies is also evident in the fact that where social justice is pursued through investment in opportunities (rather than simply supporting the non-employed on benefits) it contributes directly to economic growth. Third, businesses thrive in supportive social environments comprising networks of banks and financial institutions, community organizations, educational establishments and cultural facilities. Social capital is as important to economic performance as human, financial and physical capital (Porter 1991).

It is also important to avoid reification of the market. Markets (for labour, finance, goods and services) are not created by natural or divine forces but are the product of the values, institutions, regulations and political decisions that govern them. In short, markets are political; their structure determines their outcome (Crouch and Marquand 1993). Consider, for example, the politically contentious idea of a minimum wage. Interpreted by Conservatives solely as a cost burden on employers, it may also be viewed in a positive light for both employers and employees. While a minimum wage raises the incomes of the lowest-paid workers as well as the costs of the lowest-paying employers it may also encourage employers to invest in ways of increasing productivity in order to maintain profit levels in face of increased wage costs.

Peter Lilly, Conservative Secretary of State for Social Security in 1993, may have been correct to state that we cannot afford social justice unless we have economic growth, but a critical difference exists between the New Right and the New Left on how to engender growth. Whereas those on the Right have sought to attain economic growth by cutting labour costs and reducing social security levels, an alternative approach is to earn higher wages by producing high value-added goods using knowledge-intensive industries.

The basic principles of a social-democratic approach to urban regeneration include in *economic terms:*

1 regulation of the market in order to deliver a satisfactory combination of economic growth and social cohesion;
2 increased emphasis on the concept of value-added as a measure of success, in addition to productive efficiency;
3 labour should be seen as a resource not only a cost. While frictional unemployment is unavoidable due to changing skill requirements, long-term unemployment is regarded as a waste of human resources;
4 an economic policy cannot ignore its social consequences. Just as social cohesion has economic value, so social division has an economic cost.

In terms of *social welfare*, a realistic social-democratic perspective must acknowledge that a major change in the functions of the welfare state is necessary to reflect changed national circumstances. In the post-war years the main challenge for government was to sustain demand at a level that could support full (male) employment. The national insurance system was crucial to this objective (as well as to Beveridge's war against want). By boosting consumption in times of economic downturn (e.g. through unemployment benefit), Keynes and Beveridge used a combination of economic and social policy to maintain social harmony. A major challenge for government today is underinvestment and Britain's inability to compete successfully in world export markets. The social welfare principles needed to support an investment-led economic strategy must reflect the flexibility of (post-) modern life, characterized by a rapidly changing and insecure labour market and changing family structures. In essence, this will require the state to occupy more of an enabling role in society rather than acting simply as a safety net for the disadvantaged.

THE WAY FORWARD?

One option would be to do nothing and accept the socio-spatial consequences of market-led urban development. If, on the other hand, continuation of current social divisions within Britain's cities is considered unacceptable, then an alternative to existing urban economic and social policy must be sought. From the wide range of policies and plans examined in this book it is possible to identify a set of general principles that may inform the structure of an alternative strategy for the regeneration of Britain's cities:

1. The goal of urban regeneration must be afforded a higher priority in the list of national policy objectives.
2. To address successfully the social and economic difficulties of disadvantaged people and places in urban Britain it is necessary to acknowledge the fundamental relationship between wealth creation and wealth redistribution; the latter is dependent upon the former.
3. Government intervention in the market is essential in order to ameliorate the marked socio-spatial inequalities that result from the unrestrained operation of the capitalist development process.
4. A new form of partnership between central and local government, the private sector and local communities is required that acknowledges the different motives of partners, and enables each to achieve some of their objectives within a framework that promotes redistribution of wealth and opportunity in favour of the disadvantaged.
5. Increased public funding is required to effect the revitalization of urban areas regarded as 'no-go' areas by private capital. In some instances this may involve creation of an 'in-town' development agency comparable to the post-war New Town development corporations.
6. Urban policy needs to be reformed into a more co-ordinated framework in which social and economic dimensions are incorporated and institutional and operational divisions eliminated.
7. There should be an enhanced role for local community participation in decision-making that affects their well-being and living environment. Over-centralization of power neglects the wealth of local knowledge that can be used to inform policies and achieve maximum benefit from available investment.
8. The high level of socio-economic polarization evident in many cities between deprived areas and affluent suburbs commends the geographical concept of the city-region as an appropriate spatial framework for addressing urban inequality.
9. The increasing scale and impact of the restructuring process on Britain's cities means that idealism must be tempered with realism in formulating strategies to redress the urban crisis. While idealistic reflection has exegetical value, policy-making and planning are political activities that, if they are to be successful in tackling urban problems, must be capable of implementation in the prevailing political-economic context.
10. Finally, the multi-causal nature and complexity of the urban crisis means that no single approach is capable of resolving the social and economic difficulties stemming from the restructuring process.

An effective assault on the problems that constitute the urban crisis requires appropriate action at several scales. At the structural level, a primary requirement is recognition of the fact that the disadvantaged position of the poor is linked inextricably to the privileged position of the wealthy. This requires enacting a policy that directly addresses the fundamental question of the distribution of society's wealth. In the shorter term, easing central controls over local authorities would enable the lower tier to undertake activities of direct benefit to local groups. Local authorities might operate as land owners and developers or employ income (e.g. from council house sales) for the benefit of disadvantaged areas. Local government itself must seek to devise more effective strategies to attract private sector funding into deprived areas. This will involve both persuasion (e.g. via rates relief or flexible planning) and direction (such as linked development and leverage schemes), with the particular mix depending on the strength of the local economy. Local authorities must also accept the principle of subsidiarity. In practice, this requires the empowerment of local communities to capitalize on human resources, and the promotion of forms of aided self-help (as pioneered in community businesses and housing co-operatives). This will demand a willingness to devolve financial and political power to the neighbourhood level, with greater participation of residents in budget formulation and plan preparation for their area. Implementation of such an agenda is undoubtedly problematic in the short term, not least because it represents a challenge to the ideology of current urban policy. However, failure to achieve policies based on an integration of social and economic components will ensure that the geographies of division that characterize Britain's cities will persist well into the twenty-first century.

GUIDE TO FURTHER READING

There is a large body of literature relating to political and moral theory and to questions of social and spatial justice. An overview of government structures and political practice is provided in Hawkesworth, M. and Kogan, M. (1992) *Encyclopaedia of Government and Politics*, London: Routledge. An introduction to morality and human nature can be found in Kekes, J. (1989) *Moral Traditions and Individuality*, Princeton NJ: Princeton University Press and Poole, R. (1991) *Morality and Modernity*, London: Routledge. Heater, D. (1990) *Citizenship: The Civic Ideal in World History, Politics and Education*, London: Longman reviews the concept of citizenship, while Daly, M. (1994) *Communitarianism: A New Public Ethics* extends the idea of the public good. A classic statement on social justice is that of Rawls, J. (1971) *A Theory of Justice,* Cambridge MA: Harvard University Press, while Smith, D. (1994) *Geography and Social Justice,* Oxford: Basil Blackwell focuses on the links between geography and social justice.

REFERENCES

Barclay, P. (1995) *Inquiry into Income and Wealth*, vol. 1, York: Joseph Rowntree Foundation.

Boddy, M. and Fudge, C. (1984) *Local Socialism*, London: Macmillan.
Commission on Social Justice (1994) *Social Justice: Strategies for National Renewal*, London: Vintage.
Cooke, P. (1990) *Back to the Future*, London: Unwin Hyman.
Crouch, C. and Marquand, D. (1993) *Ethics and Markets*, Oxford: Basil Blackwell.
Gilbert, A. and Gugler, J. (1992) *Cities, Poverty and Development*, Oxford: Oxford University Press.
Glyn, A. and Miliband, D. (1994) *Paying for Inequality*, London: Rivers Oram Press.
Hasluck, C. (1987) *Urban Unemployment*, London: Longman.
Hatch, C. (1985) 'Italy's industrial renaissance', *Urban Land*, 44: 20–3.
Heidenheimer, A., Heclo, H. and Adams, B. (1983) *Comparative Public Policy*, New York: St Martins.
Hills, J. (1995) *Inquiry into Income and Wealth*, vol. 2, York: Joseph Rowntree Foundation.
Jacobs, J. (1969) *The Economy of Cities*, New York: Vintage Books.
Johannison, B. (1990) 'The Nordic perspective: self-reliant local development in four Scandinavian countries' in W. Stohr *Global Challenge and Local Response*, London: Mansell.
Keating, W. (1986) 'Linking downtown development to broader community goals', *Journal of the American Planning Association*, 52(1): 133–41.
Kraushaar, R. (1979) 'Pragmatic radicalism', *International Journal of Urban and Regional Research*, 3: 61–80.
Krumholz, N. (1982) 'A retrospective view of equity planning', *Journal of the American Planning Association*, 48(2): 163–83.
Martin, S. (1990) 'City grants, urban development grants and urban regeneration grants' in M. Campbell *Local Economic Policy*, London: Cassell.
Metcalfe, H., Pearson, R. and Martin, R. (1990) 'The charitable role of companies in job creation', *Regional Studies* 24(3): 261–8.
Morris, D. and Hess, K. (1975) *Neighbourhood Power*, Boston: Beacon Press.
Pacione, M. (1990) 'What about people? A critical analysis of urban policy in the United Kingdom', *Geography*, 75: 193–202.
——(1992) 'Citizenship, partnership and the popular restructuring of urban space', *Urban Geography*, 13: 405–21.
——(1993) 'The geography of the urban crisis: some evidence from Glasgow', *Scottish Geographical Magazine*, 109(2): 87–95.
Papworth, J. (1988) 'Non-local local government and local power', *The Ecologist*, 18(6): 213–22.
Porter, M. (1991) *The Competitive Advantage of Nations*, London: Macmillan.
Roemer, J. (1992) 'The morality and efficiency of market socialism', *Ethics*, 102: 448–64.
Short, J. (1989) *The Humane City*, Oxford: Basil Blackwell.
Squires, G. (1989) *Unequal Partnerships*, New Brunswick: Rutgers University Press.
Thake, S. and Staubach, R. (1993) *Investing in People*, York: Joseph Rowntree Foundation.
Thornley, A. (1991) *Urban Planning Under Thatcherism*, London: Routledge.
Willmott, P. (1987) *Local Government Decentralisation and Community*, London: Policy Studies Institute.

SUBJECT INDEX

PLACE INDEX